Sustainable Houses and Living in the Hot-Humid Climates of Asia

Tetsu Kubota • Hom Bahadur Rijal
Hiroto Takaguchi

Editors

Sustainable Houses and Living in the Hot-Humid Climates of Asia

 Springer

Editors
Tetsu Kubota
Graduate School for International
Development and Cooperation (IDEC)
Hiroshima University
Hiroshima, Japan

Hom Bahadur Rijal
Department of Restoration Ecology and Built
Environment
Tokyo City University
Yokohama, Japan

Hiroto Takaguchi
Department of Architecture
Waseda University
Tokyo, Japan

ISBN 978-981-10-8464-5   ISBN 978-981-10-8465-2   (eBook)
https://doi.org/10.1007/978-981-10-8465-2

Library of Congress Control Number: 2018940451

Printed on acid-free paper

This Springer imprint is published by the registered company Springer Nature Singapore Pte Ltd.
The registered company address is: 152 Beach Road, #21-01/04 Gateway East, Singapore 189721, Singapore

Preface

Today, more than half of the world's population lives in Asia, particularly in urban areas. Approximately 35–40% of the world energy is consumed in Asia, and this percentage is expected to rise further. Energy consumption has increased particularly in the residential sector in line with the rapid rise of the middle class. The majority of growing Asian cities are located in hot and humid climate regions. There is an urgent challenge for designers to provide healthy and comfortable indoor environments for occupants without consuming energy and resources excessively in growing tropical Asian cities. The aim of this book is to provide information on the latest research findings that are useful in designing sustainable houses and living in rapidly growing Asian cities.

Many parts of Asia are under the influence of monsoons, which bring rains and winds to the region. This helps to provide fertile soil and thus assists in maturing rich agricultural cultures. Countries in the Asian region therefore have much in common, mainly based on these climatic and indigenous cultural similarities. On the other hand, Asia can also be characterized as a "diverse" region. Especially after the empire era, each of the countries in Asia took its own development path and became diverse in terms of the socio-economy, culture and religion, among other concepts. In particular, the economic conditions are quite diverse among these countries. They include the emerging economies such as China, India and Indonesia, while also comprising real developed (and aged) countries such as Japan. Asia is a unique region that comprises diverse countries ranging from "emerging" to "ageing", although they have similar climatic and cultural backgrounds.

Most of the cities and towns in tropical Asia experience hot and humid climates during most of the year. In Kuala Lumpur (3°8'N 101°41'E), the capital of Malaysia, for example, the monthly average temperature ranges from 28 °C to 29 °C with an annual average of 28°C, whereas the monthly average relative humidity ranges from 78% to 84%. The tropical climates are considered one of the most challenging conditions in achieving indoor thermal comfort by non-mechanical means because both sensible and latent heat are immense. Traditional buildings in these regions are formerly used to utilize natural ventilation with proper solar shadings, as seen in an example of a wooden structure with a raised floor, which is the so-called Malay house. After WWII, most countries in this region experienced rapid urbanization along with economic growth. In most cases, they experienced Westernization in line with this urbanization and experienced dramatic changes in their building designs as well as their lifestyles. Most of the newly constructed urban houses in Southeast Asia no longer are wooden well-ventilated houses but are instead Westernized air-conditioned brick houses. The use of air-conditioning in tropical Asia is now rapidly growing and is a major concern for energy security as well as for urban and global warming.

As previously mentioned, the Asian region comprises diverse countries with different economic conditions. Although many countries are growing now, still, a large number of poor people live in the region. Those people still need improvements for their living standards and well-being. Therefore, balancing the improvement in quality of life (QoL) and energy-saving achievements in a low-carbon fashion is still a crucial question. Historically, it has been shown that nationwide energy consumption usually increases with the increase in economy, such as GDP. We therefore evaluated our QoL solely by an economic indicator such as GDP. If we do this, it is logically impossible to strike a balance between QoL and energy savings. It is very important to determine the factors (especially non-economic factors) affecting residents' QoL in developing countries while decoupling economic growth from the use of non-renewable energy, so that we can achieve low-energy and low-carbon societies without hindering the improvement of QoL.

Climate change casts a large shadow on our future. Since tropical regions experience hot and humid climates as described before, global warming will create significant impacts on indoor thermal comfort, health, and energy use. While we make efforts to mitigate further climate change through regional and/or global cooperation, it is also important to find ways to adapt to rapidly changing climates. Many cities/towns still are growing in the tropical Asian region, and therefore, it is necessary to design buildings as well as cities, and perhaps even lifestyles, that are capable of bearing these altered climates without relying on the excessive use of energy.

This book is composed of seven parts, comprising a total of 50 chapters, which were written by 55 authors from various countries, particularly from the Asian region. The three editors dealt with the different parts, depending on their specializations: T. Kubota (Parts I, V, VI and VII), H.B. Rijal (Parts II and III) and H. Takaguchi (Part IV). In Part I, vernacular houses in different Asian countries are introduced as an introduction, covering those in Indonesia, Malaysia, India, Nepal, China, Thailand and Laos. Then, in Parts II and III, indoor adaptive thermal comfort and occupants' adaptive behaviour, focusing especially on those in hot-humid climates are explored in depth. Part IV presents detailed survey results on household energy consumption in various tropical Asian cities. In Part V, we analyse the indoor thermal conditions in both traditional houses and modern houses in detail from these countries. Several actual practices of sustainable houses in Asian cities are reviewed in Part VI. Then, the vulnerability for climate change and urban heat island in Asian growing cities are discussed in the final Part VII. This book will be essential reading for anyone with an interest in sustainable house design in the growing cities of Asia.

On behalf of the editors and the authors of this book, let me pass our deepest condolences on the recent passing of Indonesian legend professor Dr. Tri Harso Karyono. He and his great works will live in our heart forever.

Hiroshima, Japan   Tetsu Kubota

Contents

List of Contributors

Mohd Hamdan Ahmad Faculty of Built Environment, Universiti Teknologi Malaysia, Skudai, Johor, Malaysia

Muhammad Nur Fajri Alfata Research Institute for Human Settlement and Housing – Ministry of Public Works, Bandung, Indonesia

Meita Tristida Arethusa PT. Hadiprana Design Consultant, Jakarta, Indonesia

Sri Nastiti N. Ekasiwi Department of Architecture, Institut Teknologi Sepuluh Nopember, Surabaya, Indonesia

J. Fergus Nicol Low Energy Architecture Research Unit, Sir John Cass Faculty of Art, Architecture and Design, London Metropolitan University, London, UK

Kazuhiro Fukuyo Graduate School of Innovation and Technology Management, Yamaguchi University, Yamaguchi, Japan

Weijun Gao Faculty of Environmental Engineering, Department of Architecture, The University of Kitakyushu, Fukuoka, Japan

I Gusti Ngurah Antaryama Department of Architecture, Institut Teknologi Sepuluh Nopember, Surabaya, Indonesia

Soichi Hata Shibaura Institute of Technology, Tokyo, Japan

Shuichi Hokoi Kyoto University, Kyoto, Japan

Southeast University, Nanjing, China

Michael A. Humphreys School of Architecture, Faculty of Technology, Design and Environment, Headington Campus, Oxford Brookes University, Oxford, UK

Ruey-Lung Hwang Department of Industrial Technology Education, National Kaohsiung Normal University, Kaohsiung, Taiwan

Pawinee Iamtrakul Faculty of Agriculture and Planning, Thammasat University, Bangkok, Thailand

Satoru Iizuka Graduate School of Environmental Studies, Department of Environmental Engineering and Architecture, Nagoya University, Nagoya, Japan

Madhavi Indraganti Department of Architecture and Urban Planning, College of Engineering, Qatar University, Doha, Qatar

Tri Harso Karyono (deceased) School of Architecture, Tanri Abeng University, Jakarta, Indonesia

Suapphong Kritsanawonghong Business Development, Renewable Energy Division, Prime Road Group, Bangkok, Thailand

Tetsu Kubota Graduate School for International Development and Cooperation (IDEC), Hiroshima University, Hiroshima, Japan

Hasanuddin Bin Lamit Faculty of Built Environment, Universiti Teknologi Malaysia, Skudai, Johor, Malaysia

Han Soo Lee Graduate School for International Development and Cooperation (IDEC), Hiroshima University, Hiroshima, Japan

Noor Hanita Abdul Majid Department of Architecture, International Islamic University Malaysia, Gombak, Malaysia

Masato Miyata Mitsubishi UFJ Research and Consulting Co.,Ltd., Nagoya, Japan

Hiroshi Mori YKK AP R&D Center, PT. YKK AP Indonesia, Tangerang, Banten, Indonesia

Yusuke Nakajima Department of Urban Design and Planning, School of Architecture, Kogakuin University, Tokyo, Japan

Miwako Nakamura MW Ecological Design, Tokyo, Japan

Takashi Nakaya Department of Architecture, Shinshu University, Nagano, Nagano, Japan

Didit Novianto Faculty of Environmental Engineering, Department of Architecture, The University of Kitakyushu, Fukuoka, Japan

Agung Murti Nugroho Department of Architecture, University of Brawijaya, Malang, Indonesia

Daisuke Oka Kyoto University, Kyoto, Japan

Eric Casimero Oliva National Housing Authority, Regional Office No. 8, Tacloban, Philippines

Ryozo Ooka Institute of Industrial Science, The University of Tokyo, Tokyo, Japan

Dilshan Remaz Ossen Department of Architecture Engineering, Kingdom University, Riffa, Bahrain

Tran Thi Thu Phuong Vietnam Institute of Urban and Rural Planning, Hanoi, Vietnam

Chanachok Pratchayawutthirat The Eco Knowledge Program, Univentures Public Company Limited, Bangkok, Thailand

Adeb Qaid Department of Architecture Engineering, Kingdom University, Riffa, Bahrain

Kavita Daryani Rao Department of Architecture, Jawaharlal Nehru Architecture and Fine Arts University, Hyderabad, India

Hom Bahadur Rijal Department of Restoration Ecology and Built Environment, Tokyo City University, Yokohama, Japan

Ikuro Shimizu School of Architecture, Shibaura Institute of Technology, Tokyo, Japan

Masanori Shukuya Department of Restoration Ecology and Built Environment, Tokyo City University, Yokohama, Japan

Daisuke Sumiyoshi Graduate School of Human-Environment Studies, Kyushu University, Fukuoka, Japan

Usep Surahman Universitas Pendidikan Indonesia (UPI), Bandung, Indonesia

Masaki Tajima School of Systems Engineering, Kochi University of Technology, Kochi, Japan

Hiroto Takaguchi Department of Architecture, Waseda University, Tokyo, Japan

Doris Hooi Chyee Toe Faculty of Built Environment, Universiti Teknologi Malaysia, Skudai, Johor, Malaysia

Andhang Rakhmat Trihamdani YKK AP R&D Center, PT. YKK AP Indonesia, Tangerang, Banten, Indonesia

Tomoko Uno Department of Architecture, Mukogawa Women's University, Nishinomiya, Hyogo, Japan

Mohammad Hussaini Wahab UTM Razak School of Engineering and Advanced Technology, UTM Kuala Lumpur, Kuala Lumpur, Malaysia

Kaede Watanabe Nikken Sekkei Civil Engineering Ltd., Tokyo, Japan

Kikuma Watanabe School of Systems Engineering, Kochi University of Technology, Kochi, Japan

Toshiyuki Watanabe Kyushu University, Fukuoka, Japan

Arif Sarwo Wibowo School of Architecture, Planning and Policy Development, Bandung Institute of Technology, Bandung, Indonesia

Takao Yamashita Japan Port Consultant Co., Ltd., Kobe, Japan

Hiroshi Yoshino Institute of Liberal Arts and Science, Tohoku University, Sendai, Japan

Mohd Azuan Zakaria Faculty of Civil and Environmental Engineering, Universiti Tun Hussein Onn Malaysia, Batu Pahat, Johor, Malaysia

Qingyuan Zhang Institute of Urban Innovation, Yokohama National University, Yokohama, Japan

Chapter 1
Introduction

Tetsu Kubota

Abstract This chapter gives an overview of Asian developing countries to determine the main focus of this book. Today, the Asian region is one of the biggest contributors in terms of energy and GHG emissions, but socio-economic conditions are very different among the countries. They experience similar paths of rapid urbanization but with different timings. Asian developing countries need to combat at least three challenges, including climate change mitigation and adaptation on top of ongoing economic growth.

Keywords Sustainability · Developing countries · Hot-humid climate ·
Urbanization · Climate change

1.1 Hot-Humid Climates of Asia

Large parts of Asia are strongly influenced by monsoons. The monsoons bring rain and winds particularly to the coastal areas. As shown in Fig. 1.1, most of the major Asian cities are located in the coastal areas and thus experience a similar climate, i.e. a hot-humid climate. The monthly average temperatures are more than 22–25 °C, whereas relative humidity exceeds 70% year-round in most of the cities of Southeast Asia (Fig. 1.2). Meanwhile, some parts of India, China and Japan also experience such a hot-humid climate during the summer months. It is worth noting that other regions such as Europe and North America rarely experience this hot-humid condition. Particularly in terms of the passive design, the local climate is one of the most important background conditions that need to be considered. Therefore, the fact that the Asian region has a hot-humid climate in common inspires us to consider the necessity of specific building designs/technologies that are suited particularly to unique Asian climatic conditions.

T. Kubota (✉)
Graduate School for International Development and Cooperation (IDEC), Hiroshima
University, Hiroshima, Japan
e-mail: tetsu@hiroshima-u.ac.jp

© Springer Nature Singapore Pte Ltd. 2018
T. Kubota et al. (eds.), *Sustainable Houses and Living in the Hot-Humid Climates of Asia*, https://doi.org/10.1007/978-981-10-8465-2_1

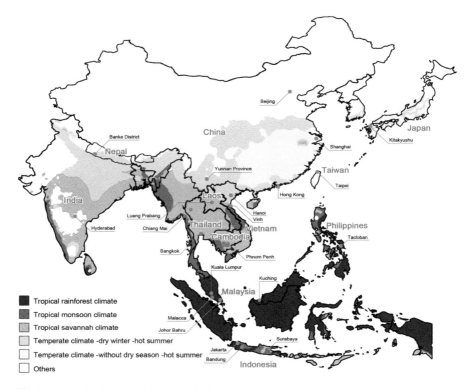

Fig. 1.1 Climates in Asia (Source: Peel et al. [1])

This is one of the motivations of this book, and this book focuses especially on Asian countries that experience these hot-humid climates.

It is worth knowing how local people coped with such a tough hot-humid condition without relying on mechanical means. Asian vernacular houses give various design hints to modern houses. Part 1 of this book explores vernacular houses and how their occupants behave in hot-humid Asia.

1.2 Rapid Urbanization in Developing Asia

The Paris Agreement came into force in November 2016. As of late 2017, a total of 172 parties have ratified the agreement [3]. Now, all the signatory parties, including developing countries, are required to declare their "nationally determined contributions" (NDCs) to put forward their best efforts to combat climate change for a sustainable low-carbon future. The Asian region, including China and India, contributes

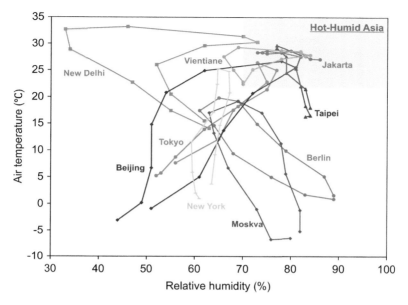

Fig. 1.2 Climograph of major cities (Source: National Astronomical Observatory of Japan [2])

approximately 48.6% of the global GHG emissions as of 2014 [4], and the region recently submitted the intended NDC to the United Nations, as shown in Table 1.1.

The Asian region is diverse. As indicated in Table 1.1, the socio-economic conditions of the countries vary. Nevertheless, most of the countries similarly experience or experienced rapid urbanization in line with economic growth (Fig. 1.3). Japan is probably the first country that accelerated urbanization after WWII in Asia. Japan achieved this prolonged economic boom until the 1990s and became the second largest economy in the world at that time. China, India and other Asian countries seemed to experience similar urbanization paths after Japan but with different timings. This, in turn, results in the unique nature of Asian socio-economic diversity, comprising real developing countries and real developed (aged) countries in the same region.

1.3 Rise of the Middle Class

Many Asian developing countries now are in the phase of the so-called population divided, which are expected to last until 2020 in South Korea, China and Thailand, and until 2020–2040 in the case of Malaysia, Indonesia, Vietnam and India. The end point of population divided means the entrance of ageing societies. This means that

Table 1.1 Key statistics of Asian countries

	Country	Population (million)	Urban population (% of total)	GDP per capita, current prices (USD)	Total final energy consumption of energy (Mtoe)	CO_2 emissions from fuel combustion ($MtCO_2$)	CO_2 emissions per capita (tCO_2/capita)	Intended Nationally Determined Contribution (INDC)	
								Target year	GHG emission reduction target
1	Singapore	5.54	100	53,629	17.1	44.4	8.0	2030	−36% of emissions intensity below 2005 levels
2	Brunei Darussalam	0.42	77.2	30,995	1.0	6.0	14.1	2035	−63% compared with BAU (total energy consumption), etc. No overall GHG reduction target.
3	Malaysia	30.72	74.7	9505	51.6	220.4	7.3	2030	−35% of emissions intensity below 2005 levels (unconditional) −45% of emissions intensity below 2005 levels (conditional)
4	Thailand	68.66	50.4	5799	98.0	247.5	3.6	2030	−20% (unconditional) and −25% (conditional), compared with BAU
5	Indonesia	258.16	53.7	3371	162.8	441.9	1.7	2030	−29% below BAU (unconditional) −41% below BAU (conditional)
6	Philippines	101.72	44.4	2866	29.6	103.9	1.0	2030	−70% below BAU (conditional)
7	Lao People's Democratic Republic	6.66	38.6	2212	–	–	–	2020–2030	Increase in renewable energy share to 30% of total, etc. (conditional) No overall GHG reduction target
8	Viet Nam	93.57	33.6	2088	58.2	168.3	1.8	2030	−8% below BAU (unconditional) −25% below BAU (conditional)

#	Country							Year	Target
9	Cambodia	15.52	20.7	1168	5.9	8.0	0.5	2030	−27% below BAU (conditional)
10	Myanmar	52.40	34.1	1147	17.7	24.4	0.5	2030	Undertaking various mitigation actions in line with its sustainable development needs (conditional). No overall GHG reduction target
11	Japan	127.98	93.5	34,493	291.4	1141.6	9.0	2030	−26% of emissions intensity below 2013 levels
12	China	1397.03	55.6	8167	1914.6	9084.6	6.6	2030	−60% to −65% of emissions intensity of GDP below 2005 levels
13	India	1309.05	32.7	1629	577.7	2066.0	1.6	2030	−33% to −35% of emissions intensity of GDP below 2005 levels
14	Nepal	28.66	18.6	751	11.6	5.6	0.2	2030	List of 14 targets (unconditional) No overall GHG reduction target

Source: UNCC [3]; IEA [4, 5]; United Nations [6]

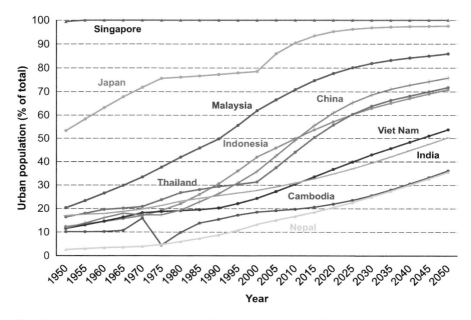

Fig. 1.3 Urbanization in Asian countries (Source: United Nations [6])

Asian countries are growing and urbanizing very quickly but will also be ageing at a fast rate, as Japan is now. Dynamic demographic change with rapid urbanization is considered to be one of Asia's most important characteristics.

The middle class now is on the rise in developing Asia. The middle-class population can be major economic consumers as well as GHG emission contributors and therefore should draw much attention from researchers and stakeholders. As illustrated in Fig. 1.4, the percentage of the middle class has been rising significantly in many Asian countries. In Indonesia (the most populated country in Southeast Asia), for example, the middle class was almost nil in 1990 but increased to 46.8% of the total until 2005 [7]. Chun [7] further projected that the middle class in Indonesia will reach 100% by 2030. It is important to provide the rapidly growing middle-class people in these Asian countries with sustainable houses that are available in the immediate future, considering the enormous size of this population.

1.4 Emerging Modern Houses

As discussed above, the development stage is very different among countries in Asia. However, to the best of the authors' knowledge, most of the newly emerging urban houses in this region are constructed of brick and/or concrete (Fig. 1.5), which seem to be very different from their original nature of vernacular houses. In

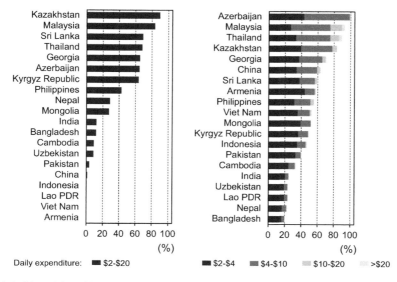

Fig. 1.4 Rise of the middle class in developing Asia (Source: Chun [7])

Fig. 1.5 Urban houses in Malaysia

Malaysia, for example, the brick houses account for approximately 91% of the total existing urban houses in which terraced houses are the most common housing type [8]. How do the local people perceive the indoor thermal environment in such modern houses? How do they adapt to hot-humid conditions? In Parts II and III, we will discuss these adaptive thermal comforts and adaptive behaviours particularly in the hot-humid climates of Asia.

A high-thermal-mass building usually has an effect of stabilizing the indoor air temperature (decreasing daytime temperature while increasing nocturnal temperature). It is still unsure whether these high-thermal-mass buildings are suitable for the hot-humid climates or not, given that natural ventilation traditionally is the norm. In fact, the ownership and usage of air-conditioning has dramatically increased in Asian developing countries as reported in [9–11]. In the case of Southeast Asia, people tend to install air-conditioners in their bedrooms rather than in the living room and, as a result, use them for a long time (approximately 7–8 hours/day) during sleep [12]. We will explore the detailed thermal conditions in several existing traditional and modern houses in Part V, after investigating household energy consumption in Asian developing countries in detail in Part IV.

1.5 Green Growth in Developing Asia

IPCC [13] predicted that the Asian region will see a temperature increase (ensemble mean changes) of approximately 1–2 °C above the baseline of 1986–2005 by 2016–2035 and approximately 2–4 °C by 2081–2100 (RCP4.5). Various negative impacts of global warming are projected in this region. One of the most serious concerns is the increase in extreme weather events [14, 15]. A large proportion of Asia's population lives in low-elevation coastal zones that are particularly at risk for climate change hazards, including sea level rise, storm surges and typhoons [14]. According to the OECD [16], the coastal flooding in Southeast Asian cities costs an estimated USD $300 million in average annual losses in 2005; even with significant investment in adaptations, the price tag could climb to USD $6 billion per year by 2050.

Urban temperature is expected to increase further not only due to global warming but also due to the urban heat island (UHI). The combined effects would be large, particularly in the case of rapidly growing cities. In fact, many Asian cities tend to propose large-scale urban development master plans to expand their cities further [17]. People living in this region already experience hot-humid climates, entirely or at least partially through the year, and therefore, such a large increase in urban temperature would create tremendous impacts on health, comfort and lifestyles. Eventually, these effects would lead to dramatic increases in energy consumption for space cooling and thus increases in GHG emissions [11, 18, 19]. In Part VII, we will predict future urban warming in several Asian growing cities.

Asian developing countries are required to combat at least three challenges ahead (Fig. 1.6). As shown, it is necessary to achieve ongoing economic development in a

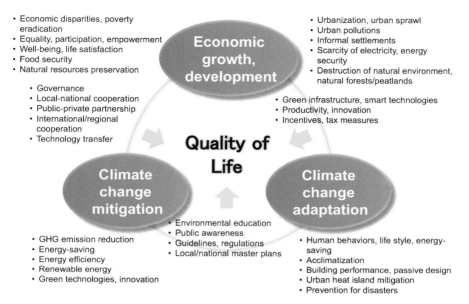

- Economic disparities, poverty eradication
- Equality, participation, empowerment
- Well-being, life satisfaction
- Food security
- Natural resources preservation

- Governance
- Local-national cooperation
- Public-private partnership
- International/regional cooperation
- Technology transfer

- Urbanization, urban sprawl
- Urban pollutions
- Informal settlements
- Scarcity of electricity, energy security
- Destruction of natural environment, natural forests/peatlands

- Green infrastructure, smart technologies
- Productivity, innovation
- Incentives, tax measures

- GHG emission reduction
- Energy-saving
- Energy efficiency
- Renewable energy
- Green technologies, innovation

- Environmental education
- Public awareness
- Guidelines, regulations
- Local/national master plans

- Human behaviors, life style, energy-saving
- Acclimatization
- Building performance, passive design
- Urban heat island mitigation
- Prevention for disasters

Fig. 1.6 Green growth in developing Asia

sustainable way under the influences of local and global warming while improving individuals' quality of life. Given that the current developed countries have achieved only one target, i.e. economic growth, in the past, the challenge ahead for the developing countries is quite literally an unprecedented challenge. As discussed internationally, technology transfer from the developed nations to the less developed nations through international cooperation is indeed one of the key components in fighting this trilemma. It is important to accumulate knowledge and share information among all stakeholders to achieve sustainable houses and living in Asia. This is one of the motivations of this book.

References

1. Peel MC, Finlayson BL, McMahon TA (2007) Updated world map of the Koppen-Geiger climate classification. Hydrol Earth Syst Sci 11:1633–1644
2. National Astronomical Observatory of Japan (2017) Chronological scientific tables. Maruzen, Tokyo
3. United Nations Framework Convention on Climate Change. http://unfccc.int/2860.php. Accessed 30 Dec 2017
4. IEA (2017) CO$_2$ emissions from fuel combustion. International Energy Agency, Paris
5. IEA (2017) World energy balances. International Energy Agency, Paris
6. United Nations. Department of Economic and Social Affairs, Population Division. http://www.un.org/en/development/desa/population/. Accessed 30 Dec 2017

7. Chun N (2010) Middle class size in the past, present, and future: a description of trends in Asia (ADB Economics Working Paper Series No. 217). Asian Development Bank

8. Department of Statistics Malaysia (2011) Population distribution and basic demographic characteristics, population and housing census of Malaysia 2010, Department of Statistics Malaysia.

9. Sahakian M (2014) Keeping cool in Southeast Asia: energy consumption and urban air-conditioning. Palgrave Macmillan, Basingstoke

10. McNeil MA, Letschert VE (2008) Future air conditioning energy consumption in developing countries and what can be done about it: the potential of efficiency in the residential sector. Lawrence Berkeley National Laboratory, Berkeley

11. Santamouris M (2016) Cooling the buildings – past, present and future. Energ Build 128:617–638

12. Kubota T, Toe DHC, Ahmad S (2009) The effects of night ventilation technique on indoor thermal environment for residential buildings in hot-humid climate of Malaysia. Energ Build 41:829–839

13. IPCC (2013) In: van Oldenborgh GJ, Collins M, Arblaster J, Christensen JH, Marotzke J, Power, SB, Rummukainen M, Zhou T (eds) Annex I: Atlas global and regional climate projections. In: Stoker TF, Qin D, Plattner G-K, Tignor M, Allen SK, Boschung J, Nauels A, Xia Y, Bex V, Midgley PM (eds) Climate change 2013: the physical science basis. Contribution of working group I to the Fifth Assessment Report of the Intergovernmental Panel on climate change. Cambridge University Press, Cambridge/New York

14. Hijioka Y, Lin E, Pereira JJ, Corlett RT, Cui X, Insarov GE, Lasco RD, Lindgren E, Surjan A (2014) Asia. In: Barros VR, Field CB, Dokken DJ, Mastrandrea MD, Mach KJ, Bilir TE, Chatterjee M, Ebi KL, Estrada YO, Genova RC, Girma B, Kissel ES, Levy AN, MacCracken S, Mastrandrea PR, White LL (eds) Climate change 2014: impacts, adaptation, and vulnerability. Part B: regional aspects. Contribution of working group II to the Fifth Assessment Report of the Intergovernmental Panel on climate change. Cambridge University Press, Cambridge/New York, pp 1327–1370

15. UN-HABITAT (2011) Cities and Climate Change: Global Report Human Settlements 2011. United Nations Human Settlements Programme (UN-HABITAT), Earthscan, London/Washington, DC

16. OECD (2014) Towards green growth in Southeast Asia, OECD green growth studies. OECD Publishing, Paris

17. Lee HS, Trihamdani AR, Kubota T, Iizuka S, Phuong TTT (2017) Impacts of land use changes from the Hanoi Master Plan 2030 on urban islands: Part 2. Influence of global warming. Sustain Cities Soc 31:95–108

18. Hekkenberg M, Moll HC, Uiterkamp AJMS (2009) Dynamic temperature dependence patterns in future energy demand models in the context of climate change. Energy 34:1797–1806

19. Isaac M, van Vuuren DP (2009) Modeling global residential sector energy demand for heating and air conditioning in the context of climate change. Energ Policy 37:507–521

Part I
Vernacular Architecture

Chapter 2
Indonesia: Dutch Colonial Buildings

Arif Sarwo Wibowo, Muhammad Nur Fajri Alfata, and Tetsu Kubota

Abstract This chapter starts with a brief history of Dutch colonial architecture in Indonesia. The arrival of Europeans in the early fifteenth century had a great impact on building construction in Indonesia. The material and spatial concepts of European buildings were completely different from those of the Indonesian indigenous people. With the passage of time, it proved that building designs imitated from existing European buildings could not be used directly in the tropical climate of Indonesia. This led to the development of buildings that were adapted to the local context. Secondly, this chapter shows the results of a field measurement conducted in a Dutch colonial building in the city of Bandung. The results showed that, overall, daytime indoor air temperatures in the building maintained relatively low values compared to the corresponding outdoor temperature mainly due to the thermal mass effect. Other passive cooling strategies found from the measurement include night ventilation, use of corridor spaces, high ceilings, and permanent openings above windows/doors.

Keywords Vernacular architecture · Colonial building · Passive cooling · Thermal comfort · Hot-humid climate · Indonesia

A. S. Wibowo
School of Architecture, Planning and Policy Development, Bandung Institute of Technology, Bandung, Indonesia

M. N. F. Alfata
Research Institute for Human Settlement and Housing – Ministry of Public Works, Bandung, Indonesia

T. Kubota (✉)
Graduate School for International Development and Cooperation (IDEC), Hiroshima University, Hiroshima, Japan
e-mail: tetsu@hiroshima-u.ac.jp

13

2.1   Introduction

The colonial building typology can be found in every European colonial town, whether in Africa, Asia, and America. Nearly everywhere in the early days of colonization, Europeans brought building technologies, materials, and designs from their home countries to the areas they colonized. Sometimes what was brought was not in accordance with local conditions, including culture and climate, and thus, over time, the Europeans adapted their technologies, materials, and designs to fit the localities, including in Indonesia. Over hundreds of years, the colonial building typology in Indonesia underwent significant development associated with architectural style and adjustment to local conditions.

2.2   A History of Dutch Colonial Buildings in Indonesia

The colonial building style in Indonesia evolved since the arrival of Europeans in the archipelago, characterized by the presence of the *Vereenigde Oostindische Compagnie* (VOC) in several port cities, such as Batavia, Semarang, and Surabaya. The initial period of the presence of colonial buildings can be associated with the construction of castle-type buildings in those cities at the beginning of the fifteenth century.

Over time, these fortress cities developed and expanded into their surrounding areas. The widening VOC areas (taken over by the Netherlands East Indies Government in 1800 due to bankruptcy of the VOC) significantly increased the number of the buildings with a European architectural style. In addition to the port cities, which were entry points for the Europeans, the government established new *gemeentes* (municipalities) in inland regions, such as Bandung, Cimahi, Magelang, and Malang.

2.2.1   Classification of Dutch Colonial Buildings

Stylistically, Dutch colonial buildings in Indonesia can be broadly classified into four periods, strongly associated with the periodization of building styles in other parts of the world, especially Europe and America (Fig. 2.1).

2.2.1.1   Before 1800s

In general, the colonial buildings in the early period mimicked those in the home country. For example, the *Stadhuis van Batavia* (City Hall of Batavia, now

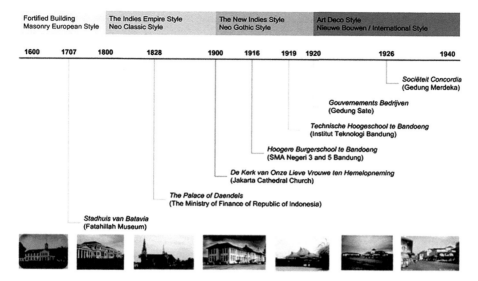

Fig. 2.1 Timeline of colonial architecture in Indonesia (Note: Photographs with permission of National Museum van Wereldculturen. Object ID: TM-60054762)

Fatahillah Museum, Jakarta), which was built in 1707–1710, closely resembles the *Koninklijk Paleis* (Royal Palace) in Amsterdam, the Netherlands.

2.2.1.2   1800s–1900s

The Indies Empire Style flourished in the early nineteenth century after the colonized areas were taken over by the Netherlands East Indies Government. This building style was inspired by the neoclassic style. The buildings in this era usually had wide front and back porches with typical Ancient Greek pillars, such as Doric, Ionic, and Corinthian. This building style was not only found in formal buildings but also in residential buildings with large yards, various applied materials, and roof shape adjustments.

2.2.1.3   1900s–1920s

The New Indies Style was developed in the early twentieth century as a form of adaptation to the tropical climate of Indonesia. In addition, the Neo-Gothic Style also evolved in this period, marked by the construction of the *Kerk van Onze Lieve Vrouwe ten Hemelopneming* (Church of Our Lady of Assumption, now Cathedral Church, Jakarta) in Batavia (completed in 1901).

2.2.1.4 1920s–1940s

The Art Deco Style spread in the 1920s and the 1930s in various colonial cities, particularly in Bandung as it was planned to become the new capital city of the Netherlands East Indies Government. The Art Deco Style itself was introduced at the *Exposition Internationale des Arts Décoratifs et Industriels Modernes* (International Exhibition of Modern Decorative and Industrial Arts) in Paris in 1925. In Bandung, this style evolved along with the *Nieuwe Bouwen*/International Style with stream-lined design characteristics. Meanwhile, in the midst of a wide variety of styles, the Neo-Gothic Style was still applied in church design in Bandung [1].

2.2.2   Dutch Colonial Buildings in Bandung

In 1906, the Netherlands East Indies Government promoted Bandung to a *gemeente* with an autonomous city government [2]. Bandung had developed as a new city in the early twentieth century. The city was one of the cities passed by *De Groote Postweg* (the Great Post Road) and is currently one of the most important colonial cities in Indonesia that still contains many colonial buildings. Several Dutch architects, such as Henri Maclaine Pont, Herman Thomas Karsten, C.P. Wolff Schoemaker, C. Citroen, and Algemeen Ingenieurs- en Architectenbureau in Batavia (A.I.A.), realized projects in Bandung. Many of their works are buildings that contributed to the development of the identity of Bandung, including the Ceremonial Hall of *Technische Hoogeschool Bandoeng* (Institut Teknologi Bandung/ITB), *Departement van Gouvernementsbedrijven* (Gedung Sate), *Sociëteit Concordia* (Gedung Merdeka), churches, schools, and others.

2.3   Field Measurement of a Dutch Colonial Building in Bandung

2.3.1   Selected Case Study

Since its establishment in 1811, many buildings, with various architectural styles, were built by the Dutch in the city of Bandung. These Dutch colonial buildings were constructed particularly during the periods of the New Indies Style (1900s–1920s) and the Art Deco Style and *Nieuwe Bouwen* Style (1920s–1940s). Approximately 1500 Dutch colonial buildings still exist in Bandung alone [3]. Due to its high altitude (approximately 760 m above sea level), Bandung has a relatively cool climate with a monthly average temperature of 23–24 °C and monthly average relative humidity of 50–80%.

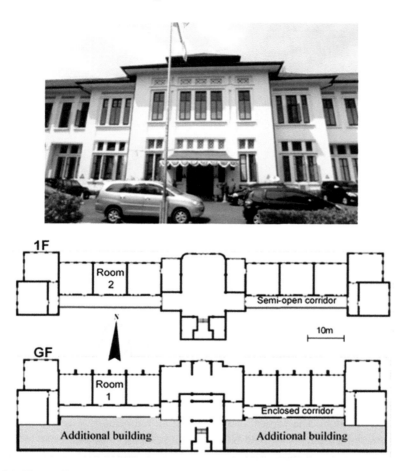

Fig. 2.2 Former *Hoogere Burgerschool* te Bandung (HBS), now is a senior high school

A detailed field measurement was conducted in one of the existing Dutch colonial buildings in Bandung in 2015 (Fig. 2.2). This building was originally constructed in 1916 as a 5-year high school (*Hogere Burgerschool*/HBS) and can be considered a typical colonial building from the period of the New Indies Style. From 1950 on, this building has been in use as a high school, accommodating approximately 500 students aged from 16 to 18 years old. The building is a two-story building constructed of timber for the main structure and brick and lime cement for the walls. The front façade faces north, whereas a large corridor (3.35 m width) is located at the southern side of the building. The corridor space on the ground floor is enclosed, while the one on the first floor is semi-open. In addition, the doors facing the corridors are permanently open (louver doors), while ventilation openings were placed above some of the windows/doors (Fig. 2.3).

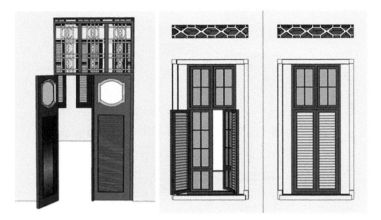

Fig. 2.3 Louver door and ventilation openings above windows/doors

The field measurement was conducted from August to October 2015, the hottest period of the dry season in Bandung. The measurements were taken in two classrooms (Rooms 1 and 2) under occupied conditions. These rooms were located at the same position between the north-facing front façade and the rear corridor but on different floors (Room 1, ground floor/GF; Room 2, first floor/1F). Room 1 is slightly smaller than Room 2 in size (80 m^2 and 91 m^2, respectively). Both rooms have high ceilings (5.5 m and 5.7 m) with large windows on both sides of the room. Major thermal parameters, i.e., air temperature, relative humidity, air speed, and globe temperature, were measured at 1.1 m above floor level in the center of the rooms and corridors. The vertical distribution of air temperature was also measured at the same points. Occupancy and window/door opening conditions were recorded. Outdoor thermal conditions were recorded by a weather station located at the measurement site (12 m above ground level).

2.3.2 Indoor Thermal Environments

Figure 2.4 presents the temporal variations of measured thermal parameters in the rooms, corridors, and outdoors, while Fig. 2.5 shows a statistical summary of the air temperature measurements under different ventilation conditions during the whole measurement period (16 days). As can be seen, outdoor air temperature ranged from 19.6 to 31.0 °C with an average of 24.7 °C, while relative humidity ranged from 31 to 88% during this period. The daily global horizontal solar radiation was measured at 15.2–24.8 MJ/m^2. Average outdoor wind speed was approximately 1.44 m/s during the daytime and 0.79 m/s at night. In both rooms, daytime ventilation (opened windows from 6:00 to 18:00) was adopted during weekdays, while

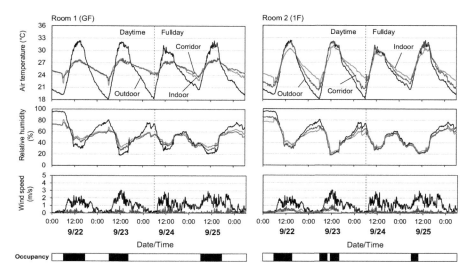

Fig. 2.4 Temporal variations of measured thermal parameters in the rooms, corridors, and outdoors

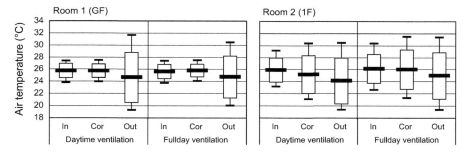

Fig. 2.5 Statistical summary of air temperature under different ventilation conditions

full-day ventilation was applied during weekends. Both of the rooms were occupied with about 30–35 students aged 16–17 during school hours (07:00–15:00).

Figures 2.4 and 2.5 show that the indoor air temperature profiles clearly differ between both floor levels (i.e., between Rooms 1 and 2) rather than because of different ventilation conditions. As can be seen, daytime indoor temperatures on the ground floor (Room 1) obtained were lower (3.3–5.0 °C lower than outdoors) than those from the first floor (Room 2) (1.7–3.2 °C lower), regardless of the ventilation conditions. Meanwhile, nocturnal indoor air temperature profiles show the opposite pattern: indoor air temperatures on the ground floor were rather higher than those on the first floor. These differences are not only due to the floor levels but also to the difference in corridor types (enclosed/semi-open corridor).

Consequently, indoor air temperatures in Room 1 adjacent to the enclosed corridor maintained narrow diurnal temperature ranges compared with outdoor

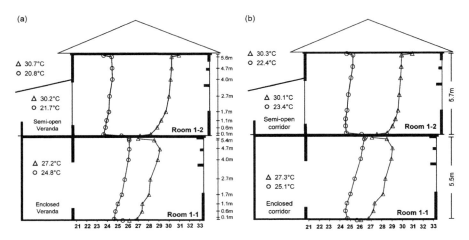

Fig. 2.6 Vertical distribution of air temperature under the different ventilation conditions (**a**) Daytime ventilation (**b**) Full-day ventilation

temperature, even when the windows/doors were open all the day (i.e., full-day ventilation). As indicated in Fig. 2.4, the measured wind speeds in the enclosed corridor and Room 1 were very low (up to 0.28 m/s) compared to those on the first floor (up to 0.50 m/s) during the daytime. It is apparent that the enclosed adjacent corridor space discourages cross ventilation in Room 1. Meanwhile, air temperatures in the semi-open corridor situated adjacent to Room 2 closely followed outdoor air temperatures (Figs. 2.4 and 2.5).

Figure 2.6 illustrates the vertical distributions of air temperature in all cases under different ventilation conditions. Overall, the ceiling surface temperatures during the daytime in Room 2 reached higher values than in Room 1 simply because the former is situated below the roof. Both roof and ceiling are not insulated, so heat is transmitted to the room below. The vertical temperature gradients are evident even in Room 1. The temperature gradients seen in Room 1 are not attributed to the transmitted heat from the ceiling but to the high ceilings (5.5 m). It should be noted that indoor air temperatures at the occupied level (around 1.1 m above floor level) maintained relatively lower values even when windows and doors were open during daytime. This implies that a high ceiling contributes to maintaining relatively low daytime air temperatures even when daytime ventilation is adopted in hot-humid outdoor conditions.

2.3.3 Thermal Comfort Evaluation

Indoor thermal comfort was indexed using operative temperature (OT) and standard effective temperature (SET*). Moreover, thermal comfort was evaluated by the adaptive comfort equation (ACE) proposed by Toe and Kubota [4] for OT. Figure 2.7 presents the evaluation results of indoor thermal comfort in Rooms

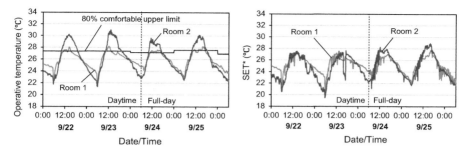

Fig. 2.7 Evaluation results of indoor thermal comfort by operative temperature and SET*

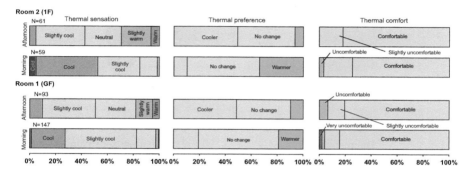

Fig. 2.8 Evaluation of indoor thermal comfort by questionnaire results

1 and 2 by the two indexes, respectively. As can be seen, over the most of the period, OT in Room 1 did not exceed the 80% upper comfortable limit regardless of ventilation conditions, while that of Room 2 exceeded this limit during most of the daytime period. Nevertheless, when evaporative cooling effects are taken into account by using SET*, the resulting SET* values in Room 2 become 1.0–2.8 °C lower than the corresponding operative temperatures. As mentioned before, Room 2 received slightly higher wind speeds during the daytime.

A brief questionnaire survey was conducted among the students in Rooms 1 and 2, respectively, during the measurement period to investigate the perceived thermal comfort using a 7-point scale of thermal sensation (from −3 for cold to 3 for hot), a 3-point scale of thermal preference (from −1 for cooler to 1 for warmer), and a 4-point scale of thermal comfort (from 1 for very uncomfortable to 4 for comfortable), etc. Questionnaire forms were distributed by the class teachers during morning class time (07:00–11:00) and afternoon class time (12:00–15:00) at the end of the respective periods. A total of 206 responses were obtained for the morning survey, while 154 responds were collected for the afternoon survey.

As shown in Fig. 2.8, there was no significant difference between the two rooms in all the answers, though there were significant differences between those for the morning period and the afternoon period in thermal sensation ($t = -6.00, p < 0.01$)

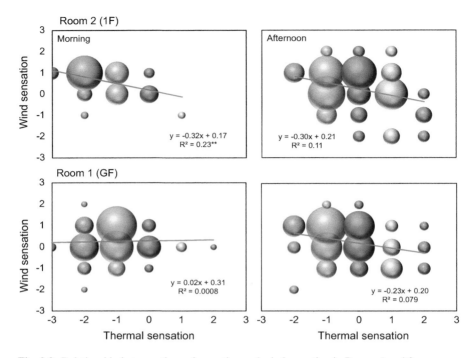

Fig. 2.9 Relationship between thermal sensation and wind sensation in Rooms 1 and 2

and thermal preference ($t = 4.61$, $p < 0.01$). This means that the relatively large differences observed between the two rooms in terms of air temperature and operative temperature (see Figs 2.4, 2.5, 2.6 and 2.7) did not cause significant differences in thermal sensation or preference. Figure 2.9 presents the relationships between the 5 scale points of wind sensation (from -2 for too still to 2 for too breezy) and the 7 scale points of thermal sensation in Rooms 1 and 2, respectively. As can be seen, during both periods, significant relationships can be seen only in Room 2 (first floor). As the wind sensation increased when stronger winds were experienced (2, too breezy; -2, too still), they tended to feel cooler (-3, cold; 3, hot). As discussed above, this result also proves that thermal comfort in Room 2 was improved by the increased wind speed primarily due to cross ventilation. The corridor space plays an important role in improving cross ventilation and thus indoor thermal comfort.

2.4 Summary

Firstly, this chapter gave a brief history of Dutch colonial buildings in Indonesia. Over the long period of colonialization, colonial buildings evolved various environmental techniques to cope with the hot-humid climate of Indonesia. Secondly, this chapter presented the results of a field measurement conducted in an existing colonial building in Bandung. The results showed that daytime indoor air temperatures in the building maintained relatively low values compared to the corresponding outdoor temperature mainly due to the thermal mass effect. Corridor spaces played an important role as thermal buffer space (if closed) as well as to encourage cross ventilation (in the case of semi-open corridor space). Increased wind speeds due to the cross ventilation improve thermal comfort in one of the rooms, as indicated by the SET* values and questionnaire results. If adjacent corridors can be designed to be closed during the daytime and to be semi-open during the nighttime, then the thermal benefits of the corridors can be obtained during both day and night. Furthermore, ceiling height also contributed in maintaining lower indoor air temperature at the occupied level (1.1 m above floor level).

Acknowledgments This research was supported by the JSPS KAKENHI (Grant No. JP 15KK0210) and the Indonesia Endowment Fund for Education (LPDP). The field measurements were conducted by the students of Hiroshima University in collaboration with the Institute of Technology Bandung. Particularly, we would like to express our sincere gratitude to Mr. Naoto Hirata and Mr. Takashi Hirose.

References

1. Voskuil RPGA et al (1996) Bandoeng, Beeld van een stad. Asia Maior
2. Kunto H (1986) Semerbak Bunga di Bandung Raya. PT. Granesia, Bandung
3. Nurfindarti E, Zulkaidi D (2015) Strategi pengelolaan cagar budaya Kota Bandung. Jurnal Perencanaan Wilayah dan Kota B 4(1):83–103
4. Toe DHC, Kubota T (2013) Development of an adaptive thermal comfort equation for naturally ventilated buildings in hot-humid climates using ASHRAE RP-884 database. Front Archit Res 2:278–291

Chapter 3
Malaysia: Malay House

Doris Hooi Chyee Toe

Abstract This chapter discusses the environmental value, in particular the indoor and outdoor living environment for thermal comfort, of a fine example of the Malaysian vernacular architecture – Malay house. Malaysia is located close to the equator with hot and humid conditions year-round at lowland areas. The Malay house is known as a well-ventilated detached building of elevated timber structure usually seen in the rural villages with many trees in its surroundings. The work presented here is based on field measurement in two traditional Malay houses in Pontian, Malaysia. The results showed that the indoor air temperatures in the front living halls were higher than the outdoor air temperatures by 1 °C during daytime under open window conditions and 2 °C at night under closed window conditions on average. The outdoor air temperature at the Malay house sites was lower than that of typical modern urban housing site by 1.7 °C on average. The passive cooling techniques emphasize shading from direct and diffuse solar radiation, maintaining a cool outdoor microclimate, and reducing the temperature of the outdoor air before entering the lightweight house for bodily cooling by cross ventilation.

Keywords Vernacular architecture · Hot-humid climate · Thermal comfort · Natural ventilation · Malay house · Malaysia

3.1 Introduction

Malaysia is a culturally diversified nation in the Southeast Asian region. In 2010 the population of Malaysia totaled 28.3 million and comprised three major ethnic groups – Malay (50%), Chinese (23%) and Indian (7%) [1]. Historically, the different ethnic groups made different living and housing arrangements that dotted the local landscape with various building forms and architecture. An illustration of these can be found in *The Encyclopedia of Malaysia* [2]. For example, traditional Malay houses

D. H. C. Toe (✉)
Faculty of Built Environment, Universiti Teknologi Malaysia, Skudai, Johor, Malaysia

© Springer Nature Singapore Pte Ltd. 2018
T. Kubota et al. (eds.), *Sustainable Houses and Living in the Hot-Humid Climates of Asia*, https://doi.org/10.1007/978-981-10-8465-2_3

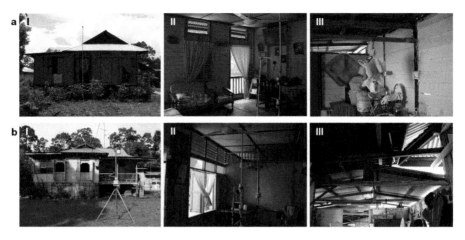

Fig. 3.1 Views of the case study Malay houses. (**a**) MH 1; (**b**) MH 2. I, front exterior view; II, front living hall interior view; III, roof construction joints (Adapted from Ref. [14])

are detached buildings arranged sparsely in rural villages (Fig. 3.1 photographs I), while traditional Chinese shophouses are narrow and deep-plan buildings situated in rows in relatively dense urban areas (see Chap. 37).

Malaysia covers a land area of about 330,803 square kilometres and consists of Peninsular Malaysia and East Malaysia which is part of Borneo Island. These two major land masses span close to the equator between latitudes 1–7°N and longitudes 100–119°E [3]. According to the Köppen-Geiger climate classification, Malaysia's climate is classified as tropical rainforest or Af climate type [4]. Since the early days, most settlements have developed at lowland coastal areas where economic activities are more vigorous. In general, the lowland areas are typified by the characteristics of Af climate. They are uniformly hot and humid year-round. There is a seasonal climatic change dominated by the monsoons which influence changes in the wind conditions and rainfalls with slight temperature variations [5, 6]. The seasons in Malaysia are recognized as the northeast monsoon (November–March), the southwest monsoon (May–September) and the two transitional or intermonsoon seasons (April and October). Historical climatic data (25–30-year period) of major cities in Malaysia show monthly mean air temperatures between 26 and 29 °C and small deviations of 1–3 °C among the months in a year [7].

The above-mentioned climatic and cultural contexts are likely the major influence on developing the traditional vernacular houses of the people. It is noteworthy that natural ventilation has been the most elementary practice by tradition for cooling in hot-humid climates.

3.2 Malay House: Overview of Its Cultural Past and Present

The Malay house can be defined as a house designed and built by the Malays in Malaysia according to their own needs and with a good understanding of nature and environment, incorporating and reflecting their way of life and culture [8]. Each house is typically composed of a main house (*rumah ibu*), kitchen (*rumah dapur*) and several transitional spaces and has a large compound without fence in the village (*kampung*) environment. Traditional Malay lifestyle is relatively informal, communal and outdoor oriented around the house compound and transitional spaces such as doorways, especially during daytime. The basic construction utilizes elevated timber post-and-beam structure. Its common building features include a large thatched roof, low walls made of lightweight materials (wood or bamboo), a raised floor, plentiful full-height operable windows with upper ventilation openings and abundant vegetation (ground cover and trees) in the surroundings. Sitting on the floor and working in the space under the raised floor are common traditionally. The spatial composition and lightweight construction provide flexibility to add spaces to the originally constructed house and thus support incremental housing. Many Malay houses are also traditionally oriented to face Mecca for religious reasons; hence, they have east-west orientation [8]. The sociocultural aspects of the Malay house are elaborated in detail elsewhere [2, 8].

Although most of the above features are still identifiable at present, the thatch for the roofing has been replaced with modern materials such as zinc for easy maintenance and other reasons. According to Markus and Morris [9], the thatched roof would give even greater cooling effect indoors than a shaded zinc roof. Part of the house is also constructed using brick and cement on ground in many cases today.

3.3 Environmental Value of Malay House: Indoor and Outdoor Living Environment for Thermal Comfort

One of the critical housing challenges faced by the world today is to house billions of people in culturally and environmentally sustainable ways. In hot developing regions such as Southeast Asia, cooling demand in residential buildings is a major concern since it is predicted to rise sharply in the coming decades, yet there is significant need for energy savings due to concern over energy security and effects of global warming [10].

Much has been written about the environmental value of the Malay house including its climatic design and indoor thermal qualities [2, 8]. As the available texts are mostly hypothetical and qualitative description, it would be important (and interesting) to analyse the traditional passive cooling techniques using scientific

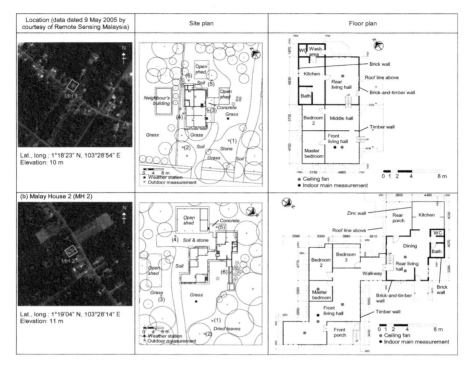

Fig. 3.2 Plans of the case study Malay houses. (**a**) MH 1; (**b**) MH 2 (Adapted from Ref. [14])

methods in order to derive principles for current use [11]. A recent study analysed the performance of natural ventilation in a Malay house during daytime (6 a.m.–6 p. m.) [12]. The study stated that its indoor air temperature was higher than the outdoors in early morning and afternoon periods. Night-time and outdoor conditions were less studied.

This section discusses the indoor and outdoor living environment of the Malay house for thermal comfort and extracts its passive cooling techniques. The work presented here is based on field measurement conducted from March to April 2011 consecutively in two selected traditional Malay houses in Pontian (Figs. 3.1 and 3.2). Pontian is a lowland coastal area located about 40 km to the west of the city of Johor Bahru in Peninsular Malaysia. Details of the measurement methods are given in [13, 14].

3.3.1 Building Features and Household Behaviours

The two Malay houses (MH 1 and MH 2) were considered typical traditional Malay houses and shared the typical Malaysian rural village setting with many trees in their surroundings (Fig. 3.2). There were few buildings and paved roads around the two sites. Thus, the average outdoor air temperature was lower than that of typical modern urban housing site by 1.7 °C on average [14].

Both houses had timber structures elevated more than 1 m above the ground for the front parts of the houses (front living hall and all bedrooms) (Fig. 3.1). The ground surface under the raised floor was soil. The rear parts of both houses were of brick-and-timber structures on the ground (rear living hall, dining, kitchen and bathroom). The roofing material was zinc without insulation. Both houses were installed with fibre cement ceiling boards, also without insulation, in the front living halls and master bedrooms. The floor-to-ceiling heights of the front living halls are 2.75 m in MH 1 and 2.95 m in MH 2. Openings in the two houses comprised full-height and half-height wooden panel windows, half-height glass louvred windows, wooden panel doors and upper ventilation openings (permanently open) above some of the windows, doors and walls.

The main measurement rooms in this study were the front living halls (Fig. 3.2). Several broadleaf trees of at least 6–7 m height were within view of each hall in addition to having low roof eaves of about 650–700 mm in depth. These trees were taller than the houses. Every external façade and internal partition had at least one operable window, door and/or fixed upper ventilation openings so that gross opening areas are 11–41% of respective wall areas. In fact, infiltration of the two houses could be increased by the porous construction joints, especially at the roof (Fig. 3.1 photographs III). An analysis of the room layout using the LT method [15] confirms that the entire floor areas are passive zone areas.

The household sizes were seven persons for MH 1 and five persons for MH 2. Both houses were occupied throughout the measurement period. Both households regularly opened windows during daytime when they were in and closed windows at night (Fig. 3.3). Ceiling fans were used.

3.3.2 Outdoor Weather Conditions

Maximum outdoor air temperatures are 31–33 °C, while outdoor relative humidity is always above 60% on the fair weather days during the measurement [14]. Daily global horizontal solar radiation ranges from about 4000 to 5900 W h/m^2. Calm conditions (outdoor wind speed ≤0.3 m/s) and very weak land breeze occur at night. During daytime westerly sea breeze (coming from the Straits of Malacca) at a mean wind speed of 0.91 m/s is seen in MH 1, while wind of varying directions and a mean speed of 0.51 m/s is seen in MH 2 probably due to the influence of the intermonsoon period [14].

3.3.3 Thermal Comfort in the Front Living Halls

In general, the indoor air temperatures in both front living halls follow the pattern of the outdoor air temperatures without time lag, as expected of the lightweight timber

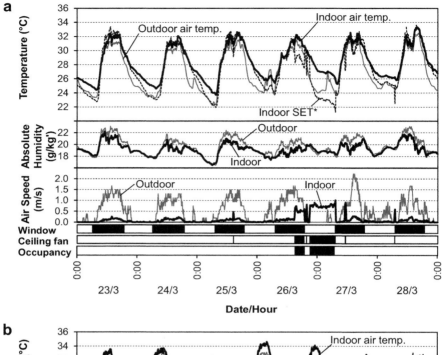

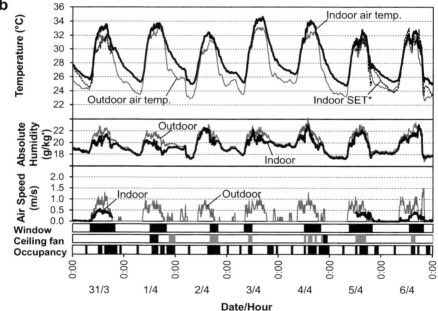

Fig. 3.3 Measured indoor and outdoor thermal environmental factors, indoor SET* and occupant's behaviour in the front living halls. (**a**) MH 1; (**b**) MH 2. Black bar indicates opened for 'window', used for 'ceiling fan' and occupied periods for 'occupancy'. Grey bar indicates usage of another ceiling fan further from the measurement point in MH 2 (Adapted from Ref. [13])

structures with low airtightness (Fig. 3.3). The indoor air temperatures are higher than the outdoor air temperatures by 1 °C during daytime under open window conditions and 2 °C at night under closed window conditions on average. Higher mean radiant temperature is seen in the front living hall of MH 2 and likely causes its slightly higher indoor air temperature compared to MH 1. Indoor air temperature elevations are higher at night as ventilation to remove or cool the heated indoor air would become slower under closed window and weak wind conditions.

The indoor absolute humidity values are lower than the outdoors by 1 g/kg' on average during daytime (Fig. 3.3). It is implied that indoor moisture production during daytime from the likely sources including occupants, moisture desorption of hygroscopic surfaces (e.g. timber and fabric) and ventilation is lower than the outdoor evapotranspiration rate. In this case, it is assumed that ventilation brings in humidity from the outdoor air.

Under the high indoor air temperature and indoor humidity ratio (>12 g/kg'), cross ventilation and/or ceiling fan were utilized by the occupants in both houses most likely to improve the indoor thermal comfort by evaporative heat loss. Cross ventilation is achieved by having openings that are perpendicular to the prevailing wind direction. It follows that the increased air speeds lower the corresponding SET* substantially for cooler thermal sensation (Fig. 3.3) [13].

3.3.4 Thermal Effects of Ceiling and Raised Floor

The vertical thermal distribution in the front living hall of MH 1 shows that the ceiling surface temperature is the highest at 34.4 °C, while the floor surface temperature is the lowest at 30.6 °C during daytime (3 p.m.) [14]. Based on heat flux measurement of the ceiling, daily maximum heat flow through the ceiling into the room averages 33.9 W/m². At night, the heat flows outwards from the room at an average of 5.7 W/m². The results signify large radiant heat gain from the roof during daytime that contributes to the increase in indoor temperature due to the high noon solar altitude in the equatorial region and low-rise building form.

The air temperature under the raised floor was investigated in a subsequent field measurement in the same houses from March 2012 to February 2013. The maximum air temperatures (95th percentile) under the raised floor are 0.6 °C and 0.8 °C lower than the outdoor air temperatures measured in open grass areas for MH 1 and MH 2, respectively [14]. The corresponding minima (5th percentile) are 0.4 °C higher than the open outdoors. The underfloor space has smaller exposure to the sun and sky for radiative heating and cooling; thus, the temperature is more stabilized. The underfloor air is a cooling source during daytime and likely maintains the relatively low floor surface temperature, but it does not act as a heat sink at night.

3.3.5   Variance in Indoor and Outdoor Air Temperatures in the Whole House

Both front living halls are among the rooms with the lowest air temperatures and were perceived by the households to be less hot than most other rooms. Rooms that were installed with ceiling, i.e. the front living halls (FL), master bedrooms (MB) and middle hall (M) of both houses, have lower maximum air temperatures compared to the other rooms without ceiling (Fig. 3.4). The presence of ceiling reduces solar heat gain through the roof and temperature in occupied rooms. The sun path analysis shows that both front living halls were shaded from direct solar radiation by low roof eaves from around noon until 4 p.m. during the measurement [13]. Further shading by tall broadleaf trees is also found to prevent direct solar radiation at lower solar altitude and could shield the building structures and indoor spaces from diffuse solar radiation.

For outdoor measurements, air temperatures under tall broadleaf trees (locations 2 of MH 1 and MH 2) have the lowest maxima (95th percentile) compared to all other locations including the open grass areas (WS) (Fig. 3.4). Relatively high maxima are seen in open paved areas (locations 4 and 5 of MH 2) and areas surrounded by low dense plants of about building height (locations 6 of MH 1 and MH 2), with differences of 2.5–3.9 °C compared to under tall broadleaf trees (Fig. 3.4). The different outdoor microclimate likely influences the indoor thermal

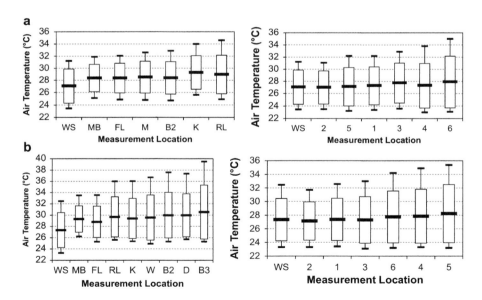

Fig. 3.4 Statistical summary (5th and 95th percentiles, mean and ±one standard deviation) of measured air temperatures. (**a**) MH 1; (**b**) MH 2. *WS* weather station, *MB* master bedroom, *FL* front living hall, *M* middle hall, *B2* bedroom 2, *K* kitchen, *RL* rear living hall, *W* walkway, *D* dining, *B3* bedroom 3 (Source from Ref. [14])

conditions of adjacent rooms. The master bedroom (MB) of MH 1 and kitchen (K) of MH 2 have relatively low indoor air temperatures despite their western locations. They receive shading and cooling effects from the immediate tall broadleaf trees.

3.3.6 Passive Cooling Techniques Derived from the Traditional Malay Houses

Figure 3.5 provides conceptual illustrations of passive cooling techniques used in the front living halls of the traditional Malay houses based on the above findings. The emphasis is on reducing the temperature of the outdoor air before entering the lightweight house for bodily cooling by cross ventilation. The techniques are:

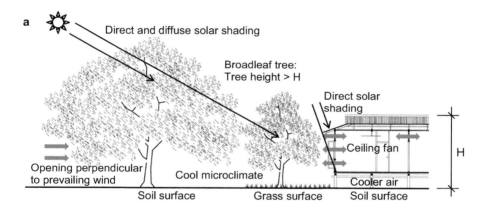

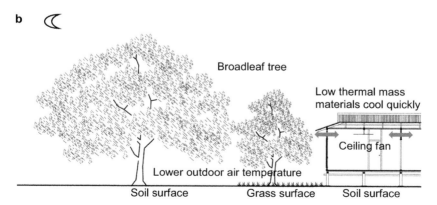

Fig. 3.5 Conceptual illustrations of passive cooling techniques used in the traditional Malay houses. (**a**) Daytime; (**b**) night (Source from Ref. [14])

3.3.6.1 Daytime

- Shading from direct solar radiation by continuous low roof eaves at approximately the window height level (650–700 mm in depth)
- Shading from direct and diffuse solar radiation by broadleaf trees that are taller than the building at strategic location
- A cool outdoor microclimate maintained by tall broadleaf trees, grass and unpaved soil
- Relatively cool air under the raised floor as a cooling source to the adjacent floor
- Cross ventilation through full-height openings that are perpendicular to the prevailing wind direction
- Use of ceiling fan to increase evaporative heat loss from the skin

3.3.6.2 Night

- Relatively low outdoor air temperature (compared to urban areas) maintained by tall broadleaf trees, grass and unpaved soil
- Use of low thermal mass materials that cool quickly as building structures
- Heat dissipation to outdoor air through fixed upper ventilation openings
- Use of ceiling fan to increase evaporative heat loss from the skin

3.4 Conclusions

The indoor and outdoor living environment afforded by the Malay houses reflects the adaptation of the traditional Malay lifestyle to the hot-humid climate. From the field measurement, the front living halls are relatively cool and comfortable compared to the other spaces in the same houses due to the implementation of the passive cooling techniques, although their indoor air temperatures are higher than the outdoor air temperatures. The relatively warm indoor environment that follows the outdoor pattern might be relatively tolerable for the residents who are closely connected with the outdoor environment. Cooling the outdoor air could be beneficial to also provide a more pleasant environment for the residents' outdoor activities in line with their livelihood and routine. The relatively cool floor surface and underfloor air would give comfort for sitting on the floor and working in the space under the raised floor. Ceiling and ceiling fans are more contemporary techniques as replacements for the thatch. Further techniques to reduce the daytime and night-time indoor air temperatures to the outdoor level at most would be roof or ceiling insulation and opening windows at night, respectively.

The future of the Malay house and similarly other types of vernacular architecture may move forwards in two ways. First is to accentuate preserving the environmental value of such buildings in situ. Second is to apply the appropriate techniques to modern local buildings as exemplified in [14]. Their environmental wisdom should not be left unused and lost.

Acknowledgements The field measurement was funded by the Asahi Glass Foundation and the G. ecbo/GELS Winter 2010 Internship Program of Hiroshima University. Special thanks are extended to the editors of this book, Professor Dr. Mohd Hamdan Ahmad of the Universiti Teknologi Malaysia, the Malay households and students who assisted in the fieldwork.

References

1. Department of Statistics Malaysia (2011) Population distribution and basic demographic characteristics. Population and Housing Census of Malaysia 2010. Department of Statistics Malaysia, Putrajaya
2. Chen VF (ed) (1998) The encyclopedia of Malaysia. Volume 5 architecture. Archipelago Press, Kuala Lumpur
3. Sani S (ed) (1998) The encyclopedia of Malaysia. Volume 1 the environment. Archipelago Press, Kuala Lumpur
4. Peel MC, Finlayson BL, McMahon TA (2007) Updated world map of the Köppen-Geiger climate classification. Hydrol Earth Syst Sci 11(5):1633–1644
5. Lim JT, Abu Samah A (2004) Weather and climate of Malaysia. University of Malaya Press, Kuala Lumpur
6. Sani S (1990/91) Urban climatology in Malaysia: an overview. Energ Build 15–16:105–117
7. World Meteorological Organization (2013) World weather information service, Malaysia. http://worldweather.wmo.int/020/m020.htm. Accessed 1 May 2013
8. Lim JY (1987) The Malay house: rediscovering Malaysia's indigenous shelter system. Institut Masyarakat, Malaysia
9. Markus TA, Morris EN (1980) Buildings, climate and energy. Pitman, London
10. Levine M, Ürge-Vorsatz D, Blok K, Geng L, Harvey D, Lang S, Levermore G, Mongameli Mehlwana A, Mirasgedis S, Novikova A, Rilling J, Yoshino H (2007) Residential and commercial buildings. In: Metz B, Davidson OR, Bosch PR, Dave R, Meyer LA (eds) Climate change 2007: mitigation. Contribution of working group III to the Fourth Assessment Report of the Intergovernmental Panel on climate change. Cambridge University Press, Cambridge/New York, pp 387–446
11. Asquith L, Vellinga M (eds) (2006) Vernacular architecture in the twenty-first century: theory, education and practice. Taylor & Francis, London/New York
12. Hassan AS, Ramli M (2010) Natural ventilation of indoor air temperature: a case study of the traditional Malay house in Penang. Am J Eng Appl Sci 3(3):521–528
13. Toe DHC, Kubota T (2013) Field measurement on thermal comfort in traditional Malay houses. AIJ J Technol Des 19(41):219–224
14. Toe DHC, Kubota T (2015) Comparative assessment of vernacular passive cooling techniques for improving indoor thermal comfort of modern terraced houses in hot-humid climate of Malaysia. Sol Energy 114:229–258
15. Baker N, Steemers K (2000) Energy and environment in architecture: a technical design guide. E & FN Spon, London/New York

Chapter 4
Malaysia: Longhouse of Sarawak

Soichi Hata and Mohammad Hussaini Wahab

Abstract Sarawak, a Malaysian state of Borneo, is covered mostly by dense tropical rainforest. The longhouse, or "Rumah Panjang" in Malay, is an architectural form in Borneo found throughout Sarawak. The traditional longhouse is home to most natives of Sarawak, including the Iban. A majority of the Iban population is concentrated around the Kapit district and along the Rajang River. The Iban are known as the Sea Dayak and are distinguished from the Bidayuh, a homologous ethnic group known as the Land Dayak. The Land Dayak also live in a longhouse similar to the Iban in an area of West Sarawak around Kuching province. Instead of constructing and settling in separate buildings, several small families live together in an elongated longhouse built along the river. Iban longhouses are found in largely "roadless" tracts of tropical rainforest, with little land access to their slash-and-burn farms or to other longhouses. Thus, every household owns a longboat, and the river is the main thoroughfare.

Keywords Longhouse · Iban · Dayak · Sarawak · Indoor thermal comfort · Sense of identity

4.1 Traditional Longhouse and Microcosmology as Expressions of Dayak

The longhouse can be conceived of as a village with inhabitants living in family apartments set neatly beside each other. Traditionally, the longhouse is built using natural resources from the surrounding areas. However, contemporary longhouse construction has lost such traditional sensitivity. Forestry development and

S. Hata (✉)
Shibaura Institute of Technology, Tokyo, Japan
e-mail: hata@sic.shibaura-it.ac.jp; sxata@jcom.home.ne.jp

M. H. Wahab
UTM Razak School of Engineering and Advanced Technology, UTM Kuala Lumpur, Kuala Lumpur, Malaysia

© Springer Nature Singapore Pte Ltd. 2018
T. Kubota et al. (eds.), *Sustainable Houses and Living in the Hot-Humid Climates of Asia*, https://doi.org/10.1007/978-981-10-8465-2_4

Fig. 4.1 Sarawak Iban longhouse in the middle of the rainforest

modernization have greatly influenced one's dwelling awareness and demand for homes. It is becoming increasingly common for households to live according to their own needs, while in the past it was common practice to take community needs as the priority instead. In addition, living itself within the longhouse has changed in light of such modernization [1]. For example, industrial materials have gradually replaced natural ones; this replacement changes the exterior of the longhouse conspicuously (Fig. 4.1). In addition, with many longhouse residents now working for timber companies, the location of their daily life has shifted to the company dormitory, further augmenting the absence of households in the longhouse. Without substantial presence or support from the principal generations, the community has been difficult to manage, and this difficulty has gradually caused the longhouse to lose much of its identity [1].

Traditional Dayak society is based on mutual benefit and cooperation among the households. It is built of social relationships and defined largely by equality among the households. The longhouse has created community and sustained community feeling by nurturing cooperativeness among the residents [2]. For example, men and women are treated equally, without discrimination, each with the same rights and obligations. For instance, all adoptions are affirmed and conducted without concern for bloodlines. Moreover, all children are treated equally, whether biological or adopted. Freeman's fieldwork differentiated this type of household group from the more general "family," or "Bilek family" [2]. This type of longhouse social structure is a unique cultural phenomenon that transcends everyday social practices.

Dayak community-style living is based on customary law known as the "Adat," a comprehensive set of beliefs and values that governs behavior. These beliefs are based on animism and emphasize relations between natural and supernatural worlds, where everything that occurs falls under the purview of this unique code of conduct [3]; it follows from this code that it is important to maintain balance and harmony in the longhouse. In other words, life and death, good and evil, and the holy and the impure are only a few of the opposed values that constitute the Adat. The Iban believe that all things and creatures have lives and souls, including the surrounding forest, rivers, and cultivated lands. Iban minds and bodies are fully committed to their beliefs, and from this perspective, faith and logic are consistent.

4.2   Longhouse Space Configuration and Construction Methods

The traditional Iban longhouse is composed of three main zones (Fig. 4.2): the household living apartments, or "Bilek"; the roofed common gallery, or "Ruai"; and an open platform called "Tanju." The Bilek are the private living quarters or apartments that separate the household units from the longhouse social group. That said, personal private space is not provided therein. The Ruai is a corridor joining all the Bilek and a means by which residents can interact with one another (Fig. 4.3). It is also an important place for entertaining visitors and for festival rituals (Fig. 4.4). The villagers also use this space for domestic activities, such as mending fishing nets and weaving mats and baskets, and as a playground for the children. The Tanju is located at the front of the compound on the river side of the longhouse; it is used both for drying harvested grain and as an open terrace for various outdoor tasks.

This kind of spatial composition and use of space, although gradually changing, still remains to this day [1]. On the other hand, since the 1970s, a special space (or "Dapor") has come to exist in the open backyard behind the Bilek. The word "Dapor" refers to kitchen in Malay. The main reason for this extension was to protect the kitchen [1]. By observing contemporary longhouse practices, we see that even a bedroom or a dining room could be called a Dapor, if it contains a fireplace. Currently, the addition of a Dapor is carried out slowly, methodically, and after the construction of the Bilek is completed. A Dapor must be built according to certain parameters set by the Bilek frontage, but this is not like the detailed prescriptions that determine the construction of the Ruai or Bilek [1].

Fig. 4.2 Longhouse space configuration

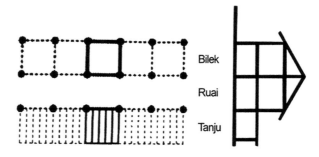

Fig. 4.3 Daily life in the Ruai

Fig. 4.4 Ruai during an Iban festival

In the case of a longhouse, construction begins on the headman's unit, and the construction method employed for it becomes the basis for configuring space throughout the entire longhouse (Fig. 4.5). What is more, the construction of the headman's unit involves all future residents and households of the longhouse. Indeed, it is through the experience of constructing the headman's unit that every member comes to share in the construction knowledge, rules, and measurements associated with every component to be built. By constructing the headman's unit, the heights of both pillars and wall plates are established, and the depths of Bileks and Ruais are determined. In addition, every household comes to share the same

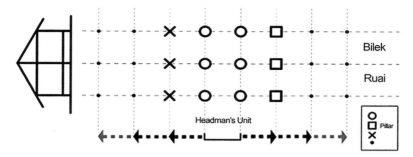

Fig. 4.5 Longhouse construction process

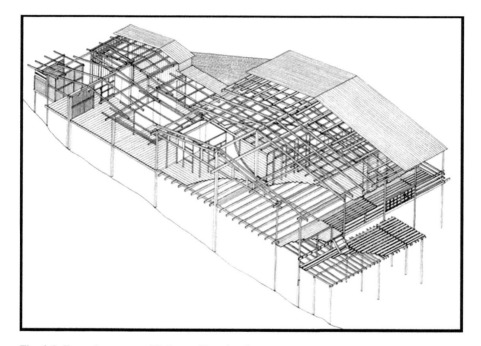

Fig. 4.6 Framed structure of Seliong village longhouse

enthusiasm for the longhouse, just as every household will come to know the work and the labor arrangements for which they are responsible [1].

With the headman's unit established as the origins, Bileks and Ruais of the same depths are built on the left and right sides of the unit. By repeating the process, the shape of longhouse is gradually formed. On the other hand, the frontage dimensions are more freely determined and can be built according to the different family configurations, labor force availabilities, and work competencies of the various households (Fig. 4.6) [1].

4.3   Longhouse Maintenance and Life Span

Constructing the Bilek, Ruai, and Tanju (including the Dapor) coupled with their maintenance and management is the responsibility of each household. In the case of the Ruai and Tanju, daily maintenance and management by the households is essential, as these are the common spaces of the longhouse. According to the Adat (Iban customary law), the most important components of Iban life are "planting and harvesting from the slash-and-burn farming, as well as building and living together in the longhouse." To live in a longhouse means to follow the Adat, to be situated cosmologically, and to satisfy one's spiritual needs. This act includes following longhouse construction and maintenance methods. In other words, it is worth emphasizing that beyond being an elongated house, a longhouse is built and maintained through strong cooperation and relationships among the households.

While a longhouse begins with the construction of a headmen's apartment unit, followed by the extensions of a new units on each end, this extension does not go on without limit. An increase in family members or a change in the family configuration of an existing household reflects a young household, with members who move out to extend a new apartment unit at the edge of the longhouse. However, the increase in long-term absences of households significantly hinders the maintenance and management of the common areas. Even if an existing household is conducting maintenance on behalf of an absent one, deterioration is inevitable. Moreover, within the same longhouse, disparities in the level of maintenance arise, depending on the number and configuration of household workers. As the longhouse ages, and maintenance becomes more sporadic and/or of lesser quality, residents then tend to leave the home, accelerating the longhouse's deterioration process.

Material for longhouse construction is chosen from among the ideal natural materials available. For timber frames, it is standard practice to choose parts that are strong and that do not rot easily. Timber is also used for the floors, a design that allows cool breezes to pass up through spaces between the planks. There are also straw mats and woven rugs made of tree bark laying on the floor. For the roof and outer walls, a combination of tree bark and sago palm leaves is used to repel water.

Ultimately, the actual life span of a longhouse varies. Essentially, it is difficult to think of living continuously in a longhouse until the life span of the timber has been reached. Living in the longhouse depends on the initiative of a headman, the longhouse's cultural significance, and its deep involvement with the Adat. Consequently, the life span of a longhouse is not only determined by the deterioration of its materials but by ominous events and misfortunes, especially those triggered by the death of a headman [1].

4.4 Longhouse Comfort Living

Most longhouses are at least 100 m in length. In Kuching province, a Bidayuh longhouse in Mujat village is 166 m long, whereas in Kapit district, the Iban longhouses of Puso village and Seliong village are 147 m and 101 m in length, respectively (Figs. 4.7, 4.8 and 4.9). Recall that the longhouse is joined by a common roofed gallery in proportion to its length, the Ruai. For the sake of this gallery, the floor must be made as flat as possible. It serves as an extended entryway for all Bilek and thus should cover the entire open space.

However, it is difficult to find a flat piece of land, especially in the upper river basin of the Rajang River, where most Iban longhouses are situated. Consequently, by adjusting the heights of the pillars that support the raised floor, a flat gallery is realized. In the case of Iban longhouse of Seliong village (located a few kilometers from Kapit Town), the gallery is up to 5 m above the ground (see Fig. 4.7). Again, the common gallery, or Ruai, is an essential place for producing and reproducing Iban communal life; as a space, it is also an important expression of a unity in the longhouse [1].

In a tropical rainforest thick with plants, the temperature is kept lower by a zephyr that passes through it. Moreover, the temperature experienced is usually lower than the true ambient air temperature. Air passing through the rainforest flows under the floor and over the shaded, now cooled ground. Thus, the floor plays a critical role as

Fig. 4.7 Seliong village view from Enchermin River

Fig. 4.8 Puso village and Seliong village longhouse. Scale 1/2000

Fig. 4.9 Section of Puso village and Seliong village longhouse

source for cooler air in the longhouse. The raised floor also serves to keep ground's moisture away from the longhouse. In addition, wind that passes beneath the floor pushes out any accumulated moisture, effectively preventing the growth of mold and bacteria [1].

In a rainforest, directly beneath the equator, the upper part of the longhouse is never shaded. As a result, the amount of heat received by the roof during the day is at a maximum. However, by using sago palm leaves and tree bark as roofing materials, the indoor air temperature is well regulated, as these materials do not absorb much solar radiation. Compare this to the corrugated iron roof often used as an alternative, with which the indoor thermal environment changes entirely. Corrugated iron roofs absorb heat, warming the attic air and heating the timber around it. Furthermore, it is difficult for the heat to escape, keeping the indoor temperature elevated and resulting in hot humid nights. In addition, while a corrugated iron roof is an industrial product with excellent durability, from the perspective of a longhouse's indoor climate, it is less beneficial than a thatched roof.

That said, a rod is used to lift up a portion of the Bilek's thatched roof, providing light to the area from the top. The Bilek is the most enclosed part of the longhouse, and beyond brightness, these openings are critical for ventilation (unlike the Ruai, lighting in the Bilek is dim even during the day, making a light-capturing device essential). In addition, the airflow below the floor increases the Bilek's air pressure and helps ventilate smoke coming from the hearth. Certainly, embodied in the construction of the longhouse is the wisdom of indigenous peoples, as they create comfortable living conditions for themselves while overcoming the difficulties of living in the rainforest.

4.5   Conclusions

Among the Iban of Sarawak, traditional relations are based on the mutual benefit and cooperation among households, creating a peaceful society and one without discrimination. For instance, in matters of adoption, every child is treated equivalently, where an adopted child is treated just as a biological one. The Adat, or customary law, plays an important role in creating this strong community and in the intimacy of its relations. Everything in Iban daily life is related to the Adat. Without the longhouse, the Iban will lose their identity. Although modernization is gradually changing the materials and structures of the longhouse, the Iban continue to live in the longhouse in an effort to maintain their identity.

References

1. Hussaini M (2007) Household composition and the longhouse cooperatively due to the formation process of Bilek, Research on the space configuration and living style of the longhouse of Sarawak. Doctor's thesis, Shibaura institute of Technology (In Japanese)
2. Freeman D (1992) The Iban of Borneo, London School of Economics; monographs on Social Anthropology No. 41, S. Abdul Majeed & Co
3. E. Hong Natives of SARAWAK 1987 Survival in Borneo's Vanishing Forests, Institut Masyarakat, Malaysia (Japanese Text by Housei University Press 1989)

Chapter 5
India: Bio-climatism in Vernacular Architecture

Madhavi Indraganti

Abstract Human beings have a natural instinct for bio-climatic home building. Vernacular architecture potentially leverages on this ability while responding to the sociocultural and economic needs of a population. India is a peninsula with a long coastline. A predominantly warm country, India has about 80% of the land under composite and warm-humid climates. This chapter navigates through the bio-climatic vernacular architecture of various climatic zones of India, emphasizing the warm-humid zones. It also provides examples that imbibed the bio-climatic spirit in creating a modern vernacular.

Keywords Indian architecture · Tropical climates · Hot-dry climate · Bio-climatic design · Warm-humid climate · Climate appropriate · Sustainability

5.1 Introduction

Climate-sensitive shelter design is inbuilt into the human knowledge. A building is often paraphrased as the dweller's third skin and clothing the second, covering the first (biological) skin. In the absence of active systems, these three together help maintain the deep body temperature at 37 °C. India's vast and diverse building traditions evolved over five millennia in response to the sociocultural, economic and thermal needs of the population. They display a remarkably sophisticated thermal adaptation [1].

Often times, it is hard to find a sharp distinction between the architecture of the vernacular and that of the refined civic buildings, such as palaces and public buildings in India. Influenced by materials availability, craftsmanship, and climate, the architectural vocabulary changes drastically from region to region. However, climate appropriateness appears to be the unifying thread in all of Indian vernacular

M. Indraganti (✉)
Department of Architecture and Urban Planning, College of Engineering, Qatar University,
Doha, Qatar
e-mail: madhavi@qu.edu.qa

© Springer Nature Singapore Pte Ltd. 2018
T. Kubota et al. (eds.), *Sustainable Houses and Living in the Hot-Humid Climates
of Asia*, https://doi.org/10.1007/978-981-10-8465-2_5

architecture (VA). With a concentration on the VA of hot-humid climates, this chapter discusses the bio-climatics in other regions as well.

5.1.1 Climate Types and Features

A predominantly warm country, India has about 80% of the land under warm-humid and composite climates. Hot-dry and cold climates form the rest. A few pockets have temperate climate [2]. The long coastal regions experience warm-humid climates with high humidity and varying levels of rainfall and temperature. The western part of Central India has Thar Desert and thus has hot-dry climate with high temperature and low humidity (Fig. 5.1).

Composite climate in Central India has a mixture of these two. It has a prominent summer season with moderate-to-low humidity. The northern, north-eastern parts close to the Himalayan ranges, and some valley/mountainous regions across the country experience cold climate, with cold winters and mild summers. Small regions

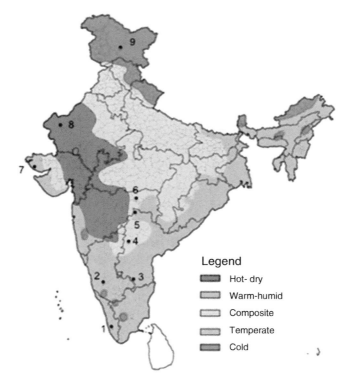

Fig. 5.1 Climatic zones of India [2]. (Legend for the numbered dots: 1. Kodianthara, 2. Chikmagalur, 3. Rishi Valley, 4. Marikal, 5. Kalashram (Tantoi), 6. Sevagram, 7. Suthari, 8. Jaisalmer, and 9. Leh

in the south also have temperate climate with mild temperature and humidity. Coastal zones of the Thar Desert experience maritime desert climate.

5.2 Warm-Humid Climate: Perforated Structure with Courtyards

5.2.1 Thermal Comfort Determinants

Warm-humid climates have little annual and diurnal temperature variation. The temperature oscillates around the skin temperature throughout. Coupled with high humidity and perpetual rain for 3–5 months, thermal comfort indoors can only be achieved by adequate ventilation (see climate charts in Figs. 5.2 and 5.3).

5.2.2 Architectural Features Modifying the Indoor Climate

A spread-out layout around a courtyard with a perforated envelope of generously sized windows offers the necessary cross-ventilation in warm-humid climate. The dwellings in Chikmagalur and Kodianthara exemplify this (Figs. 5.2 and 5.3). Deep

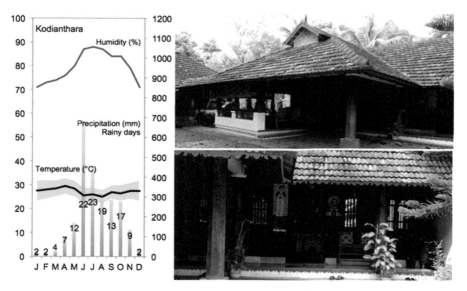

Fig. 5.2 (Clockwise from left) Climate graph, deep eaves and shaded verandahs offer protection against harsh sun and heavy rain. Perforated envelop cross-ventilates the indoors in a very humid environment at Kodianthara. A 160-year-old dwelling in Kerala (Picture credits: Nidhi Kotak Doshi)

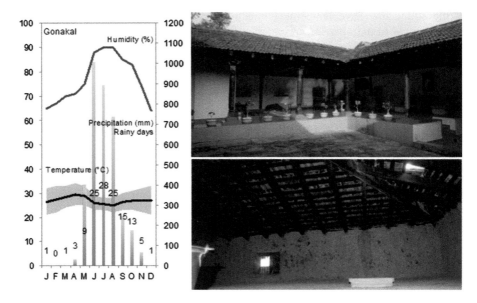

Fig. 5.3 (Clockwise from left) Climate graph, courtyard modifying the indoor environment in a warm-humid climate, ventilated loft insulating the interior space at Gonakal in Karnataka (Picture credits: Nidhi Kotak Doshi)

eaves of high-pitched roofs protect the fenestration from heavy rain and the walls from direct solar radiation. This feature forms a dominant form determinant in these climates. The *wadas* in Maharashtra on the West Coast [3] and the courtyard (*manduva*) houses on the East Coast also have similarly deep eaves, perforated envelopes, centered on courtyards.

Another common feature is the high plinth. It protects the interiors from inundation and facilitates improved ventilation. Walls are usually built with indigenous materials like mud or laterite or brick in lime or mud plaster. Light wooden roofs with clay tile roofing provide lightness with low capacitive insulation to the envelope. Oftentimes, there is no time lag between outdoor and indoor peak temperatures, if the envelope is thick and well ventilated [4]. High-level ventilators in the gable walls and roofs facilitate hot air exit, and ventilated attic spaces keep the indoor spaces insulated from overheating (Fig. 5.3).

5.2.3 Adaptive Spatial Use

Occupants chase comfort throughout the day. Thermal migration from space to space sounds a logical adaptation, considering the courtyards are 6–10 K cooler as in the traditional Nagapattinam houses featuring wind catcher (*melcaf*)-like roofs [5]. Compared to the outdoors, courtyards/semi-outdoor spaces happen to be 4–5 K cooler during the warmest period of the day, in homes in warm-humid climates of Kerala [4]. This quality encourages semi-indoor living and moving in between the open and covered spaces. As a result, everyday activities like dining and sitting tend to get enmeshed with the environmental qualities of spaces (Fig. 5.2). With air speeds as high as 0.5 m/s, the courtyard and the surrounding verandahs offer superior comfort conditions than indoors. On the contrary, during the rainy season, when humidity is near 100% outdoors, warm indoors (bedrooms with thick walls) provide comfortable conditions with about 30% lower humidity.

5.3 Warm-Humid Maritime Climate: Lightweight and Close-knit Shelter Forms

5.3.1 Thermal Comfort Determinants

High temperature, humidity, and solar radiation coupled with low precipitation and low diurnal ranges pose peculiar problems to form design in these climates (Climate graph, Fig. 5.4). While protection from heat and sun can be achieved by increasing the heat capacity of the envelope, high humidity necessitates increased ventilation.

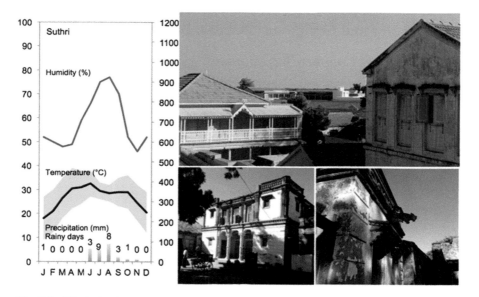

Fig. 5.4 (Clockwise from left) Climate graph, lightweight and close-knit shelter forms of Suthari, Kutch district, Gujarat. Detail of a sunshade and gargoyle on a yellow sandstone wall and "Shrinathji ni Mandir" (temple cum priest's residence) in lime-washed yellow sandstone and timber structure in Tera village in Kutch, Gujarat, in warm-humid maritime desert climate (Photo credit: Nidhi Kotak Doshi)

Providing spatial mobility within the envelope in the form of day- and night-use spaces is an ideal means to achieve thermal comfort. Smaller openings facing the inland offer better protection against the dust storms.

5.3.2 Architectural Features Modifying the Indoor Climate

Lightweight construction with large fenestration capturing the sea breeze for warmer-period usage forms an essential part of the architecture. This can be seen in the shelter forms of Kutch district of Gujarat (Fig. 5.4). Notice the large multiple sash windows with small shading and rain protection devices (*chajjas*) and balconies in the upper floors. Much similar windows with several independently operable sub-sections can also be seen in Sur, Oman [6].

On the other hand, envelopes with high capacitive insulation (high mass and specific heat) remain comfortably warm during the cold periods. These are usually provided in the lower floors. This can be seen in Tera village in Gujarat (Fig. 5.4 right). Here, high-thermal-capacity sandstone envelope with smaller openings stores heat and offers adequate comfort during the colder periods. Notably, the colder periods are usually associated with low humidity (Climate chart, Fig. 5.4).

5.4   Hot-Dry and Composite Climates: Dense Structure

5.4.1   Thermal Comfort Determinants

Hot-dry climates have high annual and diurnal temperature variation with very low humidity, rainfall, and cloud cover. Solar radiation is intense. Therefore, thermal comfort depends on protection from severe heat and hot breezes. On the other hand, composite climate usually has hot-dry summer and warm-humid monsoon seasons followed by cold-dry winter (see climate charts in Figs. 5.5, 5.6 and 5.7). Therefore, designing for hot-dry summers along with spaces for adaptive seasonal use ensures yearlong thermal comfort even in composite climates.

5.4.2   Urban Features Modifying the Macroclimate

Dense urban structure dominates these climates in contrast to the spread-out forms of warm-humid climates of coastal areas. It provides the much-needed protection from sun and hot breezes, as in Jaisalmer (26.913°N 70.915°E) in Thar Desert (Fig. 5.5) and Marikal (N16°36′ and E77°44′) in composite climate in the Telangana State [7].

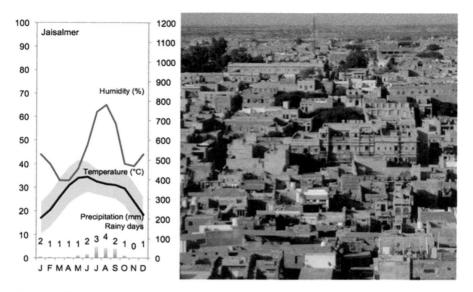

Fig. 5.5 Climate graph and dense urban structure of Jaisalmer in hot-dry climate (Picture credit: Reema Ghosh)

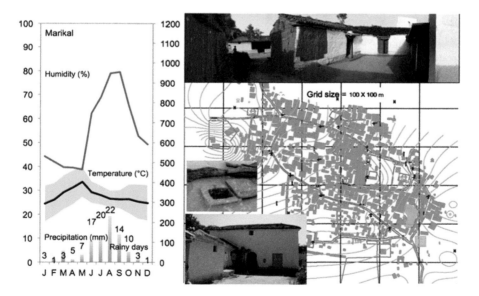

Fig. 5.6 (Clockwise left to right) Climate graph, a housing cluster, structure diagram of the settlement, a stone masonry house featuring very small openings, roof lights adaptively being closed during the midday in Marikal [7]. (Picture credit: Ravi and author)

Fig. 5.7 (Clockwise from left) Climate graph, semi-open built-form in response to the hot-dry and warm-humid seasons, detail of a clearstory window, application of mud flooring and cow-dung render at Sevagram, built by Mahatma Gandhi, in composite climate in Maharashtra

Interestingly, Marikals's 600-year-old organic built-form development and organization are akin to the traditional desert villages in Saudi Arabia and Morocco [8]. Its compact massing resulted from narrow, shaded streets and alleys, vital in the absence of cloud cover and intense solar radiation, as seen in Fig. 5.6.

Much similarly, streets and dwellings in both Jaisalmer and Marikal have strong discernible character, leaving the architectural unity of the whole undisturbed [9]. Spatial organization, vis-à-vis climatological function of all the houses big and small in Marikal, is almost the same [7]. Here, a small courtyard connects two or three small houses forming a cluster. Two or three of such clusters get organized around a slightly larger courtyard to form a unit. These are again interconnected molding it into a porous urban space.

5.4.3 Openings, Materials, Color, and Construction

Vernacular buildings in Jaisalmer and Marikal have very small openings. In Jaisalmer, the openings have intricate stone grills, and in Marikal windows are placed below the eaves. These orifices permit diffused dappled light and protect the indoors from direct sun and hot breezes, while allowing hot air exit. Small roof lights act as light shafts for inner courts. Occupants adaptively close them with stone slabs during the midday (Fig. 5.6). People often whitewash the walls. When funds are limited, they apply lime wash only on the exposed walls leaving the alley walls.

Marikal and Jaisalmer have good building stone (granite, slate, and sandstone), earth and brick in abundance to be used in walls and roofs, respectively. In Marikal, for example, locally crafted clay tiles laid to a gentle slope on wooden roofs with mud infill form the most common roofing system. Thick (450–600 mm) stone/brick/mud and reed walls and clay tile/stone roofs (200–300 mm thick) offer high capacitive insulation (time lag). It is advantageous when the diurnal range in temperature is high. Periodically residents finish smooth the roads and courts laid to the natural gradient in mud, with cow-dung paste. This material absorbs much less heat and permits usage barefoot.

5.4.4 Adaptive Spatial Use

Similar to warm-humid climates, the microclimate determines the use of a space in these climates too. For example, in Marikal, activities needing bright light, such as fishnet repair, grain cleaning, and cooking, are usually done in shaded open courts/verandahs or such threshold spaces. While choosing the optimal conditions, the inhabitants actively engage and mediate with the unbounded social spaces collectively [7].

5.4.5 Composite Climate: Semi-open Built-Form of Sevagram

Sevagram (20.736°N 78.663°E) is the hermitage of Mahatma Gandhi built in the composite climate of Maharashtra (see Fig. 5.1 and climate graph in Fig. 5.7). This envelope shown in Fig. 5.7 with low plinths and heavy mud brick walls and small orifices essentially suits the searing summer.

Clearstory openings all along the roof bearings ventilate the indoors during the humid monsoon seasons, allowing hot air exit. Barefoot users find stone paving unsuitable in these areas as it rapidly absorbs direct solar radiation with surface temperatures often crossing 50 °C. On the other hand, cow-dung render over mud plaster on the walls/floors provides cooler surfaces (Fig. 5.7). Shaded verandahs provide for alternative spaces for most of the daytime activities, like sitting, reading, napping, and cleaning the grain.

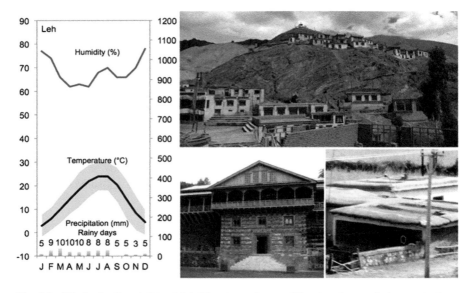

Fig. 5.8 (Clockwise from left to right) Climate graph, a traditional settlement facing south along the natural hill slopes of Leh, black-painted mud houses with flat thatch-reinforced mud roof in Udhampur, in Kashmir, and timber structure with stone infill masonry (*Dhajji Diwari* in *Kath-kuni* style [10]) in Naggar rest house in Kullu, Himachal Pradesh (Picture credits: Nidhi Kotak Doshi and Ritu Ahal)

5.5 Cold Climates: Compact Planning Along the Hill Slopes

Leh is (34.145°N 77.568°E) in cold desert climate (see Figs. 5.1 and 5.8). It has high diurnal and annual temperature ranges. Contrary to the warm-humid and hot-dry areas, the envelopes in these climates primarily need heat retention. Timber framed walls with stone masonry infills offer increased resistive insulation and the much-needed seismic resistance [10].

5.6 Experiments in the Modern Vernacular in Warm-Humid and Composite Climates

This section explains two modern examples in Andhra Pradesh that represent the vernacular in spirit and character, executed by common people, using indigenous materials. In these climates, thick stone masonry offers the required insulation for summer thermal comfort. High ceilings, gently pitched tiled roofs, and deep verandahs enhance indoor ventilation as seen in Rishi Valley (Fig. 5.9, top) in warm-humid climate. Composite climate poses some peculiar challenges, as the same building fabric needs to respond to the hot-dry summer and warm-humid monsoon

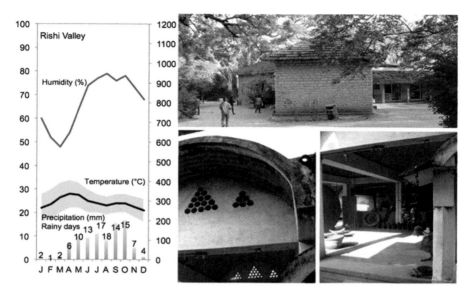

Fig. 5.9 (Clockwise from left to right) Climate graph of Rishi Valley; an elementary school at Rishi Valley, Andhra Pradesh (warm-humid climate); Kalashram, the crafts workshop in Tantoi, Adilabad, Andhra Pradesh (composite climate); and detail of the interlocking conical-shaped clay tile roof. Modern experiments imbibing the functional efficiency of the vernacular

seasons (Fig. 5.9). Responding to this, Kalashram, a crafts village in Tantoi, Adilabad district, uses interlocking potter's cones for load bearing and insulation in the roofs. Remarkably, the potter's cone inserts in gable walls give wide luminous diversity to the semi-enclosed spaces, necessary for craft activities while venting out hot air (Fig. 5.9).

5.7   Summary

In warm-humid climates, the significance of moving air in the interiors at body level cannot be overemphasized. This mandates a perforated envelope with adequate rain protection. The temperature difference between the indoors and outdoors is very little. Hence, increasing the thermal capacity of the fabric is ineffective. In such a scenario, airy semi-enclosed transition spaces offer far superior comfort conditions than the indoors.

In these climates, openings are best sized with respect to the microclimatic wind patterns. Raised plinths and wing walls to windows capture the local winds better. As thermal gradients are minimal, stack ventilation indoors may not be practical. However, ventilated roofs prevent mold growth, while contributing to convective cooling. Adequate surface drainage prevents mosquito breeding, a potential threat to

the operation of openings for ventilation. Observably, cross-ventilation is the only way to negotiate discomfort in hot-humid climes. And the occupant's engagement and migration between spaces is crucial [1].

Acknowledgments   Swarna and Murali Chanduri of Claremont, CA, USA provided the funds used for the Marikal studies through a personal grant organized via Telugu Association of Los Angeles (TELSA), USA. Nidhi Kotak Doshi, Ravi Kumar, Reema Ghosh, and Ritu Ahal from India have generously supplied some of the images. I thank all of them.

References

1. Heschong L (1979) Thermal delight in architecture. MIT Press, Cambridge, MA
2. BEE (2007) Energy Conservation Building Code, India
3. Gupta RR (2013) The Wada of Maharashtra, an Indian Courtyard House Form (PhD thesis). Cardiff University, UK/Ann Arbor, MI, Proquest LLC. http://orca.cf.ac.uk/55688/1/U584207.pdf
4. Dili AS, Naseer MA, Varghese TZ (2010) Passive environment control system of Kerala vernacular residential architecture for a comfortable indoor environment: A qualitative and quantitative analyses. Energ Build 42:917–927
5. Priya SR, Sundarraja MC, Radhakrishnan S, Vijayalakshmi L (2012) Solar passive techniques in the vernacular buildings of coastal regions in Nagapattinam, TamilNadu-India – a qualitative and quantitative analysis. Energ Build 49:50–61
6. Al-Hinai H, Batty WJ, Probert SD (1993) Vernacular Architecture of Oman: Features That Enhance Thermal Comfort Achieved Within Buildings. Appl Energ 44:233–258
7. Indraganti M (2010) Understanding the climate sensitive architecture of Marikal, a village in Telangana region in Andhra Pradesh, India. Build Environ 45(12):2709–2722
8. Saleh MAE (1999) Al-Alkhalaf: the evolution of the urban built-form of a traditional settlement in Southwestern Saudi Arabia. Build Environ 34:649–669
9. Rewal R, Veret JL, Sharma R (1985) The relevance of tradition in Indian Architecture. In: Architecture in India. Milan/Paris: Electa Moniteur, pp 12–23
10. Kumar A, Pushpalata (2013) Vernacular practices: as a basis for formulating building regulations for hilly areas. Int J Sustain Built Environ 2(12):183–192

Chapter 6
Nepal: Traditional Houses

Hom Bahadur Rijal

Abstract The traditional architecture could be one of the key issues for sustainable building design for different climates and cultures. They are well matched and adapted to the climates and cultures by using local building materials and techniques. However, traditional forms of architecture are decreasing dramatically, being replaced by artificial materials, modern designs and alien technology. We need strong policies and sound research to sustain the concepts and practicalities of traditional architecture. Thus, in this chapter, we will discuss the characteristic of the traditional vernacular houses in different climatic zones of Nepal.

Keywords Nepal · Sub-tropical climate · Temperature climate · Cold climate · traditional house · Semi-open space

6.1 Introduction

Traditional vernacular architecture is a wonderful gift from our ancestors. It has been developed over many centuries without creating many serious environmental or health problems [1]. The buildings are well matched and adapted to the climates and cultures by using local building materials and techniques. When we encounter these different kinds of traditional architecture, we feel very relaxed and comfortable in their ambient spiritual atmosphere. People travel to different parts of world to see and experience their beauty. We have to pass on this feeling to our future generations.

However, traditional forms of architecture are decreasing dramatically, being replaced by artificial materials, modern designs and alien technology. If we continue to create similar kinds of modern buildings, we will lose not only traditional wisdom and culture but also create severe environmental problems. Instead, we have to

H. B. Rijal (✉)
Department of Restoration Ecology and Built Environment, Tokyo City University
3-3-1 Ushikubo-nishi, Tsuzuki-ku, Yokohama, 224-8551 Japan
e-mail: rijal@tcu.ac.jp

© Springer Nature Singapore Pte Ltd. 2018
T. Kubota et al. (eds.), *Sustainable Houses and Living in the Hot-Humid Climates of Asia*, https://doi.org/10.1007/978-981-10-8465-2_6

redevelop the technology of traditional architecture, like our ancestors did, to suit their general and specific lifestyles. If there are problems in traditional architecture, there must be some optimal solution to fit into our modern lifestyle. We have to stop the substitution of traditional architecture by modern architecture.

We need strong policies and sound research to sustain the concepts and practicalities of traditional architecture. Of course it is important to preserve the best examples as world heritage, which have historical value. But the most important issues concern recognising its good aspects and improving its existing deficiencies for the living and working environments. We also have to identify the environmental, cultural and historical values of traditional architecture and teach these to both students and residents. Initially, we have to stop treating it as an old and valueless construct which is not suitable for modern society.

We are confident that traditional architecture could be one of the key issues for sustainable building design for different climates and cultures. It is often said that traditional buildings are cool in summer and warm in winter. Not much research has been conducted from the viewpoint of the thermal environment of traditional architecture in comparison with planning and anthropological aspects.

Although Nepal is a small country, the climate varies from sub-tropical to arctic due to the broad range in altitude (60–8848 m). Therefore different types of traditional houses, using a variety of ways to mitigate indoor thermal conditions, can be found. Some examples are (1) use of eaves or roofs cutting down sunshine, (2) designing effective ventilation arrangements, (3) creating semi-open spaces and (4) increasing thermal mass by using materials such as stone and earth. However, problems can also be found. For example, neither the windows nor doors have glass, and, as a result, excessive openness affects the winter room conditions. In addition, there is a gradual increase in the use of modern building materials, such as corrugated iron and cement, which can make the indoor environment worse.

A qualitative understanding can easily be formed from the information above. However, as we focused in part V of this book, a quantitative analysis is necessary to give the information needed to improve the indoor thermal conditions by using modification or renovation methods [1–6]. In this chapter, we will discuss the characteristic of the traditional vernacular houses in different climatic zones of Nepal.

6.2   The Areas and Climates

Investigation was carried out in six districts of Nepal (Fig. 6.1, Table 6.1). The climatic zone of Banke is sub-tropical, Bhaktapur, Dhading and Kaski are temperate, while Solukhumbu is cool climate and Mustang has a cold climate. They were chosen according to the altitude, climate, topography, ethnicity, energy and housing

Fig. 6.1 Location of the survey areas [7]

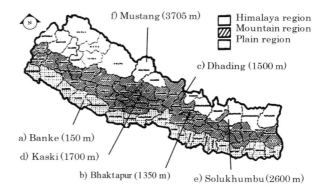

f) Mustang (3705 m)

Himalaya region
Mountain region
Plain region

c) Dhading (1500 m)

a) Banke (150 m)

d) Kaski (1700 m)

b) Bhaktapur (1350 m)   e) Solukhumbu (2600 m)

Table 6.1 Outline of the investigated districts

District	Address	Altitude [m]	Climate	T_{om} [°C] Max. (month)[a]	Min. (Jan.)[a]	Energy
(a) Banke	Mahadevpuri VDC-1 Mahadev Goun	≒150	Sub-tropical	31.4(6)	15.2	Firewood
(b) Bhaktapur	Bhaktapur Municipal-1 Tachapal Tole	≒1350	Temperate	24.0(7)	10.6	Electricity
(c) Dhading	Jyamrung VDC-1 Salle Goun	≒1500	Temperate	25.9(7)	13.3	Firewood
(d) Kaski	Dhampus VDC-5 Sude Goun	≒1700	Temperate	20.3(7)	8.9	Firewood
(e) Solukhumbu	Salleri VDC-1 Chiwang Khop Goun	≒2600	Cool	15.7(7)	4.0	Firewood
(f) Mustang	Lomangtang	≒3705	Cold	14.6(7)	−3.5	LD, shrub

T_{om}, monthly mean outdoor air temperature
[a]H.M.V. of Nepal (1995, 1997, 1999); meteorological station of each district: (a) Nepalgunj (reg. off.), (b) Kathmandu airport, (c) Dhunibesi, (d) Lumle, (e) Chialsa, (f) Lomangtang; *LD* livestock dung

conditions in Nepal. Because of the landlocked nature of the country, the climate is dry and hot in summer. In summer, the relative humidity is not too high; thus it feels pleasant in shaded areas. Nepal lies in low latitudes (26°–30° north), and therefore the solar radiation is high in winter. In Nepal, generally summer is in May and winter is in January (Fig. 6.2). Bhaktapur is an urban area and the other districts are rural. In rural areas agriculture is the main occupation and firewood or livestock dung is used for cooking and heating.

Fig. 6.2 Monthly average outdoor air temperature of each district [8]

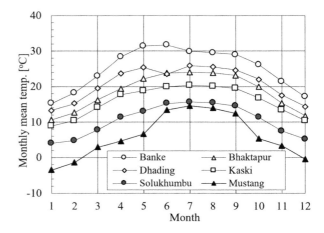

6.3 The Investigated Houses

The types of houses are one-storey mud houses in Banke; four- to five-storey brick houses in Bhaktapur; two- to three-storey stone houses in Dhading, Kaski and Solukhumbu; and two-storey sun-dried brick houses in Mustang (Figs. 6.3, 6.4, 6.5, 6.6, 6.7 and 6.8). The thickness of mud plaster, brick, stone and sun-dried brick walls are approximately 100 mm, 500 mm, 500 mm and 450 mm, respectively. The typical house spaces are as follows. (a) Banke has a kitchen, bedroom, prayer room and veranda (semi-open) on 1F (ground floor). (b) Bhaktapur has a shop, water space (an area with a water supply for washing, drawing water, laundry or shower) and toilet in 1F, bed and store room in 2F, bed and living on 3F, kitchen in 4F and store and terrace in 5F. (c) Dhading has a veranda, kitchen and bed in 1F; storage, bedroom and balcony on 2F; and store on 3F. (d) Kaski has a veranda, kitchen and bed on 1F and storage on 2F. (e) Solukhumbu has a cattle shed on 1F; kitchen, bedroom and prayer room on 2F; and store and bedroom on 3F. (f) Mustang has a cattle shed on 1F and kitchen, bedroom and prayer room on 2F. The house is constructed of natural material which is available in each area. But recently, because of modernization, the thatched roof is being replaced by a corrugated iron roof [10, 11].

In the temperate climate (Dhading and Kaski) and cold climate (some houses in Mustang), the kitchen room is used not only for cooking but also for living, dining, sleeping, etc. But in other climates, the kitchen is not used for sleeping. In the cool and cold climates, because of low temperatures in winter, people spend most of their time in the kitchen burning firewood or livestock dung.

In the Banke, Dhading and Kaski districts, semi-open spaces are also used for living, working and sleeping. Semi-open spaces are continuous with the indoor space and are connected by the external wall. It is a kind of veranda, beneath the eaves. Semi-open spaces are seen in the Banke district, but they also have open huts

 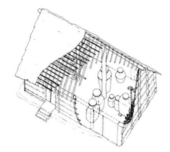

Fig. 6.3 Traditional houses in Banke

 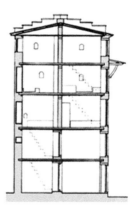

Fig. 6.4 Traditional houses in Bhaktapur (Photo by Kohei Kobata) [2, 9]

Fig. 6.5 Traditional houses in Dhading [2, 9]

Fig. 6.6 Traditional houses in Kaski [2, 9]

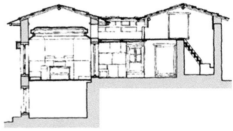

Fig. 6.7 Traditional houses in Solukhumbu (Photo by Kohei Kobata) [2, 9]

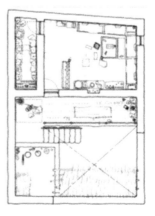

Fig. 6.8 Traditional houses in Mustang

with thatched roofs (supported by pillars without walls) and similar functionally and spatially to the semi-open spaces.

In the sub-tropical region, traditional houses are designed to exploit building elements such as earthen floors, eaves, huge mud vessels and brick walls to produce high thermal mass. Residents sleep in the semi-open spaces or in the front yard to stay cool in summer.

In the temperate climate (Dhading and Kaski), generally residents sleep in indoor space in winter and in semi-open space in summer for thermal adjustments. Thick stone walls in these areas are also effective for temperature control. Because of the low solar angle in winter, the semi-open spaces received solar radiation, and people use semi-open spaces to keep warm in the sunshine. In Bhaktapur, courtyard and roof terrace are used for the sunbathing in winter. The courtyard becomes very good shading for summer.

In cool and cold climates, the cattle are kept in the indoor space. The wall is very thick and openings are airtight. To reduce the ventilation rate in the cold climate (Mustang), the houses have few windows – these are also small and the houses are connected to each other [12]. Small holes are found in the roof for ventilation and lighting. It is interesting to note that the mud flat roof is found in Mustang due to the rain shadow of the Himalayas. Thus, the traditional buildings are well matched and adapted to the climates and cultures.

6.4 Conclusions

In this chapter, we have discussed the characteristic of the traditional vernacular houses in different climatic zones of Nepal. The houses are constructed from natural material available in each area. The traditional houses are well adapted in each climate, season and culture. The traditional architecture could be one of the key issues for sustainable building design for different climates and cultures.

Acknowledgements We would like to give thanks to Prof. Harunori Yoshida and Prof. Noriko Umemiya for their research guidance and valuable advice, to the 'Hata laboratory' for their drawings, to the investigated households for their cooperation and to our families and friends for their support. This research is supported by the Japan Society for the Promotion of Science.

References

1. Rijal HB (2012) Thermal improvements of the traditional houses in Nepal for the sustainable building design. J Human-Environ Syst 15(1):1–11
2. Rijal HB, Yoshida H (2002) Comparison of summer and winter thermal environment in traditional vernacular houses in several areas of Nepal. Adv Build Technol 2:1359–1366

3. Rijal HB, Yoshida H, Umemiya N (2002) Summer thermal environment in traditional vernacular houses in several areas of Nepal. J Archit Plann Environ Eng AIJ 557:41–48 (In Japanese with English abstract).
4. Rijal HB, Yoshida H, Umemiya N (2005) Passive cooling effects of traditional vernacular houses in the sub-tropical region of Nepal. Proceedings of the 22nd conference on passive and low energy architecture (Beirut), pp 173–178
5. Rijal HB, Yoshida H (2005) Winter thermal improvement of a traditional house in Nepal. Proceedings of the 9th International IBPSA Conference (Montréal), vol 3, pp 1035–1042
6. Rijal HB, Yoshida H (2005) Improvement of winter thermal environment in a traditional vernacular houses in a mountain area of Nepal; Investigation by simulation. J Environ Eng AIJ 594:15–22 (In Japanese with English abstract)
7. Panday RK (1995) Development disorders in the Himalayan heights: challenges and strategies for environment and development altitude geography. Ratna Pustak Bhandar, Baghbazar, Kathmandu Nepal, pp 53, 290
8. H.M.G. of Nepal (1995, 1997, 1999) Climatological records of Nepal 1987–1990, 1991–1994, 1995–1996, Ministry of Science and Technology Department of Hydrology and Meteorology Kathmandu, Nepal
9. Rijal HB, Yoshida H, Umemiya N (2010) Seasonal and regional differences in neutral temperatures in Nepalese traditional vernacular houses. Build Environ 45(12):2743–2753
10. Hanaoka S, Rijal HB, Hata S (2009) Changes and improvements to traditional vernacular houses in a mountain area of Nepal. Proceedings of the 3rd International conference on environmentally sustainable development, vol 1, pp 10–20, 16–18 August, Pakistan
11. Tiwari SR, Hanaoka S, Rijal HB, Hata S, Yoshida H (2004) Culture in development conservation of vernacular architecture. Annu J Archit (Association of Students of Architecture, Nepal) Vaastu 6:21–25
12. Rijal HB, Yoshida H (2006) Winter thermal comfort of residents in the Himalaya region of Nepal. Proceedings of International conference on comfort and energy use in buildings – getting them right (Windsor). Organised by the Network for Comfort and Energy Use in Buildings, 15 p

Chapter 7
China: Houses for Ethnic Minorities in Yunnan Province

Toshiyuki Watanabe and Daisuke Sumiyoshi

Abstract From the summer of 1990 to the winter of 1992, as the Japan Private University Promotion Foundation special research "Architectural Planning Study on Traditional Chinese Houses," at that time, the Japan-China Joint Study Group, represented by Professor Yoshimi Urano at Kyushu Sangyo University, conducted residential investigation of ethnic minorities in Yunnan Province. The study team was divided into two teams. First, the planning team gathered local hearings and housing drawings, and based on the results, the environmental group selected target residences and measured the indoor environment centered on the thermal environment.

Keywords China · Yunnan Province · Ethnic minorities · House survey · Indoor environment

7.1 Ethnic Minorities in Yunnan Province

The current total population of China is 1.37 billion, 92% of which are the Han people. The remaining 8% are regarded as consisting of 55 minorities. It is difficult to define "ethnic group." Generally speaking, sharing the same community, history, and culture is considered to be the conditions of an ethnic group. However, according to the history of mankind, the independent sense of belonging to a certain ethnic group has been more prioritized more often [1]. The case of the Han people in China is also a typical example. In the following, ethnic groups are regarded as separated from nations and races.

Fig. 7.1 shows the distributions of ethnic minorities of Yunnan Province [2]. Kunming City, the capital of Yunnan Province, spreads out near Dian Lake (1886 m above sea level) in the eastern part of Yunnan Province. In the humid

T. Watanabe
Kyushu University, Fukuoka, Japan

D. Sumiyoshi (✉)
Graduate School of Human-Environment Studies, Kyushu University, Fukuoka, Japan
e-mail: sumiyoshi@arch.kyushu-u.ac.jp

© Springer Nature Singapore Pte Ltd. 2018
T. Kubota et al. (eds.), *Sustainable Houses and Living in the Hot-Humid Climates of Asia*, https://doi.org/10.1007/978-981-10-8465-2_7

67

subtropical climate zone, it has spring-like weather throughout the year. The average temperature in January is 2.3–15.4 °C and in July is 16.9–24.0 °C. The annual mean rainfall is 1011 mm. It is also a monsoon region, having a rainy season from May to October and a dry season from November to April of the next year [3].

However, in the whole Yunnan Province, a number of mighty rivers running from mountainous areas through plateau areas create complex uneven terrains. Their climates vary from subarctic highlands to subtropical flatlands, and the fauna and flora vary accordingly. About half of the 55 ethnic groups of China live together in Yunnan Province. Each ethnic group has built their houses and communities in accordance with their history, culture, climate, and natural features in the area and has lived a self-sufficient life by gathering, growing crops, fishing, and hunting. It can be said that this very diversity is the main feature of the ethnic minorities of Yunnan Province.

7.2 Survey on Houses of Ethnic Minorities

The locations of the houses from the survey are shown in Fig. 7.1. One house from Xishuangbanna Dai Autonomous Prefecture (A), one house from Dali Bai Autonomous Prefecture (B), one house from Chuxiong Yi Autonomous Prefecture (C), and three houses from Dehong Dai Autonomous Prefecture (D–F), six in total. The

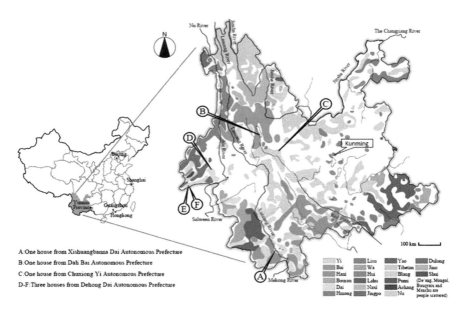

Fig. 7.1 Ethnic minorities in Yunnan Province [2] and the location of Kunming & measurement houses (**A~F**)

breakdown is three raised-floor type houses, two courtyard-type houses, and one Mu Leng house.

The houses of the ethnic minorities differ in materials, structure, and room layout, depending upon the ethnic group and its region. There is a world apart for the ethnic minorities, including the Dai people, Xishuangbanna Dai Autonomous Prefecture, along the border between Myanmar and Laos, an isolated region in the south of Yunnan Province. Xishuangbanna means rice terraces, a broad paddy field. The Lancang River runs through the center of the prefecture from north to south, and it becomes the Mekong River at the border of Myanmar. There are many raised-floor type houses with gabled, hipped roofs made of wood, bamboo, and sun-dried brick in Xishuangbanna, and among them, there are some with a floor height of about 2 m. They are more towered than raised-floor type houses, a pilotis structure so to speak. People live in the upper floor, and they keep livestock, including water buffalos, black pigs, hens, dogs, etc., on the first floor. It has been said that the purposes of raising the floor are for dehumidification, heat emission, and protection from insects and water. Certainly, because in addition to its subtropical climate, there is a fireplace indoor, and therefore heat and smoke emissions in the vertical direction would be necessary. They were nearly always barefoot and did not use footwear in 1990–1992. The earth is shared by people. The hierarchy of space can be switched by raising floor height and going up and down stairs.

Sleeping and eating areas are separated. However, the space resembling the kitchen is not clear. Occupants cook in the periphery of fireplace or in the veranda. Of course, there is no bathroom or toilet because there is no city water, which is essential for them. There is a well in the center of the village from which water is carried by human labor. Electricity is wired to each house, but there is only one bare light bulb. The females of the Dai people wear waistcloth skirts, fix their hair, and bathe. Theravada Buddhism is common there, and boy monks are often seen in the temple of the village.

7.3   Houses of Xishuangbanna Dai Autonomous Prefecture

Jinghong City, the prefectural capital, is located in the plains about 380 km toward the south-southwest of Kunming City, 550 m above sea level. The average temperature in January, the coldest month of the year, is 11.6–25.5 °C and that in June, the hottest month of the year, is 22.7–31.5 °C [3]. The average annual rainfall is 1114 mm. The wind is weak in the rainy season from May to October, and it becomes high-temperature and humid. The sunshine is strong in the dry season from November to April of the next year; however, sometimes it becomes frosty in regions of 1000 m or more above sea level. In Jinghong City, ethnic minorities including the Dai people, Hani people, Lahu people, Jino people, etc. form their own communities in flatlands and between mountains and live self-sufficient lives by growing crop and hunting.

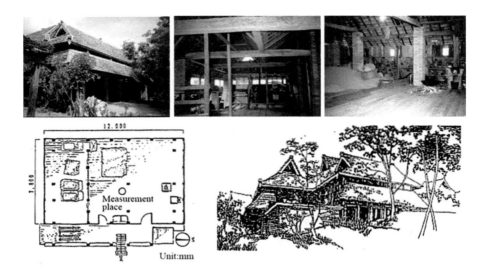

Fig. 7.2 The outside appearance, interiors, plan and sketch of Dai people house **A**

Dai people are distributed widely in Indochina and are said to be communicators of rice harvest culture. It is presumed that the rice harvest culture of Japan was also brought to northern Kyushu from the Jiangnan district of the Yangtze River Basin [4]. The total population of Dai people is approximately 65 million; 1.26 million live in China, 1.21 million in Yunnan Province, and 0.35 million in Xishuangbanna.

As the scale of the community of Dai people in Manjing Dai village, Jinghong City (565 m above sea level), there are 124 households, and its population is 630. They speak Southwestern Tai. Electricity has been supplied since 1960, and its city water system was constructed in 1985. The houses surveyed were raised-floor types of multiple-story gambrel-tiled roof structures, and all pillars are covered by bricks. The outside appearance sketch of Fig. 7.2 was drawn by Professor Rao Weichun of Yunnan Institute of Technology in 1990.

7.4 Houses in Dali Bai Autonomous Prefecture

Dali City is the capital of Dali Bai Autonomous Prefecture and is located in a basin approximately 300 km away from Kunming City toward the west-northwest, 1980 m above sea level. In the west side of this district, there is the Cangshan mountain range, famous for the production of marble. At the foot of it, there are the three pagodas of Chongsheng Temple, and in the east of it, Erhai Lake expands. It is an ancient capital where there were walled cities of Nanzhao Kingdom and Dali Kingdom. Bai people have a long history of communication with Han people, and because of this, many Han words are incorporated in the Bai language of the Sino-

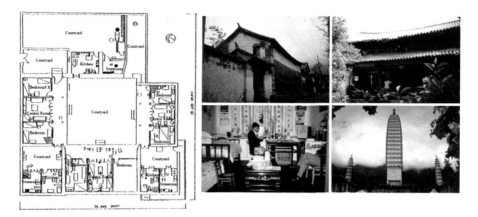

Fig. 7.3 The outside appearances, interior & plan of Bei people house **B** and three pagodas of Chongsheng Temple in Dali

Tibetan language system. The total population of Bai people is 1.93 million, 78% of them (1.5 million) live in Yunnan Province, more than 70% of them (1.1 million) live in Dali Bai Autonomous Prefecture with high population, and the rest of them live in Kunming, Nanhua, Lijiang, and Baoshan.

Many houses of the Dali Bai people are courtyard house types with courtyards surrounded by houses and screen walls (Fig. 7.3). There are those with three houses and one screen wall and those with four-sided courtyard houses with five sky-wells. In the case of three houses and one screen wall, a courtyard is surrounded by three houses and one screen wall, and the house across the courtyard from the screen wall is the main house. Each of the three houses consists of three rooms, and the central room is the main house, and those in both sides are bedrooms. Meanwhile, in the case of four-sided courtyard house with five sky-wells, four houses surround the courtyard. Sky-well means empty space through which the sky can be seen. The four leaking corners between houses become sky-wells, so that when added with the sky-well of courtyard, this structure is called a four-sided courtyard house with five sky-wells [5, 6]. The main house of courtyard house is positioned on the west side and opens eastward, facing the courtyard.

The house of Bai people in Xizhou town, Dali City, surveyed was 100 years old and had three houses and one screen wall. Hui people were living there when the survey was conducted.

7.5   Houses of Chuxiong Yi Autonomous Prefecture

Yi people migrated from the southeast of Tibet to Sichuan, Yunnan, and Guizhou Provinces. The total population is approximately 11 million. Approximately 5 million, which is the largest group, live in Yunnan Province (the largest group of minorities in the province). There is Middle Yunnan plateau in Chuxiong Yi Autonomous Prefecture, and more than 90% of the prefecture is the mountainous area. Chuxiong City, the prefectural capital, is situated almost in the middle between Kunming City and Dali City, 130 km toward the west from Kunming City. The style called Mu Leng house originated from the Lijiang Naxi people remains in the community of the Nanhua Yi people (Fig. 7.4). Mu Leng house is of a wooden masonry structure, a wall of which is taken out and three other walls are plastered with clay. Because the roof is supported by beams laid from wall to wall, the gable and tiled roof curves a little and lifts up [5, 6]. The plain composition of the courtyard house in Dali City shows the influence of dry-well Mu Leng house in Lijiang. The early Mu Leng house was a simple courtyard house. However, influenced by foreign culture later, it was transformed to the orderly form of courtyard style with three houses and one screen wall and a four-sided house with five sky-wells. Small vestiges of the courtyard house are observed in the Mu Leng houses of the Nanhua Yi people, such as the fact that the houses look toward the east, and in parts of the arrangement of outer walls.

Wujie town, Nanhua County, is situated between mountains 2280 m above sea level about 55 km toward the west-northwest from Chuxiong City, and all communities lie in steep areas, deep in the mountains in Middle Yunnan Plateau. The house of the Yi people surveyed in Wujie town was built 30 years ago. The west side of the premises abuts on a cliff about 3 m high, and the house is built facing east on flatland at the bottom of the cliff. The mountains are precipitous, and the farmlands among trees connect with each other forming rice terraces, extending nearly to the top of mountain, forming a net pattern.

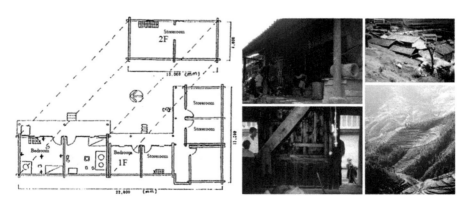

Fig. 7.4 The outside appearances, interior & plan of Yi people house **C** and the steep terraced fields spread in neighborhood

7.6   Houses of Dehong Dai and Jingpo Autonomous Prefecture

In Yunnan Province, 1.21 million Dai people live, and they are divided into seven branches. Among them, Han-Dai people, living in the periphery of Mangshi City in Dehong Dai and Jingpo Autonomous Prefecture in great number (Fig. 7.5), and Shui-Dai people, living in the periphery of Ruili City in Dehong Dai and Jingpo Autonomous Prefecture in great number (Fig. 7.6), were surveyed.

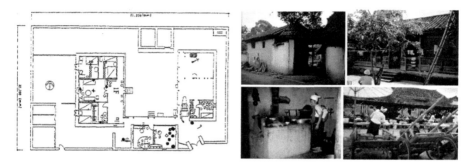

Fig. 7.5. The outside appearances, interior & plan of Han-Dai people house **D** and the market in Mangshi City

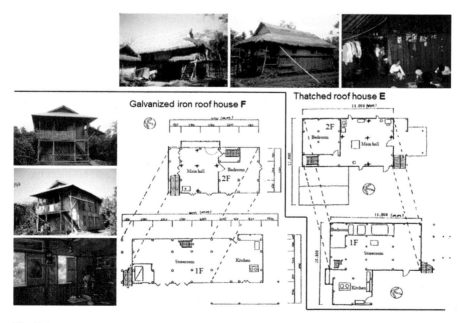

Fig. 7.6 The outside appearances, interiors & plans of Shui-Dai people houses **E**, **F** in Ruili City

The houses of Han-Dai people are of courtyard style such as the style of three or four houses surrounding a courtyard. In the case of four houses, four houses surround a courtyard, and the main house is on the west side of the courtyard. In the case of three houses, there is no house in the opposite side of the main house. The type with three houses and one wall in Dali City is considered to be the same as the type with four houses, the east house of which was converted to a wall. As for the configuration of houses, in the case of three houses, there are the main house, a kitchen on the south side of it, and a livestock barn or storeroom on the north side; in the case of four houses, there are the main house, a kitchen on the south side of it, a livestock barn or storeroom on the east side, and a grain store on the north side, and the livestock barn or storeroom is on the opposite side of the main house. The wall is mainly made of clay and bricks, and the roof is of tile and gable or straw-thatched. The doorway is at the southeast side of the courtyard (in the case of four houses, it is between the kitchen and livestock barn) [5, 6].

The houses of the Shui-Dai people are of raised-floor type, the same as Xishuangbanna. There are a storeroom, a bedroom, and a kitchen on the first floor, and there are a main hall and bedrooms on the upper floor. This means that although the living space is basically on the upper floor, part of the first floor is being converted to living space. The structure is framed by timber and bamboo and the wall is made by knitting bamboo strips. The roof used to be straw-thatched, but recently zinc roofs are increasing. Exterior stairs are installed parallel to walls. There is a side corridor or veranda at the top of the stairs [5, 6]. The raised-floor type of the Shui-Dai people is different from that of Xishuangbanna's Dai people in the following points: (1) the first floor of their raised-floor house is not a pilotis structure of open air space, but it is surrounded by walls of bamboo net, (2) part of the first floor is being converted to living space, and (3) there are exterior stairs and interior stairs.

7.7 Summary of Survey on the Houses of Ethnic Minorities

The ethnic minorities in Yunnan Province enjoy nature as it is. There are still many ethnic minorities who live almost self-sufficient lives. It is not appropriate to talk about their indoor environments uniformly based on modern environmental standards. Generally speaking, it is important that residents, designers, and builders have common values. However, in the case of Yunnan minorities, fortunately or unfortunately, there is no distinction among the three parties. From this perspective, Watanabe et al. [7] made the following points clear as a summary.

At present, the Chinese government proceeds with settlement policy. Until recently, there have been some ethnic groups changing their living locations because of slash-and-burn farming practices. Naturally, their conditions of location differ depending on their regular vocation. Some groups prefer the drafty ridges of plateaus, and some others stick to valleys near water. They may select an environment because they enjoy it. Every time when they move, they build their own houses. They can get building materials from the places near them. In principle, they use what they can easily get such as timber, bamboo, weeds, soil, stone, sun-dried brick and tile, and, recently, glass and galvanized sheet. They are doing trial and error for new materials.

The raised-floor houses in Hansha Village, Ruili City, and Dehong Dai and Jingpo Autonomous Prefecture are good examples of that. The problems of humidity, heat emission, and protections from insects and water, the reasons why the houses should be raised-floor style, have been gradually dissolved, so that the first floor also is being converted to living space. Although the use of galvanized sheet for roofing is a shift toward durable materials, new problems of overheating from solar radiation and the noise of squalls pounding the roof have also become obvious. Therefore, converting the first floor to living space is an immediate emergency measure rather than a solution.

Solar shading and cross ventilation are the base of passive cooling. In principle, the houses in Yunnan Province face the east and rooms are very dark. There are large eaves, peaks supported with the pillar built in a pit directly, and small attached annexes surrounding the living space, shielding against the high sunshine of summer. They warm themselves in semi-outdoor spaces such as verandas and courtyards during the daytime in wintertime. The traditional houses in southeastern Asia are dark inside, probably in the interest of coolness. Intrinsically, coolness is only felt in hotness, and it is a great pleasure felt when sweaty skin is exposed to a breeze. The sound of a wind chime that can be heard in a big chorus of cicadas also has the same effect.

Summer sensory temperature mitigation method by cross ventilation, *tsufu* in Japanese, has an "architectural monopoly" in the monsoon region of Asia, so to speak. There has been no idea of cross ventilation in the living habits of Westerners. Therefore, Watanabe et al. [7] hope that the English translation of *tsufu* would be *tsufu*, not cross ventilation, in the case of Japan at least. The *tsufu* of raised-floor type houses is not cross ventilation, but winds flowing both vertically and horizontally. In the summer season in particular, indoor oven heat is emitted by air flowing from under the floor upward, and it is an effective way to get *tsufu* through the floor. Many original ideas are seen in open parts including sliding doors and folding doors, and it is often the case that open parts are found in higher parts. The courtyard of courtyard houses, such as the type of four houses, is considered to be a large open part in a sense.

There are many ethnic groups who do not even control body temperature by changing clothing in wintertime [8]. Inside airtightness is poor, and they depend only upon fireplaces and braziers for warming. The reason that the houses of Bai people and Yi people have much heat quantity in the earth floor and its peripheral walls is to accumulate the radiation heat and conductive heat from fireplaces and braziers, and to keep sensory temperature high at night. They are skilled at selecting a comfortable place. It is heartwarming that they lie down near the drafty doorway in summer and near the fireplace where heat is accumulated. The introduction of passive cooling and passive heating techniques is required in the houses of the ethnic minorities in Yunnan Province.

References

1. Yamauchi M (1996) Folk problem guide. Chuko Library
2. http://www.jyfa.org/plaza/images/kihon/minzu_map.jpg
3. https://ja.wikipedia.org/wiki/%E6%98%86%E6%98%8E%E5%B8%82
4. Oketani S (ed) (1983) Visit of history and civilization in Chinese Southwest Frontier. Chinese Traditional Culture Research Center
5. Rao W (1991) Type of Yunnan people's houses. Proceedings of the symposium on vernacular houses for Yunnan minorities
6. Yunnan Institute of Technology (ed) (1983) Yunnan houses. Chinese Construction Industrial Publishing Company
7. Watanabe T, Urano Y, et al (1992) Bioclimatic design for Yunnan dwellings, Architectural Institute of Japan. Proceedings of the 22nd heat symposium
8. Song C (1990) Yunnan ethnic minorities. Japan Broadcasting Corporation

Chapter 8
Thailand: The Houses of a Khun Village in Chiang Mai

Ikuro Shimizu

Abstract The aims of this chapter are to understand the spatial characteristics of Khun houses and villages from architectural and ethnographic points of view and to consider the ways in which their houses are built to adapt to the climate. Khun is ethno-linguistically classified into the Tai-Kadai language family. They migrated from Myanmar and settled in the outskirt of Chiang Mai, the most famous old city of Northern Thailand. Khun are known for their dedicated practice of Theravada Buddhism, and active wet paddy rice cultivation is their common livelihood. The house space also has close relationship with them. This chapter shows how their houses are organized according to livelihood, religion, and other cultural features. First, the form and the spatial organization of the house will be described. Second, a typological analysis of the house from the viewpoint of gender relationships and invisible order will be made. Lastly, the most efficient way to adapt the houses to the tropical monsoon climate will be considered.

Keywords Thailand · Chiang Mai · Khun · Stilt house · Supernatural spirit · Order of the house

8.1 Northern Part of Thailand and Chiang Mai

The Kingdom of Thailand (hereinafter Thailand) is located in the northern part of the Malay Peninsula, at the center of mainland Southeast Asia, and shares its borders with Northern Laos, Eastern Cambodia, Western Myanmar, and Northern Malaysia. The country is roughly divided into four areas: the northern area, composed of plains and mountainous regions; the northeastern Khorat Plateau, called Isan in local terms; the central flat plain area, which includes Bangkok, the capital city; and the southern area, surrounded by the Andaman Sea and the Gulf of Thailand.

I. Shimizu (✉)
School of Architecture, Shibaura Institute of Technology, Tokyo, Japan
e-mail: ikuro-s@shibaura-it.ac.jp

© Springer Nature Singapore Pte Ltd. 2018
T. Kubota et al. (eds.), *Sustainable Houses and Living in the Hot-Humid Climates of Asia*, https://doi.org/10.1007/978-981-10-8465-2_8

Northern Thailand, which was dominated by the Lanna Dynasty until the beginning of the twentieth century, is famous for its multiethnic regions. Various studies on each ethnic group have been conducted over many years [1–7]. This chapter, which is based on an intensive field survey over 3 years, focuses on the Tai-Kadai language family, especially on the Tai Khun.[1] In Northern Thailand, the Khon Muang, or ethno-linguistically the Yuan, are historically the descendants of the Lanna Dynasty and the largest ethnic group. More than 100 years ago, the Khun migrated from Keng Tung, the provincial capital of Shan province in current Myanmar, and settled in the vicinity of Chiang Mai. Khun practice Theravada Buddhism, and their livelihood is commonly wet paddy rice cultivation. The aims of this chapter are to understand the spatial characteristics of Khun houses and villages from architectural and ethnographic points of view and to consider the ways in which their houses are built to adapt to the climate.

8.1.1 Climate Features of Northern Thailand

Thailand belongs to the tropical monsoon climate and is strongly influenced by monsoon patterns. One year is divided into three seasons. June to October corresponds to the rainy season, November to February is the dry season, and March to May is the hot season. In the rainy season, the air is humid and warm. Squalls may happen on a daily basis and flooding often occurs. On the contrary, the dry season has little rain and is relatively cool. The peak of the cold weather is usually in early December. In the hot season, the temperature often reaches higher than or equal to 40 °C. However, the temperature differs by region and season. While the average annual temperature across the nation is about 29 °C, in Bangkok it is 35 °C in April and 17 °C in December.

The climate of Northern Thailand is also under the influence of monsoon patterns. The rainy season and the dry season are characterized as cool seasons. In the dry season, the average temperature is about 23 °C. Moreover, there is a severe early morning chill of up to 10 °C in mountainous areas during December and January. In the rainy season, it may rain every day. Floods occur in many places and can cause great damage. March to May are the hottest months, and the daytime outdoor temperature exceeds 40 °C. It is common to require indoor cooling all day in the hot season.

---

[1]The surveys were conducted for about 10 days each year from 2014 until 2016 in T village in San Pa Tong district, Chiang Mai prefecture. The survey was carried out with physical measurement of 20 houses, house sites, and the whole village space, together with interviews with 20 villagers. In order to understand local ways of using the space, any furniture and goods in and around the house were included in the drawings. Interviews were based on a questionnaire that was prepared in advance. Examples of questionnaire topics are family structure and economic condition of households, schedule of everyday life, the use of each space inside and outside the house, remodeling history, gender relationships, and spatial perception represented in rituals.

Fig. 8.1 Map of T village

8.1.2   T Village

T village is located about 30 km south from Chiang Mai city and sits along the river Khan, which is a tributary of the river Ping. The population of the village is 661 and the number of houses is 200. Most villagers are ethnically Khun (Fig. 8.1). At the northwest of the village, there is a Buddhist temple named Donchai and an elementary school. Many elderly villagers visit the temple daily to donate food and pray. They also gather for the purpose of conversation. In this sense, the temple seems to be a place of relaxation and communication for villagers.

In the center of the village, there is an open space that is used in a variety of ways, such as a meeting place or holding the regular morning market. At the edge of the open space, there are two small huts called Sua baan, which literally means the "cloth of the village." Inside the huts, male and female guardian spirits have been worshiped since the village was established. The spirits have long been responsible

Fig. 8.2 Sections of a raised-floor house

for the prosperity and well-being of the whole village. At the southwest of the village, there is a crematorium. Outside the village, wet paddy rice fields and orchards are spread widely.

8.2 House of Khun

8.2.1 Form and Structure of the House

It is commonly known that traditional houses in Northern Thailand are stilt or raised floor and constructed with wood or bamboo. The houses (Heuan) of T village have the same characteristics. Figure 8.2 shows the typical structure of houses in T village. The house has a rigid-frame Rahmen structure, which is composed of timber posts and beams.

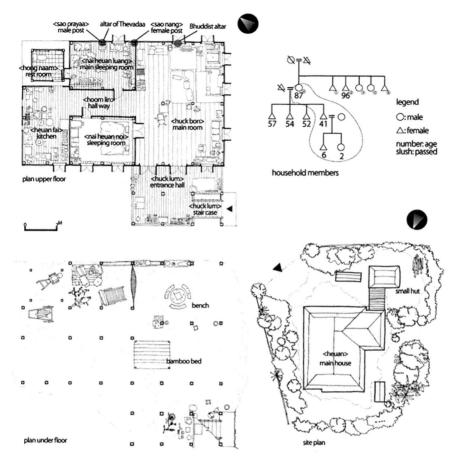

Fig. 8.3   Plan of each floor of a Stilt house and the house site

8.2.2   Spatial Organization of the House

Figure 8.3 shows a typical plan of a stilt or raised-floor house in T village. Usually, a house is composed of one or two staircases, entrance hall-like small space that is set in front of the main entrance (Huck Lum), living room (Huck Bon), main sleeping room (Nai Heuan Luang), second sleeping room (Nai Heuan Noi), hallway or corridor (Hoom Lin), kitchen (Heuan Fai), and rest room (Hong Naam).

The house site is commonly surrounded by stone wall or concrete block or, in some cases, a hedge. After entering the house site and walking toward the house, we will face to main staircase to the upper floor of the house.[2] Climbing up the stairs and

---

[2]The staircase is often placed to climb from east to west.

passing a large wooden door, there will be an entrance hall situated under the eaves. Next, to the hall, there will be a living room that can be used in many ways, such as eating, accepting guests, relaxing, taking a nap, and as a sleeping space when guests stay overnight. Between the hall and living room, there is a split-level of about 20–30 cm.

The living room is a relatively large room in the house and contains items such as TV, refrigerator, chairs, and sofa. In addition to the purposes listed above, the living room is also used for rituals and ceremonies, such as weddings, funerals, and Buddhist rituals. During the rituals, household members, relatives, villagers, and monks gather in the living room. For the purpose of these every-day activities and specific occasions, many kinds of goods and daily necessities are placed in the room. The living room seems to be a public space within the house.

On the contrary, the sleeping rooms are private and apparently used differently from the other rooms. Usually, non-household members are not allowed to enter the sleeping room. Only household members have the key to the door, and even looking inside the room will require permission from the household.

The main sleeping room is located on the east side of the house and is usually reserved for the oldest person or the oldest couple in household. The other sleeping rooms, which are placed to the west across the corridor, are for the other members of the household.

Unlike patrilineal societies, such as Japan, the societies of this language family are generally matrilineal. According to the interviews with villagers, males do not touch unmarried females. If they violate this, it will not only hurt the female herself but will also hurt her ancestral lineage. In this case, the male must compensate some amount of money as a fine for the ancestral spirit of the female side.[3]

Between the sleeping rooms, there will be a corridor that gives access to the kitchen. The kitchen is often located the furthest from the entrance. Above the corridor, the gutters run across the ridge to drain rainwater from the roof to the eaves of the long side of the house.

8.2.3 Religious Meaning Represented in the House

Inside the house, we see a variety of objects and activities related to religious beliefs. Here we discuss each of them one by one.

---

[3]Additionally, even today, males must pay what is called the "bride wealth" to the female side before the wedding.

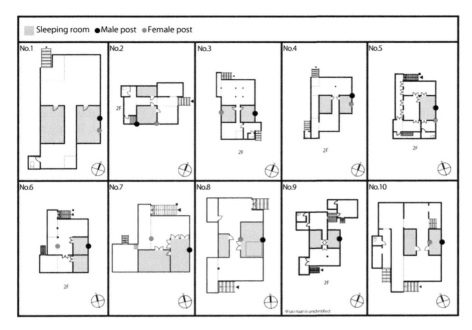

Fig. 8.4 Typological analysis of the position of the male and female posts and sleeping rooms

8.2.3.1   Sacred Post

There are two sacred posts erected in the house. These two posts are thought of as a pair. One is regarded as the male post (Sao Payaa) and the other one is regarded as the female post (Sao Naan).[4] According to the explanation given by a village carpenter, these paired posts should always be located in the main sleeping room. The male post tends to be erected in the middle of the east side of the wall, whereas the female post is in the center of the west side of the wall (Fig. 8.4).

In building a new house, short-cut banana stalks, sugar canes, coconuts, sweets made from glutinous rice, small cotton textile, and clothes belonging to the male household head are tied to the male post. On the textile, a spell called Payan is written. Clothes belonging to the oldest female and the cotton textile are tied to the female post.

The teacher or instructor of Buddhist ceremony (Achan) performs a ritual at the beginning of constructing a new house. Monks from the village temple are also invited to perform the ritual. First, at the male post, the teacher and monks pray for successful construction and for the continuity of peace in the household. After that, villagers start to erect the post.

_____

[4]Sao means post. Paya means great or large. Naan means female.

8.2.3.2 Religious Altar

There are three kinds of altar in the house. These are the altar for worship of the landowner spirit (Cao Tii), the Buddhist altar (Hing Pra), and the altar for the supernatural guardian spirit of the house (Hing Thevadaa Baan) [8].

The altar for the landowner is usually set in the front yard of the house site. The front side of the altar is often turned to face east or north. In T village, most villagers pray and make offerings to this spirit only when it is a religious event. If they need auspicious power from the spirit, they will offer a conical-shaped container (Sway Doke) made of banana leaves in which they put offerings such as incense and candles.

The Buddhist altar is, in almost all cases, set in the living room. The members of the household pray and make offerings everyday to the Buddhist saint. The altar for the supernatural spirit is located next to the male post inside the main sleeping room, and household members occasionally make various kinds of offerings and pray at this altar.

Another altar, called Tao Tan Si, is set inside the house site, usually near the entrance gate of the site.[5] This is for the worship of the plural gods, which are believed to exist in any place and in any direction. This altar is not regarded as Buddhist but as Brahman. The day before some events, such as construction of a new house, renovation of the house, celebration, and so on, household members report to the gods that they will hold the event. Materials for making this altar are mainly wood, but in some cases, the corrugated iron sheet is used.

8.2.3.3 Spatial Order of Male and Female in the House

We can understand some sort of invisible order through the concrete activities of males and females in the house, especially in sleeping and eating. Here, we concentrate on the sleeping position of males and females, specifically the orientation of each gender's head while sleeping, which is commonly observed among villagers (Fig. 8.5).

Villagers tend to turn their heads to the east or south while sleeping. Sleeping arrangement is organized by gender, with males tending to sleep on the right side, and females and children on the left side. This may come from the perceived association between maleness and the right side, and, in contrast, femaleness and the left side. An example of this is also demonstrated in the positions in which the bride and groom sit during the marriage ceremony. The groom is arranged to sit on the right side of the house, whereas the bride sits on the left side.

Funeral rites also show another example of spatial order. The monks sit at the back end of the wall that divides the living room and sleeping rooms. Males sit near the monks. Females sit near the entrance, which is furthest from the monks.

---

[5]Tao means Thevadaa. Tan means above. Si means four directions of north, south, east, and west.

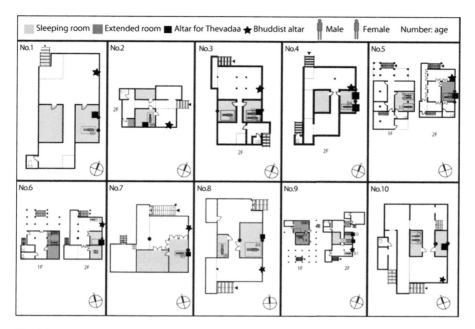

Fig. 8.5 Altars installed in the house

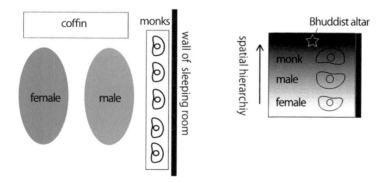

Fig. 8.6 Sitting order in various rituals and ceremonies

According to Buddhist practice, females should not be in contact with the monks. Figure 8.6 shows the model of the spatial order.

There are a lot of religious rituals performed by monks inside the house, such as Suup Chata. This is a kind of curing rite performed when illness or injury occurs in the household.[6] Monks preach and pour holy water over each household member,

---

[6]Suup means inherit or to extend the life. Chata implies the fate or life of human being.

followed by the house itself. After that, monks wind white threads around the wrist of the sick or injured household member.

When religious rituals are performed with invited monks, the ritual is usually held in the living room, and the monks sit near the Hing Pra (the place where Buddhist altar is set). This place is thought of as the highest level in the spatial hierarchy inside the house. From the altar to the peripheral space, monks, males, and females sit in the same order as we see in funeral rites (Fig. 8.6).

8.3 Climate Control Method Seen in and Around the House

The most popular way of adapting the house to the hot and humid climate, as observed by the author, is with the use of raised floors. In the tropical monsoon climate, which is clearly separated into the rainy season and the dry season, most of the year's rainfall occurs in the rainy season. In the rainy season, swollen river water often causes flooding. One advantage of the raised-floor house is that people can remain safe and secure on the upper floor if flooding occurs.

Throughout the year, people use the space under the floor in many ways. During the rainy season, it is possible to avoid the rain while under the floor. On the other hand, in the dry and hot season, the wind passes through and provides a cool space. In fact, people practice a variety of activities under the floor. Common activities include weaving, accepting guests, taking a nap, or watching TV on a bed. Cooking usually takes place inside the cooking hut or at the hearth in the main house, but occasionally meals are cooked and eaten at a table under the floor. This space under the floor is multipurpose and accommodates a variety of activities.

The windows of the house are well designed to fit the climate. Normally, the windows are set directly from the floor. The window is commonly composed of two parts, and each one has a double door. The lattice of the iron is fitted in the lower layer (Fig. 8.7).

During the day in the hot season, both layers of the window will be opened. At night, only the lower part will be opened to draw the cool air indoors. The window frame that separates the upper and lower layers will be put at the height of the elder woman's elbow when she is sitting by the window. In the village, a lot of women spend time sitting by the window and looking outside.

Originally, Khun houses were not built with a ceiling. If the entire window is wide open, the air passes through the entire indoor space. An opening or a gap created in the roof gable also helps to create a cooler indoor climate.

Fig. 8.7   Window of a Khun
house

8.4   Conclusions

Whenever staying at a Khun house, the author did not feel inconvenienced. In the hot season, many residents spend daytime under the floor. For this study, interview with villagers were also carried out under the floor. In T village, many houses have changed roofing materials from traditional grass to tiles or slates. Although the quality of such material is still good, it is too hot in the hot season to stay under the roof, even if the windows are opened to draw in outside air. Currently, in T village, residents are trying to restore and preserve traditional houses, and what to use as roofing materials is an important question for such activities.

As mentioned earlier, a lot of work is carried out under the floor in the rainy season. In addition to the space under the floor, under the eaves on the floor is a space for residents to stay comfortable. In Khun houses, the building materials and structure are gradually changing in the modern context, but the traditional way of living and elements such as the windows, the roof gradient, and the raised floor still remain with important meaning and function.

References

1. Rhum MR (1994) The ancestral lords, gender, descent, and spirits in a northern Thai Village. Northern Illinois University Center for South East Asian Studies
2. Shimizu I (2005) An ethnography of interaction between house and people. Fukyosha (in Japanese)
3. Shimizu I et al (2015) On characteristic of the spatial organization of the village: study of spatial characteristics in Lowland Society of Northern Thailand, Part 1. In: Summaries of technical papers of annual meeting 2015. Architectural Institute of Japan, pp 1197–1198 (in Japanese)
4. Shimizu I et al (2015) A study of the use of the space from the viewpoint of gender relationship: study of spatial characteristics in Lowland Society of Northern Thailand, Part 2. In: Summaries of technical papers of annual meeting 2015. Architectural Institute of Japan, pp 1199–1200 (in Japanese)
5. Shimizu I et al (2014) Characteristics of the House of Tai Yai in Northern Thailand: Integrated study of man-lived-space at the wetland area of Mainland South-East Asia part 4. In: Summaries of technical papers of annual meeting 2014. Architectural Institute of Japan, pp 963–964 (in Japanese)
6. Izikowitz KG et al (eds) (1982) The house in east and Southeast Asia: anthropological and architectural aspects. Curzon Press, London
7. Sparks S, Howell S (eds) (2003) The house in Southeast Asia: a changing social, economic and political domain. Routledge Curzon, London
8. Waterson R (1990) The living house: an anthropology of architecture in South-East Asia. Oxford University Press, Oxford

Chapter 9
Laos: Indigenous Houses of a Lue Village in Luang Prabang

Ikuro Shimizu

Abstract This chapter discusses the villages and houses of the Lue ethnic group, which is one of the ethnic minority groups in the peripheral area of Luang Prabang, Laos. Lue originally lived in Sip Song Pan Na, a former kingdom of Dai in Yunnan province of current China. Today, Lue are spread throughout Yunnan, Shan State of Northern Myanmar, and Northern Thailand. They are organized into nuclear families and are dedicated practitioners of Theravada Buddhism. This chapter describes the remarkable characteristics of their village and houses, from architectural and ethnographic perspectives. The spatial structure of the village will be described. This chapter also examines the physical structure, construction process, and spatial organization of Lue houses.

Keywords Laos · Lue · Luang Prabang

9.1 Lue in Luang Prabang, Laos

9.1.1 Laos

Lao People's Democratic Republic (hereinafter, Laos) is a landlocked country surrounded by Myanmar, China, Vietnam, Cambodia, and Thailand. Located in the center of the Indochina peninsular, the country has a population of approximately 6,000,000, and tropical monsoon climate covers the whole region. The most popular subsistence is agriculture, especially paddy rice cultivation, which is seen everywhere in the country. Furthermore, people's everyday lives are in close proximity to the water. They do active fishery not only in rivers and lakes but also in paddy rice fields. In the paddy rice fields, many kinds of natural resources, such as trees, bamboos, and grasses, are growing, and people are producing building materials from them. The man-made habitat greatly relies on the ecological environment.

I. Shimizu (✉)
School of Architecture, Shibaura Institute of Technology, Tokyo, Japan
e-mail: ikuro-s@shibaura-it.ac.jp

© Springer Nature Singapore Pte Ltd. 2018
T. Kubota et al. (eds.), *Sustainable Houses and Living in the Hot-Humid Climates of Asia*, https://doi.org/10.1007/978-981-10-8465-2_9

Laos is also known as a multiethnic country, having more than 50 ethnic minority groups in the country. Until recent years, the characteristics of each ethnic group were handed down from ancestors to descendants in every society [1–5]. However, since the introduction of the market economy in 1986, the socioeconomic situation has been changing significantly.

This chapter focuses on the village (Baan) and the house (Heuan) of a Lue community living in N village of Nam Bak district, Luang Prabang province.[1]

9.1.2 Climate Condition of the Research Site

Laos belongs to the climate zone of the tropical monsoon. This climate is changed by the wind direction of the monsoon. Across the Indochina peninsula, including Laos, during almost half the year from May to October, there is a moist southwest wind blowing. This period is equivalent to the rainy season. Around 80–90% of the annual precipitation concentrates in this season. Although the average annual rainfall is about 1834 mm (2011), precipitation in the rainy season depends on the region and the altitude.

On the other hand, from November to April, there is little rainfall and this time is equivalent to the dry season. The temperature is lowest from December to January and is often less than or equal to 15 °C in the early morning or midnight. Furthermore, at high altitude mountainous areas, the temperature is lower than lowland areas, and there may be frost in some places. However, during the cool season, people will be released from agricultural work to participate in other kinds of business, such as trade, house construction, or important social activities such as marriage. From February, the temperature gradually rises across the country. In particular, April is the hottest season of the year.

Luang Prabang is also influenced by the tropical wet and dry climate. Although the city is generally warm throughout the year, it is noticeably cooler during December and January. The city receives approximately 1450 mm of precipitation annually.

9.2 N Village

9.2.1 Luang Prabang

Luang Prabang province, with a population of about 400,000, lays approximately 350 km north of the capital city, Vientiane. The city of Luang Prabang is located on a

---

[1]Field surveys were conducted in 2012, 2013, 2015, and 2016, for a total of 40 days. We got the measurement of 40 houses and conducted 42 interviews with villagers, together with the measurement of each house site and the whole village. After the measurement of the house, we drew plan (1/50), site plan (1/200), and whole village map (1/500).

confluence of the River Khan and the River Mekong.[2] The population of the city is estimated to be approximately 56,000. King Fa Ngum first settled the city in 1353 as the capital of Lan Xang Kingdom. There was a royal palace until the communist takeover in 1975; it is now used as the national museum.

9.2.2   N Village and the Lue Ethnic Group

N village is located about 140 km north from Luang Prabang city. The population is 548, including 248 males and 300 females, with 118 houses. The village society is organized into headman, deputy headman, association of the elderly, women's alliance, youth alliance, police, and the guardians. Several male elders are responsible for the ritual practices related to Buddhism and animistic belief. Electricity has been laid since 2006. On the contrary, the road to the village has not been paved.

The Lue of Laos originally migrated southward or eastward from Sip Song Pan Na, a former kingdom of Dai established in Yunnan province, China. The domestic population was 123,000 in 2005. Lue language is similar to both Thai and Lao languages. Social features such as the nuclear family as the basic unit of social structure, maternal residence after marriage, property inheritance, and dedicated practice of Theravada Buddhism are similar to another Tai-speaking group such as Shan, Yuan, Lao, and so on. The ancestors of N village used to work as caretakers of horses raised by the Lan Xang Kingdom and established N village around 650 years ago.

About 10 years ago, N village was registered as a "cultural village" by the local government. Cultural villages are usually certified according to national policy.[3] The chief livelihood of the village is paddy rice cultivation on wetland. Cultivation of cash crops is widespread in the village periphery.[4]

Stilt or raised-floor houses have been traditionally popular among the Lue. As mentioned earlier, in 1986, the Laos government introduced and promoted the market economy mechanism. As a consequence, the socioeconomic situation of the country changed considerably [6]. These changes had a great impact not only on the capital city but also on rural areas and villages. House style has also been impacted by these changes. Today, it is less common to see houses built in the traditional form across the country.

---

[2]The town is well known for its numerous Buddhist temples. Every morning, hundreds of monks from the various monasteries walk through the streets collecting alms.

[3]Selection criteria for the cultural village is as follows: good health management, village security, does not use slash-and-burn agriculture and deforestation, good reservation of the water source, and observes the traditional customs of the village.

[4]Although there was slash-and-burn agriculture in the village in the early 1970s, villagers discovered that rice crops are sufficient without the slash-and-burn. Slash-and-burn has not been carried out at all since then.

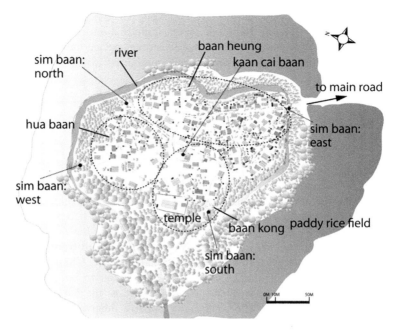

Fig. 9.1 Whole map of N village

9.3 Spatial Characteristics of the Village

9.3.1 Cardinal Point and Center of the Village

N village is spatially divided into three areas according to the history of expansion (Fig. 9.1): highland area (Baan Kong), north side area (Hua Baan), and the recently developed area (Baan Heung). At the highland and north side areas, there are old houses and the village temple. This suggests that these areas were the first to be pioneered. The recently developed area was once forest and believed to contain many spirits.

At the center of the village, there is a stone (Kaan Cai Baan) that marks the center of the village. Marking the center by stone or wooden post is commonly seen among Tai-speaking groups. In this case, at the establishment of the village, they decided on the central point of the village and put the stone on the ground.

Additionally, at four points marking the outer edge of the village, a stone (Sim Baan or Sii Che Baan) is placed. Each point corresponds to the directions of north, south, east, and west. At the purification rite, villagers chased the evil spirit out from inside the village. At that time, villagers made a great number of bamboo crafts and put them around the stones to prevent the spirit from entering the village and disturbing the villagers.

The river flows around the village, and the river water is used to supply daily life and the paddy rice fields. Today, they have an irrigation system to direct water from the river during the dry season, making double cropping available. In the dry season, paddy fields are also used for planting other cash crops and fruits, such as tobacco, green onions, garlic, coriander, mango, and so on. The forests surrounding the village are composed of protected area and free-use area. Villagers cooperatively manage the protected area.

9.4 Lue Houses

9.4.1 Household Situations

The average house site is approximately 573 m^2. The widest is 1140 m^2 and the narrowest is 214 m^2. Electricity, water supply, and sewage systems are not supplied to all households, and only one household uses propane gas. Electrical appliances are owned by some households: 15 have a telephone, 17 have a refrigerator, 37 have a TV, and 8 have an audio player. Furthermore, 39 households have a bicycle, 8 have a motorbike, and 18 have a tiller. Most households possess a mobile phone.

The average number of residents per house is four people. The most popular occupation is agricultural work. However, some households earn money by weaving or teaching. In 2012, there was only one household that was running a shop alongside part-time agriculture. However, villagers who are running the shop are now increasing. For agriculture, the main crops are sticky rice and tobacco. Women tend not to engage in agriculture and, in many cases, do weaving and household chores instead. Fields for planting fruits or vegetables are owned by 12 households. Thirty-six households possess paddy rice fields with an average area of about 1.37 hectares. The most popular livestock is 20 to 30 chickens and ducks. There are also seven households that have pigs and two that have water buffalos.

All households practice Theravada Buddhism, and several households also have faith in Pam, the traditional animistic belief. The average annual income per family is about 12,950,000 kiip[5]. The highest income is 126,000,000 kiip and the smallest is 30,000 kiip. Households that gain over 10,000,000 kiip per year do not always have a lot of agricultural land or farmland. The difference in annual income may be attributed to crops that are more profitable, or sideline businesses such as weaving.

The average age of the houses in the village is estimated to be about 22 years (Fig. 9.2). The oldest house was built in 1972 and the newest was built in 2009. There were few houses built during the 1990s, perhaps because of the Asian economic crisis.

---

[5]1 US dollar is nearly equal to 8300 kiip in 2017.

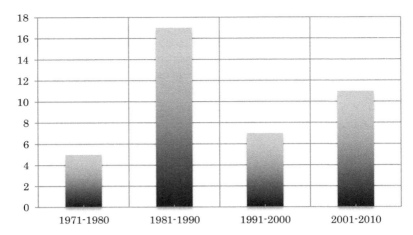

Fig. 9.2 Number of village houses built in the past four decades

9.4.2 Material and Structure of the Houses

There are 79 stilt or raised-floor houses in the village. An additional 25 houses are made of concrete and all are two-story. The remaining 14 are one-story houses. Currently, two-story and one-story houses are increasing, and traditional stilt houses are decreasing.

The most traditional and popular material for house construction is timber. Even in houses that use concrete block or red brick to make the walls, the house frame is built with timber. Besides timber, bamboo, concrete block, concrete, slate, brick, and plaster are commonly used as building materials.

Although concrete is a relatively new material, bamboo has long been a popular material for villagers because of its easy processing. Bamboo has been used in walls, roofs, and floors. In traditional houses, the roof was usually thatched with many flattened bamboo panels. Visually, the row of such roofs makes the village landscape remarkable. Bamboo material must be replaced every 8–10 years. In recent years, an increasing number of houses have changed roofing material from bamboo to slate or corrugated iron sheet. Slate is a popular alternative roofing material because its lifetime is longer than that of bamboo or other natural materials, and it is less likely to leak. Bricks have been used as the outer wall of the house. Stucco has also been used to cover the surface of the outer wall of some houses.[6]

Figure 9.3 shows the procedure of traditional house construction, as explained in an interview with a village carpenter who had engaged in constructing village houses as a master carpenter. On the first day, the team of carpenters places the footing stones on the ground. After checking that the stones are level, they erect all posts on

---

[6]It has also been used on the outer wall and posts of village temple.

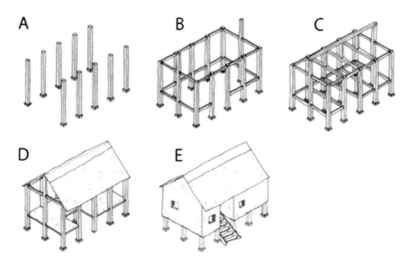

Fig. 9.3 The procedure of construction

them (A). To erect the posts, they use strings to ensure accurate positioning. Next, they join beams and cross beams to the posts. After that, they erect center posts, which support the ridgepole at both gable sides of the house. Floor beams are also erected on the same day (B). On the second day, they put the ridgepole at the top of the center posts. Then, they create the floor and roof (CD). To thatch the roof, half cleaved bamboo is used as rafters, which are put on the ridgepole and beams. The rafters are tied to the beams with bamboo string. On the rafters, they thatch panels of flattened bamboo mat. On the third day, they construct the walls all around the house (E). On the fourth day, they build the space for the kitchen.

The structure of the house is the timber framework. In order to join the posts and beams to each other, they use tenon and mortise joints. Cotter is also put in the mortise to make the structure rigid.

Figure 9.4 shows each element of the raised-floor house, with a brick-walled room under the floor.

9.4.3 Spatial Features

Figure 9.5 shows the spatial organization of a typical raised-floor house [7–9]. The house is divided into two buildings. One is the cooking hut (Hong Khua) and the other is the main house (Heuan), which includes a large main room (Kaan Heuan). The main room is used for many purposes, including sleeping, eating, and accepting guests. The staircase is set at the front side of the house. In this case, it is set between the cooking hut and the main house.

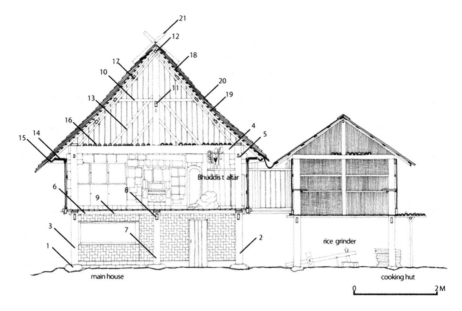

Fig. 9.4 The sectional plan of typical Lue house
1. Footing stone / 2. Male post (Sao Tao) / 3. Female post (Sao Naan) / 4. Beam / 5. Floor girder /
6. Floor beam / 7. Bunch / 8. Floor girder / 9. Floor / 10. Balanced beam / 11. Roof girder / 12. Ridge
pole / 13. Angle brace / 14. Arm to support rafters / 15. Bamboo to receive rafters / 16. Half bamboo
sprit to create storage area / 17. Climbing beam / 18. Rafter / 19. Purlin / 20. Bamboo panel/ 21.
Ornament of the ridge

Traditionally, there was a hearth for cooking inside the main room. Furthermore, they had another hearth for performing childbirth rituals inside the main room, which was usually hidden under the bamboo mat. This hearth was used for the mother's recovery and caring for infants. Today, the childbirth hearth has almost disappeared.

The use of the main room depends on the daily activities of residents. The side nearest to the cooking hut is used for taking meals. At mealtime, residents sit around a round table set on the floor. At the deepest side of main room, a bamboo wall divides the main room and the sleeping room (Suam).[7] All household members share the room in sleeping as shown in Fig. 9.5. The front side of the wall is reserved for accepting guests or daily relaxation.

The space under the floor is mainly used for storage of firewood, motorcycles, motorbikes, agricultural equipment, fishing equipment, and so on. In addition, it is used as a workplace for blacksmithing, making bamboo crafts, and maintenance of fishery gear. The weaving machine is also usually set under the floor. Weaving textiles is

---

[7]Entering into the sleeping room without permission from the house owner or members of the house is strongly prohibited in the village.

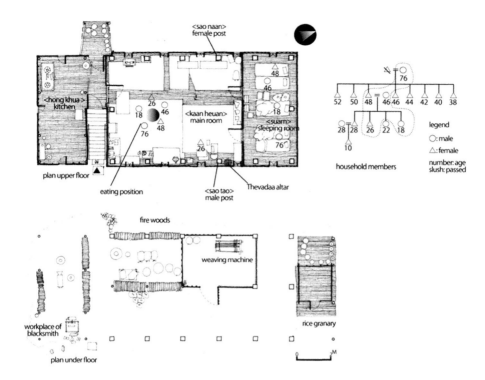

Fig. 9.5 Typical plan of a Lue raised-floor house in N village

a woman's activity and is an important way to earn cash income for the household. The rice granary is placed under the eaves at the backside of the space under the floor.

Two of the most important posts, the male post (Sao Tao) and the female post (Sao Naan), are usually erected near the sleeping room. At either of these posts, there is space for the altar to worship the supernatural spirit called Thevadaa, which is believed to protect the household. The sleeping room is usually on the north and deepest side of the house, whereas the cooking space or hut is often placed on the south or southeast and most front side.

9.4.4 Control of Humidity and Temperature

The village ground has not been paved and remains as soil. The ground becomes muddy in the rainy season and dusty in the dry season. Lue houses are raised from the ground to make life comfortable during the rainy season. When the ground is made muddy by the rain, villagers take off their shoes before going up the stairs.

They do not bring muddy and wet shoes inside the house. In addition, the space under the floor can be used in many ways even in the rainy season.

Inside the house, because there is no ceiling, the space between the floor and the roof is quite high. Lue usually like to sit directly on the floor inside the house with a large space spread overhead. Although the timber plate covers both sides of the gable, there is a gap at the edge where it touches the roof. The girder receives the rafters that are descended from the ridgepole, and the roof is placed on the rafters. The wind blows through this gap to help ventilate the house.

Double windows are installed on the walls of both longitudinal sides. The humidity and temperature inside the house can be adjusted by the use of these windows. Even if all windows are not opened, the air can be circulated through the gap between the girder and the roof. In the hottest season, between March and April, everyday life on the raised floor is more comfortable than we expected. Even in the coldest season, around January, it is also common to adjust the indoor environment by using the windows.

9.5   Conclusions

Among Lue villages throughout Laos, N village has the most traditional houses. Inside the traditional house is cool and comfortable, even in the hot season, with stilt form, high-slope roofs, and many windows. In the rainy season, the stilt form allows residents to live a life away from the muddy ground.

Currently, such traditional houses have been reduced to less than 20 in N village. The influx of factory-made building materials has been rapid. It is not only the physical and material characteristics of the houses that are changing; it is also the lives of the residents. It will be necessary to actively and urgently promote maintenance and preservation of these traditional houses.

References

1. Ikuro S (2014) Spatial characteristics seen among the Phuan in Vientiane province, Lao P.D.R., especially focusing on the house and village. J Fine Arts 2557(1):21–66 (CMU)
2. Izikowitz KG, Sorensen P (eds) (1982) The house in East and Southeast Asia: anthropological and architectural aspects. Curzon Press, London
3. Charpentier S (1982) The Lao house: Vientiane and Louang Prabang. In: Izikowitz KG, Sorensen P (eds) The house in East and Southeast Asia: anthropological and architectural aspects. Curzon Press, London
4. Clément S (1982) The spatial formation of the Lao house. In: Izikowitz KG, Sorensen P (eds) The house in east and Southeast Asia: anthropological and architectural aspects. Curzon Press, London
5. Clément S (1982) The Lao house among the Thai houses: a comparative survey and a preliminary classification. In: Izikowitz KG, Sorensen P (eds) The house in east and Southeast Asia: anthropological and architectural aspects. Curzon Press, London

6. Ikuro S (2009) Houses. In: Akimichi T (ed) An illustrated eco-history of the Mekong River Basin. White Lotus, Bangkok, pp 74–77
7. Ikuro S et al (2015) A study on religious space in and around the house: Ethnographic-Architectural Study of the House of Mainland South-East Asian Society, Part 2. In: Summaries of technical papers of annual meeting. Architectural Institute of Japan, pp 1207–1208 (in Japanese)
8. Ikuro S et al (2015) Symbolic analysis of the ritual process at the house among Lue in Lao P.D. R.: Ethnographic-Architectural Study of the House of Mainland South-East Asian Society, Part 1. In: Summaries of technical papers of annual meeting 2015. Architectural Institute of Japan, pp 1205–1206 (in Japanese)
9. Ikuro S (2013) Characteristics of the house of Lue in Northern part of Lao P.D.R.: integrated study of man-lived-space at the wetland area of mainland South-East Asia, part 3. In: Summaries of technical papers of annual meeting 2013. Architectural Institute of Japan, pp 1307–1308 (in Japanese)

Part II
Adaptive Thermal Comfort

Chapter 10
Principles of Adaptive Thermal Comfort

Michael A. Humphreys and J. Fergus Nicol

Abstract The adaptive approach to thermal comfort recognises that people are not passive with regard to their thermal environment, but actively control it to secure comfort. Thermal comfort can thus be seen as a self-regulating system, incorporating not only the heat exchange between the person and the environment but also the physiological, behavioural and psychological responses of the person and the control opportunities afforded by the design and construction of the building. This chapter considers the more important means of adaptation and draws attention to factors that tend to inhibit effective adaptation.

Keywords Thermal comfort · Adaptation · Clothing · Fieldwork

10.1 Thermal Comfort Research

There are two kinds of thermal comfort research

1. Climate chamber research, where the environmental conditions can be accurately controlled and where comprehensive measurements may be taken, both of the environment and of its occupants. It can be very accurate and it enables controlled experimentation, but may seem far from everyday life.
2. The field study, which takes place in the normal settings of daily life, at home or at work. It has the advantage that peoples' reported responses are those in daily

---

This chapter is developed from a lecture given in the International Exchange Committee of the Kinki Branch of the Society of Heating, Air-Conditioning and Sanitary Engineers of Japan, 17 October 2008, Kyoto, Japan.

M. A. Humphreys (✉)
School of Architecture, Faculty of Technology, Design and Environment, Headington Campus, Oxford Brookes University, Oxford, UK
e-mail: mahumphreys@brookes.ac.uk

J. Fergus Nicol
Low Energy Architecture Research Unit, Sir John Cass Faculty of Art, Architecture and Design, London Metropolitan University, London, UK
e-mail: f.nicol@londonmet.ac.uk

© Springer Nature Singapore Pte Ltd. 2018
T. Kubota et al. (eds.), *Sustainable Houses and Living in the Hot-Humid Climates of Asia*, https://doi.org/10.1007/978-981-10-8465-2_10

life and that the environments can be fairly typical of the normal building stock. The drawbacks of field research are two: comprehensive measurement either of the people or their environments would be intrusive and unpractical and that the field study does not lend itself to controlled experimentation. Adaptive thermal comfort necessarily rests on field research because adaptive behaviour is necessarily studied in the normal habitat.

Early field studies date from the 1930s. The pattern was laid down by Thomas Bedford, who in 1936 published his report *The warmth factor in comfort at work* [1]. It concerned the thermal comfort of people working in light industry in England and was conducted during the heating season. He surveyed 12 factories and conducted almost 4000 interviews. Bedford started his interviews by asking: 'Do you feel comfortably warm?' From the varied answers to his structured follow-up questions, he built a seven-category ranked scale that has become known as the 'Bedford Scale':

Much too warm
Too warm
Comfortably warm
Comfortable
Comfortably cool
Too cool
Much too cool

He measured the temperature of the hand, the foot and the forehead and of the surface of the clothing of each participant. From his measurements of the environment at the workstation, he calculated the temperature of the air (Ta), the mean radiant temperature of the surrounding surfaces (Tw), the air speed (v) and its humidity. Bedford was the first to use multivariate statistical analysis in a thermal comfort survey. All the calculations were done by hand or by using mechanical adding machines. He produced histograms, gave means and standard errors, calculated Pearson correlation coefficients and performed multiple regression analyses. Using multiple regression analysis, he derived an 'equivalent temperature' (ET):

$$ET = 0.522Ta + 0.478Tw - 0.01474\sqrt{v}(100 - Ta) \qquad (10.1)$$

(The temperatures are in degrees Fahrenheit and the air speed is in feet/minute.)
He found that the best ET for wintertime comfort was 65 °F (18 °C).

Others followed Bedford's lead, and many field studies of thermal comfort were conducted worldwide in the following years. Few were as comprehensive, and few as thoroughly analysed. Charles Webb, who perhaps should be regarded as the originator of the adaptive approach to thermal comfort, was a physicist at UK government Building Research Station (later the Building Research Establishment, BRE). We were both young researchers in his unit. Charles obtained field data from Singapore [2], Baghdad (Iraq), Roorkee (North India) [3] and Watford (near London, UK) [4]. He was the first to apply electronic data logging and computer processing to comfort surveys (in 1966). Charles noticed that his respondents were

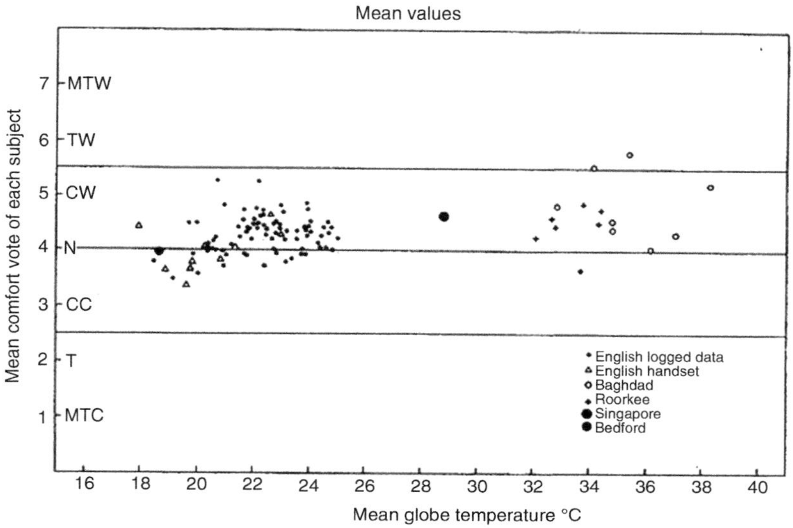

Fig. 10.1 Monthly mean comfort votes including data from various sources (Source: Humphreys and Nicol [4]) (Copyright BRE, reproduced by permission)

comfortable close to the mean conditions they had experienced, whether in Singapore, North India, Iraq or England. This suggested that they had adapted to their disparate mean conditions. Figure 10.1 shows the relationship between the new English data and Charles's other sets of data, together with Bedford's result. Notice how little the mean warmth sensation depends on the mean room temperature. Rather it depended on the departure from the mean temperature.

10.2   Thermal Comfort Meta-analysis

We thought long and hard about this result and drew a flow diagram showing thermal comfort as a part of a self-regulating adaptive system (Fig. 10.2). It includes both physiological and behavioural adaptation as feedback loops [5]. But did the total evidence from all available field studies support this interpretation? What if we collected together all their results? We found nearly 40 studies from countries all over the world, published between 1936 and 1974 and representing more than 200,000 'comfort votes' given by respondents [6]. From most of these studies, it was possible to extract the optimum temperature for comfort and the sensitivity of the respondents to temperature changes. If people had adapted to their normal indoor environment, the optimum temperature for comfort should be highly correlated with the mean temperature they had experienced. Figure 10.3 shows this to be true. And the mean 'comfort vote' should change little over a wide range of mean indoor temperatures. This was found to be so (The two outliers were from surveys

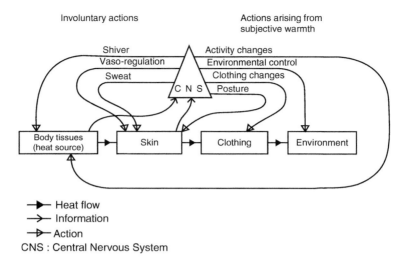

Fig. 10.2 The thermal regulatory system (Source: Nicol and Humphreys [5]) (Copyright BRE, reproduced by permission)

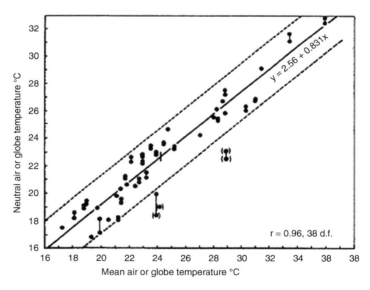

Fig. 10.3 Scatter diagram of mean temperature and neutral temperature (Source: Humphreys [6]) (Copyright BRE, reproduced by permission)

conducted when the researchers chose atypical conditions, so helping to confirm the adaptive hypothesis.) The range of comfort temperatures was too wide to be explained by new PMV eq. [7]. Next the comfort temperatures were analysed in relation to the monthly outdoor temperatures [8], these being obtained from

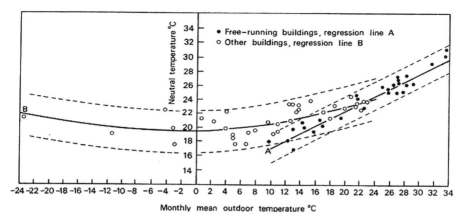

Fig. 10.4 Scatter diagram of mean temperature and neutral temperature (Source: Humphreys [8]) (Copyright BRE, reproduced by permission)

published world meteorological tables. We found the optimum temperatures most strongly related to the corresponding mean outdoor temperatures (Fig. 10.4). The strongest relation was for the 'free-running' mode of operation (no heating or cooling in use).

Changing the clothing is the most obvious behavioural adaptation to temperature. So studying clothing change should tell us more about how people adapt to their indoor environment. At the BRE field studies (1969–1977) on adaptation by clothing-change were conducted, for children at school, for people out shopping and at a zoo and for bed clothing during sleep. We found that there was little adaptive clothing change during the day, more change from day to day and more still from week to week. Clothing changes lagged behind temperature changes. People sometimes 'traded' thermal comfort for fashion (social comfort) [9].

After the publication of the meta-analysis, other researchers explored comfort from an adaptive perspective: Ian Griffiths (UK and European surveys), John Busch (surveys in Bangkok, Thailand), Andris Auliciems and Richard de Dear (Australian surveys), Gail Schiller (Brager) and team (USA surveys) [10]. Many of these surveys included an assessment of the thermal insulation of the clothing and the metabolic rate of the respondents. This enabled the result to be compared definitively with the predicted mean vote (PMV). These researchers found adaptation to be taking place, sometimes to an extent inexplicable on the PMV/PPD model. Andris Auliciems suggested that energy could be saved by using a 'thermobile' rather than a 'thermostat' [11]. The thermobile would adjust the indoor temperature set-point according to the prevailing outdoor temperature. Systematic discrepancies were found, particularly in warm and hot indoor temperatures.

The need for an adaptive approach to thermal comfort became more pressing in the 1990s. Awareness of global warming gave thermal comfort standards a new importance and urgency. Troubles with air-conditioned buildings (discomfort, high energy use, 'Sick Building Syndrome') made alternatives seem attractive. Post-

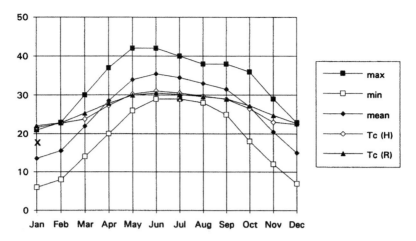

Fig. 10.5 Nicol graph for Multan, Pakistan (Source: Nicol [3])

Occupancy Evaluation demonstrated that simple buildings often performed better. There was a renewed interest in vernacular architecture, and there was a movement to encourage passive and low energy architecture (PLEA). These factors favoured the adaptive approach to comfort, seeing thermal comfort as part of a broad social response to climate and culture rather than a matter only of the physics and physiology heat exchange. Gail Brager and Richard de Dear encouraged ASHRAE to fund field studies, leading to the ASHRAE world database of field experiments and an ASHRAE session dedicated to adaptation in 1998 [12].

The Oxford Brookes Thermal Comfort Unit conducted surveys to establish adaptive comfort standards for Pakistan and followed this work by similar studies in the UK. Nicol and collaborating teams – a cooperative venture by five countries – conducted European surveys on the same pattern. The pattern comprised once-a-month surveys all year round (the transverse data), with a subset of respondents giving four reports per day for as long as they were willing to do so (the longitudinal data). Automatic environmental monitoring took place, and there was a good variety of types of office building. Figure 10.5 shows a Nicol graph for Multan, Pakistan. The graph compares the comfort temperature with the outdoor maximum and minimum temperatures, indicating which months of the year would not normally require the use of heating or cooling.

There was conceptual advance too. Nick Baker and Mark Standeven working in Cambridge, UK, linked comfort to the available means of thermal adaptation – the 'adaptive opportunity' [13]. If there was little adaptive opportunity, discomfort was likely to occur. Bordass and Leaman (working in the UK) developed protocols for the 'Post-Occupancy Evaluation' (POE) of buildings. Their results seemed to show that people who had control over their environment were more tolerant of it. They called this the 'Forgiveness Factor' [14]. If the occupants could control their environment, discomfort was less likely. Accepting the adaptive hypothesis,

Elizabeth Shove argued that comfort is a 'social construction'. Different societies, historically and geographically, have had very different comfort temperatures [15]. This suggested that in a particular society, adaptation may be limited or controlled by social and cultural factors.

We have already mentioned the ASHRAE database. It comprised some 20,000 sets of observations, each with a subjective warmth vote, corresponding thermal environmental measurements and clothing and activity records. The data are from 9 countries, 160 buildings and a variety of climates. de Dear and Brager did a meta-analysis of these data, and their results broadly confirmed the findings of the meta-analyses of 1978–1981. The 2004 revision of ASHRAE Standard 55 used this result to provide a graphical relation between comfort indoors and the outdoor mean temperature. The adaptive approach has since been included in other standards and guides for thermal comfort: CIBSE Guide Section A1 2006, 2015 (UK), CEN Standard prENrev 15,251:2007 (European Union).

10.3 Explaining the Adaptive Model

We will now explain the basic principles of the adaptive model of thermal comfort and illustrate some main features. Fundamental is the 'adaptive principle'. *If a change occurs that produces discomfort, people tend to act to restore their comfort.* The return towards comfort is pleasurable – if you are too hot, cooling down feels wonderful! People are not passive receptors of their thermal environment, but continually interact with it. So except in extreme climates, people will become adjusted to the conditions they normally experience. Discomfort arises from insufficient adaptive opportunity. Thermal comfort is an example of a 'complex adaptive system' – a multitude of interacting and possibly nonlinear variables. People must therefore be studied in their everyday habitat in all its complexity, or important variables will be overlooked. Two properties of complex adaptive systems are mathematical intractability and multiple equilibria – if the system is disturbed, it may settle at a different equilibrium position. (Other examples of complex adaptive systems are the world climate and the world economy.) So there is no such thing as a fixed optimum temperature for comfort – rather there are numerous discrete and temporary optima, each depending on its particular circumstances, social, physical and cultural.

10.4 Adaptations

Adaptation may be physiological, behavioural or psychological. Physiological adaptations to coldness include vasoconstriction, shivering, eating more and perhaps some acclimatisation to cold, though that is less certain. To warmness they include vasodilatation, sweating, eating less and acclimatisation to heat. Behavioural adaptation applies equally to adults and children and is widespread in the animal and plant worlds. The list of behavioural adaptations is endless. Responses to coldness include increased activity, increased clothing, a closed posture, cuddling up, heating the room, finding a warmer place, closing windows, avoiding draughts, modifying the building and emigrating to a more congenial climate. A parallel list could be constructed of responses to warmth. Psychological adaptations are not yet well understood. They may include expecting a range of conditions, accepting a range of sensations, enjoying variety of sensation, accepting behavioural adaptations and accepting responsibility for control.

10.5 Constraints

If the combined effect of the available actions is sufficient, comfort will be achieved, otherwise it will not. There may be insufficient opportunity for adaptive action to be fully effective. It may be *constrained* by, for example, culture and fashion, by poorly designed buildings, by work requirements or by personality. We consider some of these constraints.

Clothing fulfils a social function as well as a thermal function. It is for display as well as for comfort. This double function may constrain its use for thermal adaptation, as people might have to balance the social against the thermal. Posture may be dictated by the task that is being undertaken rather than by any thermal consideration. Posture and room temperature may both be constrained by the task, as in cooking by a hot stove. The activity level also may be dictated by the task, rather than freely chosen. Window opening may be constrained by noise and fumes, as in a building overlooking a busy street. A person working in a single office, with openable windows, controllable blinds, adequate controls for the heating, the use of a fan when needed and a liberal dress code is unlikely to suffer discomfort. There is very little constraint on adaptation; there is adequate adaptive opportunity. The same is not true of a person at work in a deep open-plan office, well away from the window, with no control over the temperature and in a company imposing a strict dress code. Again, consider a person sitting for a portrait; he or she cannot adjust the clothing, the posture, the activity or the room temperature – so discomfort is likely.

The importance of clothing needs to be fully appreciated. By means of suitable clothing, societies have achieved comfort temperatures anywhere in the range from 15 °C to 35 °C, the latter assuming substantial air movement and moderate to low humidity. In a world where energy is to be used frugally, perhaps we need to recapture the art of being fashionable in highly insulating clothing and of being smart in clothing of very light insulation. Earlier civilisations have done so – and we could do so too.

10.6   Adaptation and Building Design

It follows that the architect and engineer can do much to ensure that people will be comfortable in a building, without specifying exact comfort conditions. Factors to consider include the provision of a variety of environments, their controllability, their stability and their predictability. The thermal conditions should be capable of adjustment to be within the range considered normal in that society at that time of year, or, alternatively, people should have access to a variety of environments. Thus building design and construction should be related to the climate of the region. Seen overall, the design and construction of the building is itself an adaptive action set in the context of climate and society.

For example, consider a traditional courtyard house in northern India, where the climate is characterised by hot summers and cold winters. There are spaces for winter and summer, spaces with massive construction, spaces with light construction, sunny spaces and shady spaces, open spaces and enclosed spaces. The result is that there is always a space that is comfortable, whatever the season of the year and the time of day. Alternatively, consider an eco-house in Oxford, UK, designed by Professor Sue Roaf. The UK weather is erratic: very variable from day to day. The summers are usually moderate, but can be hot, while winters are usually moderate, but can be cold. Its heavyweight construction gives stable indoor temperatures; the high insulation minimises heating requirements; solar gain is available when needed from the glazed conservatory; the photovoltaic roof exports energy to offset electricity consumption. This house has for 20 years proved to be very comfortable to live in despite the very variable climate.

Perhaps the most difficult climate to design for without resort to air-conditioning is one that has hot humid summers and cold winters. Much can be learned from traditional design and construction and from a study of the clothing worn in the different seasons. Comfortable adaptive solutions can be found – but it is necessary to live in sympathy with the natural rhythms of the thermal environment, rather than insisting on constant conditions throughout the year.

10.7   Concluding Comments

At the beginning of the chapter, we introduced the self-regulating system that lies at the heart of the adaptive approach. During the chapter we have come to see that it must be set within the context of climate and culture (Fig. 10.6).

The adaptive model shows that comfort temperatures are diverse and variable rather than single and fixed. We have seen populations comfortable at temperatures as low as 15 °C and as high as 35 °C. Comfort temperatures in the free-running mode depend strongly on the outdoor temperature, suggesting that a society could vary its comfort temperatures to minimise fuel use.

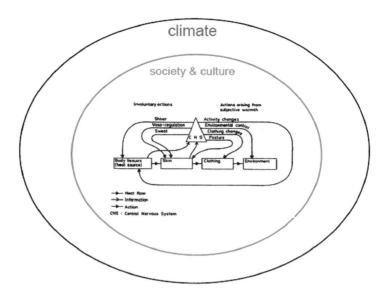

Fig. 10.6 The context of the self-regulating system (Nicol and Humphreys [5])

The comfort temperature can be seen as the current equilibrium position of a complex adaptive system. Modifying the pattern of constraints acting on the system will modify its equilibrium position. Gradual changes in the indoor thermal environments are unlikely to produce discomfort, provided there is adequate adaptive opportunity. Too much constraint will produce discomfort. The judicious use of constraints could lead to comfort solutions that are frugal in their demands on energy.

Thermal physiology and heat exchange are components within the adaptive model. But even if we had a perfect heat-exchange model, it would not be sufficient for understanding thermal comfort; contextual factors are also needed. The adaptive model does not fit easily into the current ways of expressing standards for thermal comfort. Its focus is entirely different and will lead to differently expressed standards.

References

1. Bedford T (1936) The warmth factor in comfort at work. Medical Research Council Report no. 76, HMSO, London (110pp)
2. Webb CG (1959) An analysis of some observations of thermal comfort in an equatorial climate. Br J Ind Med 16:297–310
3. Nicol JF (1974) An analysis of some observations of thermal comfort in Roorkee, India and Baghdad, Iraq. Ann Hum Biol 1(4):411–426
4. Humphreys MA, Nicol JF (1970) An investigation into thermal comfort of office workers. J Inst Heating Vent Eng 38:181–189

5. Nicol JF, Humphreys MA (1973) Thermal comfort as part of a self-regulating system. Building Res Pract (J CIB) 6(3):191–197
6. Humphreys MA (1975) Field studies of thermal comfort compared and applied. Department of the Environment, Building Research Establishment Current Paper CP 76/75 (reissued in: Journal of the Institution of Heating and Ventilation Engineers 44:5–27, 1976)
7. Fanger PO (1970) Thermal comfort. Danish Technical Press, Copenhagen
8. Humphreys MA (1978) Outdoor temperatures and comfort indoors. Build Res Pract 6 (2):92–105
9. Humphreys MA (1979) The influence of season and ambient temperature on human clothing behaviour. In: Fanger PO, Valbjørn O (eds) Indoor climate. Danish Building Research Institute, Copenhagen, pp 699–713
10. de Dear R, Brager G, Cooper D (1997) Developing an adaptive model of thermal comfort and preference. Final report on RP-884. Macquarie University, Sydney, Australia (296pp)
11. Auliciems A (1981) Towards a psycho-physiological model of thermal perception. Int J Biometeorol 25:109–122
12. ASHRAE (1998) Technical Bulletin 14(1): Field studies of thermal comfort and adaptation
13. Standeven MA, Baker NV (1995) Comfort conditions in PASCOOL surveys. In: Nicol F, Humphreys M, Sykes O, Roaf S (eds) Standards for thermal comfort. E & FN Spon/Chapman & Hall, London, pp 161–168
14. Bordass WT, Bromley AKR, Leaman AJ (1993) Are you in control? Build Serv 15(4):30–32
15. Shove E (2003) Comfort, cleanliness and convenience. Berg Publishers, Oxford

Chapter 11
Thermal Comfort in Indonesia

Tri Harso Karyono

Abstract The general thermal comfort theory assumed that there are only six parameters which affect the thermal sensation of a person. However, some studies show that a different group of people might be comfortable at different thermal conditions. Thermal comfort researchers predict that there might be other factors, such as adaptation, which affect human thermal sensations. Adaptation can be a matter of bodily adjustment to the prevailing climate that people experienced within a particular period, or it can be affected by local culture and local foods, which may affect the human thermoregulatory system and its thermal preference. As a country which possess thousands of islands, having various types of thermal conditions, cultures, languages, and foods, Indonesia have a very limited number of thermal comfort research. This chapter discusses the possible factors, which could affect the different comfort temperatures of people in this country due to the different outdoor temperature they have experienced and perhaps different cultures. A history of thermal comfort study in this country is also introduced in this chapter.

Keywords Adaptation · Climate · Comfort temperature · Culture · Indonesia · Thermal comfort study

11.1 Introduction

Indonesia, lies in the equatorial belt between 6° north and 11° south latitudes, is a country with various cultures. There are more than 300 ethnic groups with their own cultural identities. Not only they speak different languages but also they have their own distinctive cultures, including their own particular clothes, which may relate to the need of different thermal comfort. In every ethnic group, people have their own type of local foods, although most of the Indonesians today consume rice as their main food. Along with the rice, people take meat, poultry, seafood, tofu, tempeh, and

T. H. Karyono (deceased)
School of Architecture, Tanri Abeng University, Jakarta, Indonesia

© Springer Nature Singapore Pte Ltd. 2018

T. Kubota et al. (eds.), *Sustainable Houses and Living in the Hot-Humid Climates of Asia*, https://doi.org/10.1007/978-981-10-8465-2_11

vegetables. Most of people take rice about three times a day for their daily meal, including breakfast. Hot chili are quite often consumed along with the meal, while pork and alcohol are forbidden for the Moslems. The variety of food consumed by people in this country may also affect the different thermal comfort required by them since particular ingredients such as pepper, chili, and alcohol would give a sense of warmth to the human body as it may increase the metabolic rate.

Unlike in the four-season countries in which ambient temperature changes significantly between the seasons, Indonesia with its warm and humid tropical climate, on the other hand, has a similar ambient temperature between its seasons, wet and dry seasons, throughout the year. The temperature variations between dry and rainy seasons in most places in this country are very small, which doesn't affect to the most of people activities and the clothing ensembles that people wore between these two seasons.

Since thermal comfort directly corresponds to a number of factors, such as air temperature, radiant temperature, relative humidity, air velocity, human activity, and the clothing insulation, any changes of these factors would change the degrees of human thermal comfort.

Clothing ensemble which affects the thermal insulation values is mainly correspondent to the weather, predominantly the air temperature. The higher the air temperature, the thinner the clothes would be worn by people, and, vice versa, the lower the air temperature, the thicker the clothes would be worn by people. The changing of clothes in a country with four or more seasons was noticeable; there would be a summer, winter, and other seasonal clothing types within a year. This is because the different seasons have different air temperatures and people tend to adjust their clothing to respond to a particular temperature so that thermal comfort can be achieved.

In Indonesia, on the other hand, air temperature has a very small variation throughout the year, so, unlike people in the four-season climate, the Indonesians have no particular seasoning clothes. On the other hand, the Indonesian traditional people tend to wear different types of clothes as they have different cultures and beliefs. As there are hundreds of ethnic groups with their own cultures, the traditional clothes that people wore tend to be corresponded to the local climate, cultures, and beliefs of traditional people. Moreover, the more developed the culture, the more varieties the clothing ensembles would be for the traditional people. For example, the Papuans in Papua wear less clothes than the Dayaks in Kalimantan, while Javanese and Balinese wear more clothes in terms of layers and parts of the body covering the clothes.

There is a tendency that people living in the lowland and coastal areas with a higher outdoor temperature would wear fewer layer clothes than those living in the highland and mountainous areas with a lower outdoor temperature. Sarong is one of the Indonesian traditional clothes which is used by many ethnic groups, particularly in the lowland areas, such as Jakarta. Sarong is used as the main cloth in addition to shirt; however, in the highland areas, Sarong may be used as an addition, such as for blanket.

Since the nation has been developed and the country's economy is growing, there has been a change on people's cultures. People started to adopt a modern way of life. The modern Indonesians have changed the daily clothes they are wearing along with strong influence of the Western culture in this country. They left the traditional clothes and started to wear modern clothes such as shirt, trouser, blouse, skirt, and suits as their fellow in the west do.

To respond to the prevailing climate, the modern Indonesians today tend to wear light tropical clothes. Men going to work usually wear short or long sleeved cotton shirts and cotton trousers, in addition to shoes and socks. Suits are relatively rarely worn by the ordinary workers in ordinary situations. More suits are usually worn by people who hold higher positions, or those who are attending formal meetings. For women, blouses (either short or long sleeved) and medium or short skirts without stockings are commonly worn, in addition to shoes without socks.

There were not much complaining about the indoor thermal environment in the past, when most of the cities had not been developed yet and when cities were mainly still covered by green areas with not many modern buildings, let alone air conditioners installed in the buildings. Many people still wore their traditional clothes as the clothes were designed to adjust to the local climate and culture. Thermal comfort science might not be developed in the past since most of the people had never paid any intention to their thermal environment as it did not create any problem to them.

The changes in clothes from traditional to modern and also the changes in the built environment of modern Indonesia had started to create a number of problems in building indoor thermal environments. People started to feel that their indoor thermal environments were no longer comfortable as before. The massive development of air-conditioner industry and triggered by the country's economic development in the 1970s, many people began to use air-conditioners in their rooms and buildings. People started to be more aware of their indoor thermal environments, and thermal comfort science has slightly become more popular, in which a number of thermal comfort studies have been carried out in this country.

11.2   Thermal Comfort Studies in Indonesia

When the Dutch colonized Indonesia before 1945, their people were hardly adjusted to the tropical climate with its warm and humid conditions. Thermal comfort was one thing that they had to consider in designing the building and cities in their colonial countries, such as Indonesia.

It was Mom, a Dutch scientist, who firstly conducted a thermal comfort study with his colleagues in the Technische Hoogeschool Bandung (currently the Bandung Institute of Technology, ITB) in Bandung, Indonesia, between 1936 and 1940 [1].

This is the first recorded comfort study, which has been carried out in Indonesia. In this comfort study, there were three different groups of subjects: Europeans and Chinese who lived in Bandung and Indonesians had been involved. The last group was divided into two subgroups: educated and uneducated subjects. There was no information concerning the number in each group. The Indonesians wore tropical (thin) clothes; the other subjects wore suits. There was no information about subjects' clothing values. A five-point scale of the thermal comfort vote was used: uncomfortable cold, comfortable cool, comfort or optimum comfort, comfortable warm, and uncomfortable hot. The air velocities during the measurement were taken to be between 0.1 m/s and 0.2 m/s. The neutral temperature of the Indonesian subjects was 24.4 °C ET or about 26 °C T_a (air temperature), with the comfort range being 20.5 °C ET or about 23 °C T_a (with 50% RH) to 27.1 °C ET or about 31 °C T_a (with 60% RH).

In the 1970s Indonesia has developed its economy: cities were developed, wide roads were built, middle and high-rise buildings were erected, and new houses were built, occupying most of the existing green areas. The massive development of the cities has been leaving only small spaces of the open green areas in the cities. Cities are becoming warmer, while people started to be familiar with the use of air conditioners everywhere in their houses, offices, classrooms, and public buildings. People started to be aware about thermal comfort in their working spaces and homes, and the science of thermal comfort has started to be discussed slightly in the academic level.

Since then, a second thermal comfort study was recorded in 1993. This study was carried out in the office buildings in Jakarta in 1993 [2], in which 596 office workers in seven multistory buildings participated. The mean monthly outdoor temperature was about 28 °C. Thermal responses, based on ASHRAE seven-point scale [3], were collected from all the subjects. Indoor climatic parameters, i.e., air temperature (T_a) and relative humidity (RH), were recorded along with the questionnaire by using Thermal Comfort Meter BK 1212 and a thermo-hygrometer. All the measurements were carried out during office hours between 10:00 a.m. and 4:00 p.m. The comfort temperature of the whole subjects was 26.4 °C T_a [2].

The next recorded comfort study in Indonesia was carried out by Feriadi and Wong [4] between 2000 and 2001. This study was carried out in the low-income settlement in the town of Yogyakarta. There were 525 comfort votes collected from inhabitants living in this settlement. The average daily temperature in this town was 29 °C, and the comfort temperature of the subjects was 29.1 °C T_a.

The following thermal comfort study has been done by Karyono [5] in the city of Bandung in 2005. The study involved 20 students of 10 males and 10 females in the longitudinal survey. The subjects' comfort temperature was found to be 24.7 °C in terms of air temperature, or 25.7 °C in terms of globe temperature.

Some more further studies have been carried out by Alfata et al. [6] in 2012. These studies were conducted in four different office buildings in four Indonesian major cities: Medan (111 subjects), Jakarta (169 subjects), Surabaya (110 subjects), and Makassar (109 subjects). All of these cities are located in the coastal area, having temperature ranges of between 24 °C and 34 °C, with an average of 28 and 28.5 °C. Comfort temperature of subjects in Medan was 27.9 °C, while in Jakarta was 26.6 °C, Surabaya was 28.9 °C, and Makassar was 27.7 °C, all in terms of air temperature.

Recent thermal comfort studies have been done by Karyono et al. [7] in Tarumanagara University (Untar) West Jakarta with an average monthly outdoor temperature of 28.5 °C. Using a classroom, which was slightly modified to resemble a climate chamber, the longitudinal surveys, involving 54 subjects of 30 males and 24 female students, have provided a subjects' comfort temperature of 24.1 °C [7]. Furthermore, a comfort study was conducted by Karyono et al. [7] in Mercu Buana University. This study also uses a classroom as a climate chamber, as it was in Untar. There were 36 undergraduate students of 20 males and 16 females involved in this study, applying a longitudinal survey. Analyzing the data, subjects were comfortable at 24.9 °C air temperature [7].

A series of thermal comfort studies were conducted by Sri, Sulistiawan, Triswanti, and Karyono in free-running public buildings of a Chatedral (70 subjects), a museum (77 subjects), and a traditional market (72 subjects). Analyzing the data of the three buildings, subjects were comfortable at 27.7 °C, 27.7 °C, and 27.3 °C in terms of air temperature, respectively [8].

11.3 Predicting Comfort Temperature in Indonesia

By plotting all the comfort temperatures of subjects from all the above studies, except of Mom et al. [1] studies, on the mean monthly or daily outdoor temperatures of related venues where the study was taken place, a regression equation of the predicted comfort temperature (T_c) on the mean daily outdoor temperature was found to be [9]

$$T_c = 0.749 T_d + 5.953 \tag{11.1}$$

T_d is the average daily or monthly outdoor temperature. The coefficient determination ($R^2 = 0.38$) and the regression coefficient are significant at a 95% confidence level.

In general, the variations of the average daily temperatures from the previous studies' locations were limited at 24 °C (Bandung), 28 °C (South and West Jakarta), 28.5 °C (Central and North Jakarta), and 29 °C (Makassar, Surabaya, Yogya). Therefore, the equation generated in this study tends to be limited in its application. The T_c tends to be reliable to be used for locations with a range of average daily temperature of between 24 °C and 29 °C. Although it is unlikely that there is a town in Indonesia with an average daily temperature higher than 29 °C, it is not yet known whether the proposed T_c equation would work properly for people living in a lower temperature than 24 °C.

Some areas of Indonesia possess average daily temperatures lower than 24 °C (Bandung). As with some villages in the Dieng Plateau, Central Java, they have quite low average daily temperatures of about 13–15 °C [10]. There would be a question regarding the comfort temperature of these people living in this area. Since adaptation involves the way people wear their clothes, a lower temperature would make people wear thicker clothes. In addition to the metabolic rate adaptation, clothing adjustment would help people to be comfortable at a lower temperature.

Using the comfort equation of $T_c = 0.749\ T_d + 5.953$, a comfort temperature for any given location with any given average daily temperature of between 24 °C and 29 °C can be predicted.

On the other hand, in some cities in the highland areas, such as Bandung, where the average daily temperature is 24 °C (the range of daily temperatures is between 20 °C and 28 °C, or 18 and 30 °C), the comfort temperature would be 23.9 °C. Although the application of the initial T_c would be limited to a particular range of temperatures, between 24 °C and 29 °C, this comfort predictor is likely still a better tool in predicting a comfort temperature in a certain location than the current Indonesia comfort standard [11]. The current Indonesian standard was derived from the previous standard, which was purely adopted from the old American standard [3] by adding the value from 24 °C to 25.5 °C, considering the wide range of critics that Indonesians should have a higher comfort temperature than stated in the American standard. This standard is mainly addressed to office buildings built on the lowland or coastal cities, such as Jakarta and Surabaya.

The proposed comfort temperature predictor, however, does not take into account the effects of humidity since most of the Indonesian regions have no significant difference in their humidity levels, regardless of the location, whether it is in the coastal or mountainous areas. The only climatic parameters, with regard to thermal environment, which are different between locations are air temperature and slightly the air velocity. Therefore, using the daily or monthly average outdoor temperatures as the single variable to predict comfort temperature in any location in this country would be sufficient.

11.4   Conclusions

The prevailing climate in most of the Indonesian regions tends to be fairly comfortable for the most of the Indonesian people, since the outdoor temperatures are relatively close to people's comfort temperatures. In the past, when people were living modestly, with a simple culture, based on the traditional way of life, people tend to accept everything that is provided by nature. In this moderate environment, thermal comfort had never been discussed as people tend to be comfortable in any given climate environment where they live. Moreover, the traditional buildings were simply designed to respond to the local climate, which made the buildings and houses fairly comfortable for them. When the Dutch colonized Indonesia, the Dutch had to adjust to the tropical climate with its warm and humid conditions. Thermal comfort was started to be discussed in the building design.

Later on, after Indonesia proclaimed its independence and developed its economy quite well, people have started to change their lifestyle to be more modern, while the use of air conditioning in the buildings was unavoidable. People started to feel uncomfortable in the naturally ventilated building since they have already adjusted to the lower indoor temperature, created by air-conditioning system. This change has created a highly dependency of building to be air conditioned. People started to recognize about thermal condition in their rooms. A number of thermal comfort studies have been carried out since then in a number of lowland cities of Indonesia. Result from these studies, however, showed that comfort temperatures of subjects found in these studies were likely higher than the comfort temperature suggested by the current Indonesian standard. Since there are some climatic variations between cities in the lowland and highland in Indonesia, which could lead to the difference on the people's comfort temperature due to bodily adaptation, there should be a new adaptive standard introduced in this country to provide a more appropriate guidance for building to make buildings' occupants be more comfortable, without consuming a lot of energy. A standard in which comfort temperature of people in a particular location could be determined by corresponding it with its daily or monthly outdoor temperature.

References

1. Mom CPP, Wiesebron JA, Courtice R, Kip CJ (1947) The application of the effective temperature scheme to the comfort zone in the Netherlands indies (Indonesia). Chron Nat 103:19–31
2. Karyono TH (2000) Report of thermal comfort and building energy studies in Jakarta, Indonesia. Build Environ 35:77–90
3. ASHRAE, Standard 55-1992 (2013) Thermal environmental conditions for human occupancy. American Society of Heating, Refrigerating and Air-Conditioning Engineers (ASHRAE), Atlanta
4. Feriadi H, Wong NH (2004) Thermal comfort for naturally ventilated houses in Indonesia. Energ Buildings 36:614–626
5. Karyono TH (2008) Bandung thermal comfort study: assessing the applicability of an adaptive model in Indonesia. Archit Sci Rev 51:60–65
6. Alfata MNF, Sujatmiko W, Widyahantari R (2012) Thermal comfort study in the office buildings in Medan, Jakarta, Surabaya and Makassar. Final report of innovation research: the effect of air movement on thermal comfort in some office buildings in some big cities in Indonesia. Unpublished annual report. Indonesian Ministry of Public Works, Jakarta
7. Karyono TH, Heryanto S, Faridah I (2010) Air conditioning and the neutral temperature of the Indonesian university students. Archit Sci Rev 2015, 58:174–183. 33. Shahin H. Coping with nature, ten years thermal comfort studies in Iran. In Proceedings of conference: adapting to change: new thinking on comfort. Cumberland lodge, Windsor, UK, 9–11 April 2010, pp 59–80
8. Karyono TH, Sri E, Sulistiawan JG, Triswanti Y (2015) Thermal comfort studies in naturally ventilated buildings in Jakarta, Indonesia. Buildings 5:917–932
9. Karyono TH (2015) Predicting comfort temperature in Indonesia, an initial step to reduce cooling energy consumption. Buildings 5:802–813
10. Dieng Plateau. Available online: http://en.wikipedia.org/wiki/Dieng_Plateau. Accessed 22 Feb 2015
11. Badan Standardisasi Nasional (BSN) (2011) Standar Nasional Indonesia (Indonesian National Standardization), SNI 03-6572-2001. Tata Cara Perancangan Sistem Ventilasi dan Pengkondisian Udara pada Bangunan Gedung (Desain procedure for Building Ventilation and Air Conditioning Sytems)

Chapter 12
Exergetic Aspect of Human Thermal Comfort and Adaptation

M. Shukuya

Abstract This chapter describes why passive-technology measures such as better thermal insulation of exterior building envelope systems and more efficient solar control of window systems come to the top priority, from the findings available from a series of human-body exergy research. The human-body exergy consumption rate varies very much with the change in mean radiant temperature, and there is a value of mean radiant temperature giving the smallest possible human-body exergy consumption rate under a chosen set of air velocity and air temperature indoors. In such a case, where room air temperature is higher than the conventional target value in mechanical air cooling, the smallest possible human-body exergy consumption rate emerges with the mean radiant temperature ranging between 26 °C and 29 °C.

This result suggests that the first priority for summer seasons especially in hot and humid climate regions is to reduce solar heat gain from windows by external shading and from electric lighting and others in order to make the mean radiant temperature stay sufficiently low, but not too low. Such control of indoor radiant environmental condition, which is primarily to be done by passive-technology measures, should allow building occupants to take rational adaptive behaviour such as window opening and thereby to perform natural ventilation. Provided that the rational passive-technology measures are taken, then right-sized radiant cooling systems should also function effectively with the use of natural exergy resources.

Keywords Human body · Exergy consumption · Radiant temperature · Air velocity · Thermal comfort

M. Shukuya (✉)
Department of Restoration Ecology and Built Environment, Tokyo City University, Yokohama, Japan
e-mail: shukuya@tcu.ac.jp

© Springer Nature Singapore Pte Ltd. 2018
T. Kubota et al. (eds.), *Sustainable Houses and Living in the Hot-Humid Climates of Asia*, https://doi.org/10.1007/978-981-10-8465-2_12

12.1 Introduction

Research on the built environment with exergetic viewpoint has been grown to the present status since early 1990s. In due course, the understanding of exergy concept itself has advanced and been sharpened to a large extent so that it can be fully applied to the field of building science, in particular, indoor thermal environmental science [1, 2].

Exergy analysis of the built environment equipped with space heating and cooling systems, whether they are passive technology based or active technology based, articulates how much and where exergy is consumed in the whole process from its supply and consumption to the resulting entropy generation and disposal [1, 2].

Space heating and cooling systems themselves are physical systems at their own right, but their purpose is to control the built environmental condition within a certain range that allows the building occupants to take rational adaptive behaviour being healthy and comfortable with the lowest possible and rational human-body exergy consumption rate.

The present chapter first reviews briefly, without mathematical formality, the fundamentals of thermodynamics, with a focus in particular on how the "thermal energy dispersion" occurs in nature and on what the "consumption" implies. Then, a couple of new findings obtained from a series of sensitivity analysis focussing on the effects of mean radiant temperature, air temperature, and air velocity indoors under hot and humid conditions are described.

12.2 Exergy Balance Equations in General

Generally speaking, a thermodynamic system to be investigated, whether it is a human body, a building wall, a heat pump, or others, is regarded to be surrounded by its environmental space. Any working system works feeding on some energy and matter, while at the same time storing their portions and/or giving off the remainders. In due course, the whole amount of energy is necessarily conserved, while, on the other hand, an amount of entropy is necessarily generated. The generated entropy may be stored for a while not to be discarded into the environment, but sooner or later it must be discarded so that the homeostatic condition of the system remains unchanged within a certain required range.

A portion of energy, whose associated temperature, pressure, and chemical potentials are in equilibrium with their corresponding values in the environment, has no capability of dispersion. The other portion of energy, which has not yet dispersed, has a capability of dispersion. That is exactly "exergy", which is the driving agent for any workable system. As can be seen in the left-hand side of Fig. 12.1, a working system feeds on exergy from a source and dumps the generated

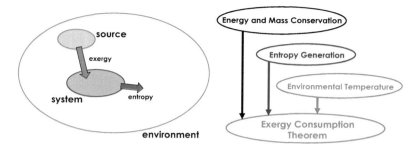

Fig. 12.1 A system and its environment. The system performs "exergy-entropy" process. The exergy balance equation is derived from the concepts of energy and entropy together with the environmental temperature [2]

entropy, which is proportional to the exergy consumption within the system, into the environment. Such a process is called "exergy-entropy process" [2].

Exergy balance equation for the system can be set up in a general form as

$$[\text{Exergy input}] - [\text{Exergy consumed}] = [\text{Exergy stored}] + [\text{Exergy output}]. \quad (12.1)$$

In order to set up the detailed form of Eq. (12.1), we first set up an energy balance equation according to "the law of energy conservation (the 1st law)" and then the corresponding entropy balance equation according to "the law of entropy generation (the 2nd law)". Extraction of the product of the entropy balance equation and the environmental temperature from the energy balance equation brings about the "exergy" balance equation. The whole procedure is as schematically shown in the right-hand side of Fig. 12.1.

12.3 The Human-Body Exergy Balance

A set of thermal exergy balance equations for human body based on the two-node model of body core and body shell was developed by combining the energy and entropy balance equations together with the environmental temperature. The detailed formulae of the respective terms are thoroughly given in [1–3].

Knowing the values of exergy input, stored, and output with the calculated results of body core and body shell temperatures, the exergy consumed can be determined so that the equality in Eq. (12.1) is true.

The exergy consumption emerges due to three kinds of dispersion: one is the absorption of radiant exergy incident on the body surface; another is heat transfer caused by the temperature difference between the body core being almost constant at 37 °C and the body shell and the clothing surface ranging from 20 °C to 36 °C; and the last is the dispersion of liquid water into water vapour, free expansion of water molecules into the surrounding space.

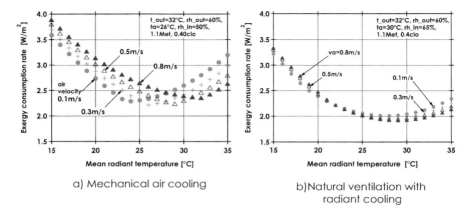

a) Mechanical air cooling

b)Natural ventilation with radiant cooling

Fig. 12.2 Two examples of the relationships between human-body exergy consumption rate and mean radiant temperature with average air velocity as parameters. In these charts, outdoor air temperature and relative humidity are assumed to be 32 °C and 60%, respectively [3, 4]

The exergy output plays important roles in disposing of the generated entropy due to the exergy consumption within the human body. This process of outgoing exergy flow together with exergy consumption influence very much on human thermal well-being.

12.4 Mean Radiant Temperature, Air Velocity, and Exergy Consumption Rate

A series of analyses were made for outdoor condition of air temperature and relative humidity of 32 °C and 60%, assuming that metabolic energy emission rate and clothing insulation are 1.1 Met and 0.4 clo, respectively, in order to investigate further and thoroughly the relative importance of air velocity, which is realized by natural ventilation or the use of fans together with the relative importance of mean radiant temperature that are likely to occur in a variety of built environment under hot and humid climate regions [3, 4].

Figures 12.2 and 12.3 show their results. Either of Fig. 12.2 or Fig. 12.3, there are two charts: the left one represents a case of mechanical air cooling with the target air temperature and relative humidity of 26 °C and 50% and the right one a case of air temperature and relative humidity being assumed to be 30 °C and 65%. The room air temperature and relative humidity assumed in the latter case may be considered to represent a condition achievable by moderate radiant cooling either with nocturnal ventilation for cool-exergy harvest in the following daytime within the building envelopes or that with thermally activated building envelope system with a mechanical radiant cooling system.

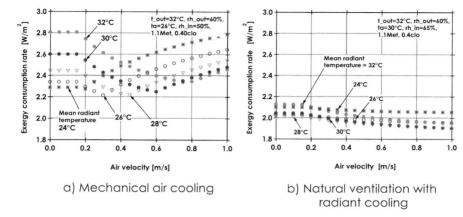

Fig. 12.3 Two examples of the relationships between human-body exergy consumption rate and average air velocity with mean radiant temperature as parameters. In these charts, outdoor air temperature and relative humidity are assumed to be 32 °C and 60%, respectively [3, 4]

Looking at Fig.12.2a, we find that there are values of mean radiant temperature providing the human body with the smallest exergy consumption rate; e.g., in the case of average air velocity of 0.1 m/s, the mean radiant temperature of 24 °C gives the smallest value of human-body exergy consumption rate at 2.3 W/m².

As the average air velocity becomes higher, the mean radiant temperature, which results in the lowest exergy consumption rate, shifts from lower to higher values. The lowest human-body exergy consumption rate emerges at the mean radiant temperature of 26 °C in the case of air velocity at 0.3 m/s.

In Fig.12.2b, as mentioned above, the air temperature is assumed to be 30 °C, that is, 4 °C higher than in the case of mechanical air cooling, whose target air temperature is 26 °C. Mainly for this reason, the lowest human-body exergy consumption rate in all cases of Fig.12.2b stays from 1.9 to 2.0 W/m², while, on the other hand, that in Fig.12.2a from 2.2 to 2.3 W/m². If the mean radiant temperature is kept within the range from 21 °C to 24 °C, the change in air velocity does not affect the human-body exergy consumption rate at all. This is due to the skin temperature being lowered mainly by radiant cooling rather than air cooling.

In Fig.12.2b, similarly to Fig.12.2a, there are the values of mean radiant temperature giving the lowest human-body exergy consumption rate; they are all a little higher than the cases (b) shown in Fig.12.2a, that is, from 26 °C to 29°C.

In Fig.12.2b, for the range of mean radiant temperature from 20 °C to 25°C, any changes in average air velocity do not affect the values of human-body exergy consumption rate. But, for the range of mean radiant temperature above 26 °C, as the mean radiant temperature rises, the influence of a change in average air velocity becomes gradually significant. This is due to the skin layer exposed more to long-wavelength radiation in the cases of higher mean radiant temperature than those of lower mean radiant temperature.

The lowest possible human-body exergy consumption rate being 1.9 W/m^2 in the cases of air velocity at 0.8 m/s by natural ventilation with the mean radiant temperature from 28 °C to 30 °C is about 20% smaller in the case of mechanical air cooling shown in Fig.12.2a.

What we can see from Fig.12.2a and b suggests that lowering the mean radiant temperature being sufficiently low, let's say, down to 30 °C, should be achievable not by mechanical cooling immediately but first with appropriate thermal insulation of building envelopes together with the installation of external shading devices over the windows.

The charts shown in Fig.12.3a and b demonstrate the same results as those in Fig.12.2, but shed light from another angle. These charts are the relationships between human-body exergy consumption rate and average air velocity with mean radiant temperature as parameters.

In the conditions of mechanical air cooling shown in Fig.12.3a, average air velocity is required to be lower than 0.15 m/s for avoiding draught. In such cases, lowering the mean radiant temperature from 32 °C to 24 °C always results in smaller human-body exergy consumption rate. But for the conditions of air velocity higher than 0.15 m/s, the lowest human-body exergy consumption rate occurs at different air-velocity values depending on the values of mean radiant temperature.

For such a condition of very high mean radiant temperature such as 30 or 32°C, it is necessary to make the air velocity up to 0.6 or 0.8 m/s; this is why people start fanning themselves under hot conditions. But, for the mean radiant temperature of 28 or 26 °C, the value of air velocity decreases down to 0.45 m/s or 0.3 m/s, respectively, and thereby the resulting human-body exergy consumption rate also becomes smaller.

In the cases of higher air temperature as shown in Fig.12.3b, if the air velocity remains low, the lowest human-body exergy consumption rate is achieved in the case of 28 °C. In other words, either higher or lower than 28 °C of mean radiant temperature results in larger human-body exergy consumption rate.

Increasing the air velocity with low mean radiant temperature does not bring about a large change in human-body exergy consumption rate. But, if the mean radiant temperature is 28 or 30 °C, it is possible to make the human-body exergy consumption rate as small as 1.9 W/m^2 by enhancing the average air velocity up to 1.0 m/s.

Radiant cooling systems, which require moderately low surface temperature, should be attractive, because of the necessity of smaller cool-exergy input that should be available from a variety of immediate natural sources.

12.5 Conclusions

This chapter has demonstrated the exergetic characteristics of human body indoors under hot and humid summer conditions. Discussion on the human-body exergy consumption rate in relation to mean radiant temperature and air velocity has led to the following knowledge:

1. The human-body exergy consumption rate, which is a kind of thermal-stress indicator, is very sensitive to the changes in mean radiant temperature and air velocity.
2. The human-body exergy consumption rate becomes smaller with a combination of moderately high air and low mean radiant temperatures than a combination of low air and high mean radiant temperatures.
3. In the case of moderately low mean radiant temperature with rather high air temperature, the enhancement of air velocity necessarily brings about smaller human-body exergy consumption rate.

Therefore as summarized in 2. and 3., the implementation of thermally well-insulated building envelope together with well-solar-controlled windows and also the reduction of internal heat generation as much as possible are the top priority in the pursuit of low-exergy buildings under hot and humid outdoor climatic conditions. Such measures should enable the building occupants to take rational adaptive behaviour to maximize thermal comfort with the least exergy consumption of fossil fuels by optimizing the use of natural exergy sources to be found in the immediate environmental space.

References

1. Shukuya M (2009) Exergy concept and its application to the built environment. Build Environ 44:1545–1550
2. Shukuya M (2013) Exergy – theory and applications in the built environment. Springer, London
3. Shukuya M, Iwamatsu T, Asada H (2012) Development of human-body exergy balance model for a better understanding of thermal comfort in the built environment. Int J Exergy 11 (4):493–507
4. Shukuya M (2015) Indoor-environmental requirement for the optimization of human-body exergy balance under hot/humid summer climate. PLEA2015, Bologna, 9–11 September 2015

Chapter 13
Thermal Sensation and Comfort in Hot and Humid Climate of Indonesia

Tomoko Uno, Daisuke Oka, Shuichi Hokoi, Sri Nastiti N. Ekasiwi, and Noor Hanita Abdul Majid

Abstract In conventional air-conditioning design, the comfortable range is considered to be between temperatures of 25 and 27 °C and relative humidity of 40 and 60%; these numbers vary only slightly based on a person's race and country. However, several studies conducted in regions with hot and humid climates show that the observed thermal comfort requirements often do not agree with those obtained from European-based studies. In this chapter, the results from the questionnaire surveys and measurements in residences in the hot and humid regions of Indonesia and Malaysia are shown. It was found that the residents use air conditioning at low temperatures for sleeping and that many people consider cold condition as "comfortable" and desire "cooler" conditions even if they feel "cold" in thermal sensation.

Keywords Field survey · Thermal environment · Sleeping · Comfort temperature · Thermal sensation · Comfort sensation

T. Uno (✉)
Department of Architecture, Mukogawa Women's University, Nishinomiya, Hyogo, Japan
e-mail: uno_tomo@mukogawa-u.ac.jp

D. Oka
Kyoto University, Kyoto, Japan

S. Hokoi
Kyoto University, Kyoto, Japan

Southeast University, Nanjing, China

S. N. N. Ekasiwi
Department of Architecture, Institut Teknologi Sepuluh Nopember, Surabaya, Indonesia

N. H. A. Majid
Department of Architecture, International Islamic University Malaysia, Gombak, Malaysia

© Springer Nature Singapore Pte Ltd. 2018
T. Kubota et al. (eds.), *Sustainable Houses and Living in the Hot-Humid Climates of Asia*, https://doi.org/10.1007/978-981-10-8465-2_13

13.1   Introduction

In conventional air-conditioning design, the comfortable range of temperature is considered to be 25–27 °C and relative humidity levels as 40–60%, as defined by American Society of Heating, Refrigerating, and Air-Conditioning Engineers (ASHRAE) [1]; variation in the abovementioned numbers is negligible based on person's race and country [2]. The predicted mean vote (PMV) and the predicted percentage of dissatisfaction (PPD) are based on these concepts, and they are widely used as the international thermal comfort standard ISO7730 for air-conditioning design. Because the comfort equation is based on experiments using North Americans as subjects, the differences from the results obtained in the regions with hot and humid climates have been discussed. Several studies on thermal comfort in tropical climates have been conducted [3–7]. These studies show that the observed thermal comfort requirement in hot climates is not consistent with that dictated by the thermal comfort equation. It has also been observed that the neutral temperature in naturally ventilated buildings is higher than that in the air-conditioned buildings because of the adaptation to outdoor conditions [8].

de Dear et al. [9, 10] conducted an experiment in Singapore in which subjects in a climate chamber adjusted the temperature according to their wish. On observing the preferred temperature, it was found that the thermal sensation was significantly lower than neutral. This indicated that subjects prefer to select cool conditions. The observed "higher neutral temperature" and "preference for cool condition" seem to be conflicting. The reasons for this conflict could be attributed to the differences between the climate chamber experiment and field survey or whether use of air conditioning is prevailing.

Therefore, based on the results of a questionnaire survey and measurements of indoor thermal conditions in residences, this chapter aims to clarify the temperature and humidity level preferred by the people in these regions to feel comfortable.

13.2   Field Survey on Air Conditioner Usage

13.2.1   Target and Outline of the Survey

Surabaya, Indonesia (SB, 7°S 112°E), and Kuala Lumpur, Malaysia (KL, 3°N 102°E), which are characterized by the humid, tropical climate of Southeast Asia, were chosen for the survey. The daily maximum temperature is above 32 °C throughout the year, and relative humidity is high (Fig. 13.1). The first survey was conducted once in September 2009, and the second one was conducted twice in August and September 2010. The measurements of indoor thermal environment were taken in April, May, November, and December 2010.

Fig. 13.1 Climate in
Surabaya and Kuala Lumpur

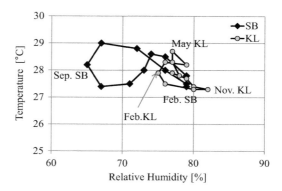

Table 13.1 Questionnaire items

1st survey
Living condition: Family structure, house condition (layout, construction, surrounding environment), income, owned electrical appliances, windows opening pattern, electric fees, sleeping condition (hours, usage of AC, clothes and usage of blanket)
Condition during AC was used: AC manufacturer, year AC installed, price, usage hour of AC, set temperature, thermal sensation, comfortable sensation, problem of health
2nd survey
Condition during AC was used: Necessity of AC, consciousness for AC use, consciousness for energy saving under AC use, conditions in bedroom with AC (occupants, set temperature, thermal sensation, comfortable sensation, prefer temperature, behavior when they feel cold in the bed room, electric cost)

Residential buildings were chosen for the survey because the occupants have a higher chance of being able to choose their preferred temperature setting for the air conditioner (AC).

The questionnaire surveys for the air-conditioning usage and housing conditions in Table 13.1 were conducted for students in the Architecture Department of the Institut Teknologi Sepuluh Nopember (ITS), Indonesia, and in the Architecture Department of the International Islamic University, Malaysia (IIUM). The questionnaires were handed out during a class, and students were directed to fill them out at home. They were then collected in a later class. Since the families of university students usually belong to the wealthier class, the results of the questionnaire might be biased against an average Indonesian or Malaysian sample. Results from 56 houses (SB) and 108 houses (KL) in 2009 (1st survey) and 37 houses (SB) and 50 houses (KL) in 2010 (2nd survey) were obtained. In the questions about the set temperature of ACs and the resident's sensations in the bedroom in both first and second surveys, the several residents in a house answered. So the number of the response was more than the number of households.

13.3 Results of Survey

13.3.1 Classification of the Targets

The number of family members ranged from 1 to 11 with an average of 4.7 (SB) and from 2 to 9 with an average of 5.1 (KL), which included the domestic help. The monthly income of 60% of the respondents ranged from 3,000,000 IDR to 6,000,000 IDR (SB) and had an average of 5,500,000 IDR (610 USD, 1 USD = 9000 IDR, April 2010). In Kuala Lumpur, 26% of the respondents had annual incomes less than 20,000 MYR, while 20% had more than 100,000 MYR. The average monthly income was 4500 MYR (1400 USD, 1 USD = 3.2 MYR, April 2010). Because the price of a new AC ranged from 2,500,000 IDR to 18,000,000 IDR (SB) [11], a low-cost air conditioner could be obtained with half the monthly income of the respondents.

On an average, 2.1 (SB) and 3.0 (KL) ACs were installed per family, which was almost the same as 2.3 in Japan in 2010. According to studies conducted in 2009, 70% of the ACs were installed after 2005, while 17% were installed between 2000 and 2004, which shows the rapid rise in the number of installations in residences.

13.3.2 Results for Use of Air Conditioner

In both cities, ACs were placed almost exclusively in bedrooms or living rooms. In particular, the installation rate of ACs in bedrooms was very high, i.e., 92% (SB) and 77% (KL), while that in the living rooms was 8.3% (SB) and 18% (KL). This implies that they gave priority to the thermal environment during sleeping.

The longest daily operating time of ACs was 24 h in both cities (Fig. 13.2); the average use was 10 h (SB) and 9.1 h (KL). These are very long periods compared to typical usage patterns in Japan [13].

As shown in Fig. 13.3, the temperature settings of ACs widely varied from 16 to 28 °C in both cities, and two peaks were observed—one between 18 and 20 °C and another around 25 °C. The average temperature settings were 21.6 °C (SB) and 22.0 °C (KL). These values were rather low for ambient temperature setting for sleeping compared with the average value in Japan, which is approximately 25.8 °C [12].

Figure 13.4 gives information regarding the thermal sensations that the respondents reported while sleeping (1 Cold, 2 Cool, 3 Neutral, 4 Warm, and 5 Hot). It can be seen that there was a clear difference between the respondents in two cites. The percentage of respondents in Surabaya who reported a thermal sensation of 1 was approximately 40%, which is higher than the respondents who reported 3 (neutral) as the thermal sensation, while in Kuala Lumpur, the percentages of 1 (cold) and 3 (neutral) are almost same. Anyhow, the votes for 1 (cold) and 2 (cool) accounted

Fig. 13.2 Operating hours (first questionnaire) [15]

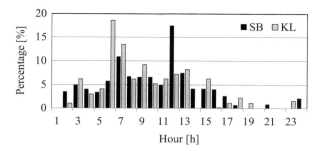

Fig. 13.3 Set-point temperature (first questionnaire) [15]

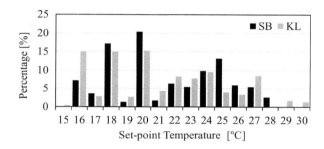

Fig. 13.4 Thermal sensations (second questionnaire) [15]

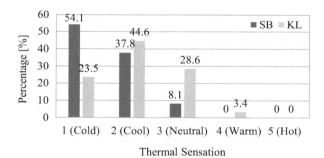

for more than 95% in Surabaya and 75% in Kuala Lumpur of the total votes. The average values were 1.7 (SB) and 2.0 (KL). The ACs were set to temperatures that produced "Cool" or "Cold" thermal sensations, especially in Surabaya.

From responses to the question about clothing worn during sleeping, which was asked only in Surabaya, it was found that most of the respondents wore T-shirts and light shorts or *Kebaya*, a traditional Muslim clothing, whose insulation values are approximately 0.3 clo. In addition, the bedclothes for half of the respondents were light cotton blanket, and 20% of the respondents used nothing.

Because ACs were mainly installed in the bedrooms and the people used them for long hours at the temperature settings noted above, they felt cold during their sleep that negatively affected their health.

In both SB and KL, 46% and 27% of the respondents, respectively, were aware of health problems such as drying (25%), feeling chill, and feeling ill. Some respondents reported other symptoms such as headache, stomach ache, pain in eyes, dehydration, and absence of sweating. It was clear from the survey that using ACs at low-temperature settings caused health problems.

13.4 Thermal Sensation

Here, the thermal sensation, comfort sensations, comfort temperature, and their relations based on the second survey are discussed. Figures 13.5 and 13.6 are the breakdowns of comfort sensation and thermal preference in each thermal sensation, respectively. The numbers in the figures are the numbers of respondents.

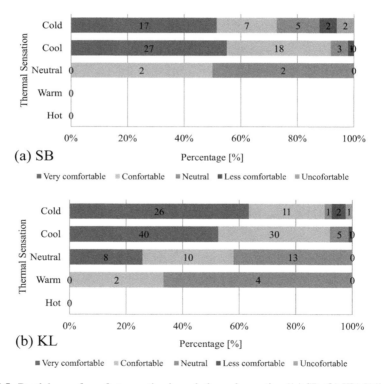

Fig. 13.5 Breakdown of comfort sensation in each thermal sensation ((**a**) SB, (**b**) KL) [14]

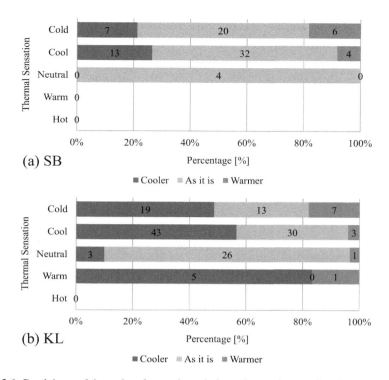

(a) SB

(b) KL

Fig. 13.6 Breakdown of thermal preference in each thermal sensation ((**a**) SB, (**b**) KL) [14]

13.4.1 Relation Between Thermal Sensation and Comfort Sensation

In both cities, with regard the respondents who reported a "cold" or "cool," the percentage of a "very comfortable" or "comfortable" was quite high, i.e., 80% (SB) and 69% (KL) of all respondents, and 84% (SB) and 91% (KL) of the respondents who reported "cold" or "cool" (Fig. 13.5). With regard to the respondents who reported "neutral," the percentage of "very comfortable" or "comfortable" is lower, i.e., 50% (SB) and 58% (KL). This shows that the residents felt comfortable because of the sensation of cold and they preferred cooler conditions rather than the neutral condition.

In addition, in the respondents who answered "cold" in the thermal sensation, the percentage of respondents who reported "less comfortable" and "uncomfortable" were low at 12% (SB) and 8% (KL), respectively. Thus, the people do not have a negative impression about "cold."

13.4.2   Relation Between Thermal Sensation and Thermal Preference

Twenty-five percent (SB) and 50% (KL) of respondents who reported "cold" or "cool" desired a cooler environment, and these numbers are much higher than those who preferred "warmer" (Fig. 13.6). Therefore, even if the people in Surabaya or Kuala Lumpur experienced cold or cool conditions, they did not feel "uncomfortable" and desired the "cooler" condition.

13.4.3   Relation Between Comfort Sensation and Thermal Preference During Cold Thermal Sensation

With regard to respondents who reported "very comfortable" in the comfort sensation and "cold" in the thermal sensation, the rate of the respondents who reported "as it is" on the questions of the thermal preference accounted for 59% (SB) and that of the respondents who reported "cooler" accounted for 72% (KL), as shown in Fig. 13.7. Especially in Kuala Lumpur, respondents who felt the cold condition tended to desire "cooler" condition.

In both cities, the percentage of respondents who preferred "cooler" among the respondents who reported "very comfortable" or "comfortable," and those who reported "very comfortable" was higher than that of "comfortable." While for each type of comfort sensation, the percentage of respondents who reported "warmer" among the respondents who reported "uncomfortable" was higher rather than that of "comfortable." From these results, we can say that there are two attitudes toward "comfortable" under "cold" thermal sensation: (1) a person who prefers the cool or cold condition and desires a colder condition and (2) a person who feels uncomfortable about the cold condition and desires a warmer condition.

13.4.4   Relation Between the Comfort Sensation and Thermal Preference

Among respondents who reported "cold" and "cooler," all of the respondents answered "very comfortable" or "comfortable," as can be observed in Fig. 13.8. This percentage was higher in the group of the respondents who reported their satisfaction level as "very comfortable" with regard to the comfort sensation.

In addition, considering those respondents who reported "very comfortable" (Fig.13.8), the percentage of respondents who preferred "cooler" was higher than that of respondents who perceived the current condition as the most suitable for

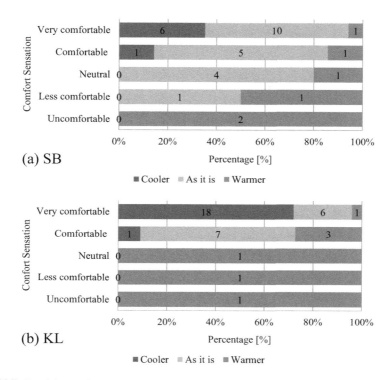

Fig. 13.7 Breakdown of thermal preference in each comfort sensation with regard to respondents who reported "cold" in the thermal sensation ((**a**) SB, (**b**) KL)

them. Furthermore, there were 33% (SB) and 57% (KL) of respondents who reported "very comfortable" or "comfortable" in the group who reported "cold" and "warmer." This showed that there were people who felt "cold" while not feeling uncomfortable and desired a warmer condition. Conversely, the 50% (SB) and 29% (KL) of the respondents who reported "warmer" and "cold" felt "less comfortable" or "uncomfortable." Thus, we can say that there are two attitudes toward the "cold" condition.

13.4.5   Thermal Sensation and Temperature Setting of Air Conditioner

Figure 13.9 shows the breakdown of the thermal sensation in each group of the temperature setting of AC. Each result obtained from either city is divided into three groups: below 18 °C (29.4%), from 19 to 21 °C (22.4%), and above 22 °C (48.2%) in Surabaya and from 15 to 19 °C (28.3%), from 20 to 24 °C (39.2%), and from 25 to

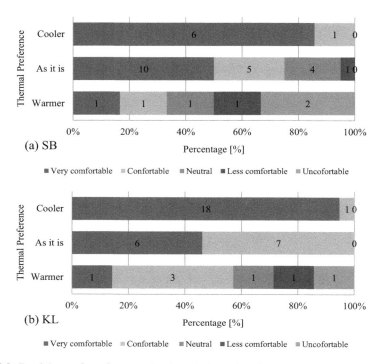

Fig. 13.8 Breakdown of comfort sensation in each thermal preference with regard to respondents who reported "cold" in the thermal sensation ((**a**) SB, (**b**) KL)

30 °C (32.5%) in Kuala Lumpur. Because we have assumed the control range of local AC in Kuala Lumpur to be lower than 30 °C, responses of over 31 °C have been omitted.

In Surabaya, 75% of the respondents in low-temperature setting group (under 18 °C) reported "cold" as the thermal sensation. The number of respondents, which is higher than that of the group that selected the high-temperature setting, considered the low-temperature condition as "cold." In Kuala Lumpur, the respondents who reported "cold" typically selected lower set temperatures. In both cities, the percentage of respondents who reported "very comfortable" was the highest among those who selected the low-temperature setting, while the percentage of the respondents who reported "very comfortable" and "comfortable" are over 80% in each group.

This indicated that there were two peaks when the set temperature was plotted: one around 18 °C and another around 24 °C, as shown in Fig. 13.4. The people who selected the low-temperature setting preferred the cooler condition, and the others who selected the higher temperature setting choose the suitable temperature depending on the situation.

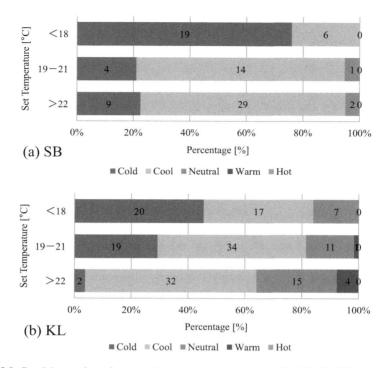

Fig. 13.9 Breakdown of comfort sensation in each set temperature ((**a**) SB, (**b**) KL)

13.4.6 Thermal Environment in Residences

According to the measurements of the thermal environment in the bedrooms of the 13 semidetached houses, the minimum room temperature occurred during time of sleep (21:00 and 6:00), and the average minimum temperature was 23.9 °C. Because the room temperature continued to decrease in most of the houses, it can be said that the residents desire lower-temperature environment; however, because of non-optimum values of thermal performances such as insulation or airtightness and the lack of capacity of ACs, the room temperature cannot be realized at the desired level.

An appropriate temperature setting is imperative from the viewpoint of the residents' health. Furthermore, the thermal performance of the houses should be improved from an energy-saving viewpoint. Improving the airtightness of air-conditioned rooms can be especially effective in this regard [16].

13.5 Comparison with Previous Studies on Thermal Sensation

There are two types of comforts: active and passive [17–20]. According to Nagano et al. [18], active comfort refers to an active and vibrant condition such as "a desire for development," while passive comfort refers to a physiologically easy condition such as "a desire for deficiency." de Dear et al. [7, 9, 10] conducted an experiment and a field study in Singapore to calculate a "neutral temperature" based on the subjects' reports of thermal sensations. The neutral temperature found was 28.5 °C [7]. Conversely, in an experiment wherein subjects chose their own preferred temperature, the value chosen was 25.4 °C [9]. The difference of 3 °C in the two experiments points to the result in which the low-temperature condition causing pleasant cold comfort sensation based on "active comfort" selected in the latter experiment.

Figure 13.10 shows the percentage of respondents who reported "uncomfortable" or "less comfortable" in each thermal sensation. The percentage of "uncomfortable" was lower than 10% among those who chose the sensation of "cold," "cool," or "neutral," while it was 67% in the sensation of "warm." The cold–hot asymmetry indicated that people desired or preferred cooler condition. Further, this showed that although $-0.5 < \text{PMV} < 0.5$ and PPD $<10\%$ conditions are recommended by the 7-point ASHRAE scale as the international standard for thermal environment, it is not necessarily relevant in regions with hot and humid climates. A similar result was obtained from the field survey of a hot dry climate in Oman [14].

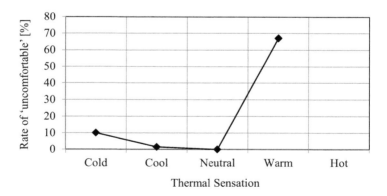

Fig. 13.10 Rate of "uncomfortable" in each thermal sensation (This figure shows the rate of the respondents who reported "uncomfortable" in the comfort sensation to each thermal sensation. The total number of the respondents were 86 (SB) and 154 (KL))

13.6 Conclusions

In this chapter, the current thermal environment, residents' thermal sensation, and comfort sensations were surveyed in two cities—Surabaya, Indonesia, and Kuala Lumpur, Malaysia. Based on these results, differences in the existing thermal sensation studies are examined.

From the results of the field survey on the thermal environment and the usage of ACs in Surabaya and Kuala Lumpur, it was shown that ACs are used for long hours at low-temperature settings during sleeping and people tended to select a low-temperature setting, which they felt was "cold." However, some of the respondents reported health problems with use of the low-temperature condition.

Many people selected "cold" in the thermal sensation, "very comfortable" in the comfort sensation, and "cooler" in the thermal preference at the same time. It was shown that there are people who desire for the "cooler" condition as comfortable.

It can be said that most of the people belonging to regions with hot and humid climates evaluate their thermal environment based on "active comfort." The selected temperature differences in the two experiments to obtain the neutral temperature and preferred temperature can be ascribed to the difference in the response of the subjects—whether based on passive comfort or active comfort. Moreover, the cold–hot asymmetry of the PMV and PPD is cleared, and thus the careful use of the international standard ISO 7730 is necessary in the design of air conditioning in regions with hot and humid climate.

References

1. ASHRAE (2005) Chapter 8: Thermal comfort. In: ASHRAE handbook, fundamentals. ASHRAE
2. Fanger PO (1970) Thermal comfort analysis and applications in environmental engineering. McGraw-Hill, New York
3. Nicol JF (1973) An analysis of some observations of thermal comfort in Roorkee, India and Baghdad, Iraq. Ann Hum Biol 1(4):411–426
4. Busch J (1992) A tale of two populations: thermal comfort in air-conditioned and naturally ventilated offices in Thailand. Energ Buildings 18:235–249
5. Sharma MR, Ali S (1986) Tropical summer index – a study of thermal comfort in Indian subjects. Build Environ 21(1):11–24
6. Nicol F (2004) Adaptive thermal comfort standards in the hot–humid tropics. Energ Buildings 36:628–637
7. de Dear RJ, Leow KG (1990) Indoor climate and thermal comfort in high-rise public housing in an equatorial climate: a field–study in Singapore. Atmos Environ Part B 24(2):313–320
8. ASHRAE (2010) Thermal environmental conditions for human occupancy. ASHRAE Standard 55-2010
9. de Dear RJ, Leow KG, Ameen A (1991) Thermal comfort in the humid tropics. Part I. Climate chamber experiments on temperature preference in Singapore. ASHRAE Trans 97(1):874–879
10. de Dear RJ, Leow KG, Ameen A (1991) Thermal comfort in the humid tropics. Part II. Climate chamber experiments on temperature acceptability in Singapore. ASHRAE Trans 97(1):880–886

11. http://www.harganya.com. Accessed June 2016
12. KIRIN (2008) Consciousness research on "way to spend in summer", report vol. 9. http://www.kirinholdings.co.jp/news/2008/0725_01.html. Accessed June 2016
13. Bogaki K, Sawachi T, Yoshino H, Suzuki K, Akabayashi S, Inoue T, Ohno H, Matsubara N, Hayashi T, Morita D (1998) Study on the living room temperature and it's regional differences in summer & winter season: study of energy consumption in residential buildings from the viewpoint of life style, on the basis of national scale surveys Part 2. J Archit Plann 63 (505):23–30
14. Madjid NHA, Takagi N, Hokoi S, Ekasiwi SNN, Uno T (2014) Field survey of air conditioner temperature settings in a hot, dry climate (Oman). HVAC&R Res 20(7):751–759
15. Ekasiwi SNN, Madjid NHA, Hokoi S, Oka D, Takagi N, Uno T (2013) Field survey of air conditioner temperature setting in hot, humid climates, Part 1: questionnaire results on use of air conditioners in houses during sleep. J Asian Archit Build Eng 12(1):141–148
16. Uno T, Hokoi S, Ekasiwi SNN (2003) Survey on thermal environment in residences in Surabaya, Indonesia: use of air conditioner. J Asian Archit Build Eng 2(2):15–21
17. Kuno S (1989) A study on comfort. Summaries of technical papers of annual meeting, AIJ, pp 71–72
18. Horikoshi T (1991) Investigation on thermal comfort and thermal sensation. Summaries of technical papers of annual meeting, AIJ, pp 707–708
19. Nagano K, Matsubara N, Kurazumi Y, Narumi D, Gassho A, Ito K (1997) Psychological and physiological influence of combined environment on non-specific evaluation: considerations based on the concept of positive and negative comfort. AIJ (Kiniki), pp 69–72
20. Seo H, Bogaki K (1995) Fundamental study on the structure of comfort and pleasantness. J Archit Plann Environ Eng 60(475):75–83

Chapter 14
Development of an Adaptive Thermal Comfort Equation for Naturally Ventilated Buildings in Hot and Humid Climates

Doris Hooi Chyee Toe

Abstract The objective of this study was to develop an adaptive thermal comfort equation for naturally ventilated buildings in hot-humid climates. The study employed statistical meta-analysis of the ASHRAE RP-884 database, which covered several climatic zones. The data were carefully sorted into three climate groups including hot-humid, hot-dry, and moderate and were analysed separately. The results revealed that the adaptive equations for hot-humid and hot-dry climates were analogous with approximate regression coefficients of 0.6, which were nearly twice those of ASHRAE Standard 55 and EN15251, respectively. Acceptable comfort ranges showed asymmetry and leaned towards operative temperatures below thermal neutrality for all climates. In the hot-humid climate, a lower comfort limit was not observed for naturally ventilated buildings, and the adaptive equation was influenced by indoor air speed rather than indoor relative humidity. The new equation developed in this study can be applied to tropical climates and hot-humid summer seasons of temperate climates.

Keywords Thermal comfort · Adaptive model · Hot-humid climate · Natural ventilation · ASHRAE RP-884

14.1 Introduction

The adaptive model of thermal comfort is used in ASHRAE Standard 55 [1] as the code for naturally conditioned spaces and in EN15251 [2] for buildings without mechanical cooling systems. The adaptive model investigates the dynamic relationship between occupants and their general environments based on the principle that *if a change occurs such as to produce discomfort, people react in ways that tend to restore their comfort* [3]. Such adaptation encompasses physiological, psychological, and behavioural adjustments simultaneously. Therefore, the adaptive model provides greater flexibility in matching optimal indoor temperatures with outdoor climate,

D. H. C. Toe (✉)
Faculty of Built Environment, Universiti Teknologi Malaysia, Skudai, Johor, Malaysia

© Springer Nature Singapore Pte Ltd. 2018   145
T. Kubota et al. (eds.), *Sustainable Houses and Living in the Hot-Humid Climates of Asia*, https://doi.org/10.1007/978-981-10-8465-2_14

Table 14.1 Classification of the ASHRAE RP-884 database for naturally ventilated buildings according to climate

Climate	Number of observations	
	Original database	Refined database
Hot-humid	1682	1673
Hot-dry	4339	2776
Moderate	4044	3213
All	10,065	7662

particularly in naturally ventilated buildings [4, 5]. Adaptive models are thus considered more appropriate for supporting comfort in low-energy buildings [4, 6].

Because climatic context is a primary consideration in the adaptive model, it is imperative to evaluate the comfort requirements of people worldwide, particularly in tropical regions that lack comprehensive standards [7, 8]. This study examines the thermal adaptation of occupants and develops an adaptive thermal comfort equation to be used as a standard for naturally ventilated buildings in the hot-humid climate [9]. It employs statistical meta-analysis of the ASHRAE RP-884 database [10, 11]. It is hypothesized that reanalysis of the ASHRAE RP-884 database according to climate would clarify any differences in thermal adaptation among climates.

14.2 Meta-analysis Method

The data files supplied in the ASHRAE RP-884 database [10] were classified into one of three climate groups including hot-humid, hot-dry, and moderate according to survey locations and seasons. Table 14.1 shows that of the 10,065 observations for naturally ventilated buildings in the database, 1682 represent hot-humid climate while 4339 represent hot-dry climate. The remaining 4044 observations apply to moderate climate. Both residential buildings and offices were surveyed in each climate. In the hot-humid climate, 583 observations were gathered from subjects in residential buildings and 1099 observations were from offices. Overall, the database contained 2209 observations in residential buildings and 3657 observations in offices solely, while the other 4199 observations followed subjects in their houses and offices and were a mix of both building types [11]. The latter involved the hot-dry and moderate climates.

The classified data were then checked for the consistency of each variable and refined where necessary. In particular, the outdoor temperatures for all observations were standardized by using the daily (24-h) mean outdoor air temperature for each exact survey date and station in the survey location. These data were obtained from Global Surface Summary of Day Data Version 7 by NOAA [12]. The final refined database for analysis consisted of 7662 observations (Table 14.1).

Linear and probit regressions using the least-squares method were employed in the data analysis. Analyses of both regression models were conducted at the individual observation level with raw data used as a single unit. All transverse and longitudinal surveys in the database were treated similarly. A complete outline of the method is detailed in [9].

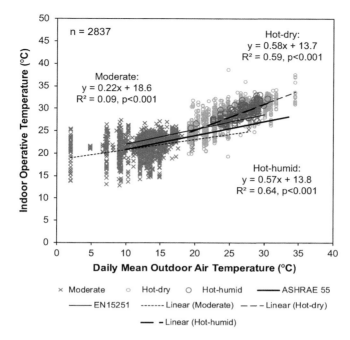

Fig. 14.1 Scatter diagram of indoor operative temperatures at thermal neutrality and the daily mean outdoor air temperatures. The discontinuous lines are linear regression models of this study and represent adaptive equations for predicting neutral temperatures (Source from [9])

14.3   Results and Discussion

14.3.1   Adaptive Thermal Comfort Equation

Figure 14.1 presents a scatter diagram of observed indoor operative temperatures at thermal neutrality and the corresponding daily mean outdoor air temperatures together with the adaptive thermal comfort equations (linear regression lines). It is clear that data for each climate have a distinguishable range of daily mean outdoor air temperatures. It is noteworthy that the adaptive comfort equations underlying ASHRAE [1] and EN15251 [2] standards are

$$T_{comfop} = 0.31\,T_{outpm} + 17.8, \qquad (14.1)$$
$$T_{comfop} = 0.33\,T_{outrm} + 18.8, \qquad (14.2)$$

respectively, where T_{comfop} is indoor comfort operative temperature (°C), T_{outpm} is prevailing mean outdoor air temperature (°C), and T_{outrm} is running mean outdoor air temperature (°C).

The regression lines for hot-humid, hot-dry, and moderate climates are defined by the following equations, respectively:

$$T_{neutop} = 0.57\,T_{outdm} + 13.8 \qquad (14.3)$$
$$T_{neutop} = 0.58\,T_{outdm} + 13.7 \qquad (14.4)$$
$$T_{neutop} = 0.22\,T_{outdm} + 18.6 \qquad (14.5)$$

where T_{neutop} is indoor neutral operative temperature (°C) and T_{outdm} is daily mean outdoor air temperature (°C). All regression coefficients are significant at the 0.1% level. As indicated, these regression lines differ among themselves and from those of the standards in terms of their gradients and the outdoor temperature ranges (Fig. 14.1). Compared with the ASHRAE adaptive equation, the regression lines for hot-humid and hot-dry climates are nearly twice as steep with regression coefficients close to 0.6.

This result supports our hypothesis such that climate is a major influence on the thermal adaptation of occupants in naturally ventilated buildings. It also implies that people living in hot climates, particularly regions with daily mean outdoor air temperatures higher than 20 °C, adapt to a wider and higher range of indoor operative temperatures relative to the same magnitude of outdoor air temperature increases than those living in colder climates.

Further, when the outdoor air temperature in Eq. (14.3) is characterized as monthly mean, prevailing mean [1], and running mean [5], respectively, similar neutral operative temperatures are predicted for hot-humid climate. The adaptive equation based on the daily mean shows the highest coefficient of determination (R^2) and predicts at least 10% more variability in the neutral operative temperature compared with the other outdoor temperature characterizations [9]. The result implies the above climate classification that considers season sufficiently distinguishes the data to explain adaptation to thermal history in this climate even without considering the previous outdoor temperatures. The dependence on the previous days' outdoor temperatures, hence acclimatization in the time frames of a month or a week, in the hot-humid climate is likely negligible due to the small changes in its daily outdoor weather conditions over the entire year.

14.3.2 Acceptable Comfort Limits

An acceptable range of temperature deviation from the predicted neutral operative temperature (Eqs. (14.3, 14.4 and 14.5)) for each climate is analysed in Fig. 14.2 by using probit models in consideration of the thermal sensation votes. Figure 14.2a shows that the proportion of occupants voting "neutral" does not exceed 30% and peaks at 0.7 °C lower than the neutral operative temperature for hot-humid climate. The probit line for "comfortable" thermal sensation, which includes "slightly cool," "neutral," and "slightly warm," or −1, 0, +1, respectively, is one-tailed and has no symmetry within the observed temperature range (Fig. 14.2a). The proportion of

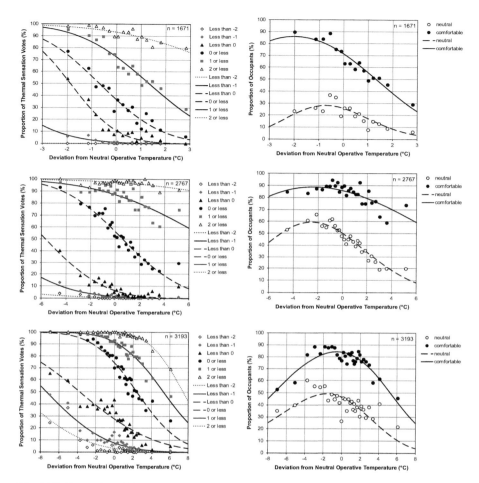

Fig. 14.2   Proportion of thermal sensation votes (left) and proportion of occupants voting "neutral" (0) and "comfortable" (± 1) (right) as a function of deviation from the predicted neutral operative temperature. (**a**) Hot-humid climate; (**b**) hot-dry climate; (**c**) moderate climate. Lines indicate probability predicted by probit regression models. Points represent observed values for equal bins of the temperature deviation (102–105 data per bin). In the left figure, dashed lines and black points represent "neutral" votes (0); continuous lines and grey points represent "comfortable" votes (± 1) (Source from [9])

occupants who voted "comfortable" increases from 30% at 2.5 °C higher than the predicted neutral temperature to 86% at 2 °C below the predicted neutral temperature. Eighty percent "comfortable" votes are predicted at 0.7 °C less than the neutral temperature for hot-humid climate.

In comparison, at least 80% of the occupants voting "comfortable" appear within temperature deviations of approximately 2 °C above and 6 °C below the predicted neutral temperature for hot-dry climate (Fig. 14.2b) and approximately 1.5 °C above and 2.5 °C below that for moderate climate (Fig. 14.2c). The comfortable temperature range is largest for hot-dry climate, at 8 °C for 80% of "comfortable" votes, likely because adapting to a wider temperature range is easier when humidity is low.

The analysis implies the upper and lower comfort limits must be considered separately for each climate, as determined in Fig. 14.2 for the respective percentages of "comfortable" votes. In particular, a lower comfort limit is not observed for naturally ventilated buildings in hot-humid climate. The upper comfort limit for this climate is recommended to not exceed 0.7 °C below the predicted neutral operative temperature so that at least 80% of the occupants would be in comfort.

14.3.3 Effects of Indoor Air Speed and Humidity

As discussed in Sect. 14.3.1, the adaptive thermal comfort equations for hot-humid and hot-dry climates are steeper than that for moderate climate. The indoor air speed and indoor humidity levels are two possible factors affecting the thermal adaptation in hot-humid and hot-dry climates.

The effects of indoor air speed on the adaptive equations are analysed in Fig. 14.3. In the figure, the data are categorized into three groups of indoor air speeds including low (<0.3 m/s), moderate (0.3 to <0.65 m/s), and high (≥0.65 m/s). Figure 14.3a shows similar linear regression lines for low and moderate air speeds that maintain regression coefficients at 0.57 and 0.54, respectively, for hot-humid climate. These regression lines predict that moderate air speed has little to no effect on neutral temperatures compared with low air speed. Still air conditions do not generally occur in naturally ventilated buildings in hot-humid climate. The regression line for high air speed (0.80) is steeper and higher than that for low air speed (0.57) by up to approximately 2 °C at 29 °C daily mean outdoor air temperature. The analysis of variance reveals a significant mean difference of $F_{(2, 309)} = 4.52$, $p < 0.05$. These results imply that air movement is likely a possible factor for increasing the gradient of the adaptive equation for hot-humid climate.

For hot-dry climate, the regression lines predict no constant increase in indoor neutral operative temperature at moderate and high air speeds when compared with low air speed (Fig. 14.3b). The thermal adaptation processes of occupants in dry air conditions differ in humid air conditions at high temperatures. Increased air speed allowance is not applicable to hot-dry climate.

In terms of humidity, the regression line for low relative humidity (<60%) predicts higher neutral operative temperatures than for high relative humidity (≥60%) by 0.6–1.7 °C for hot-dry climate (Fig. 14.4b). The analysis of variance shows a significant mean difference at $F_{(1, 1045)} = 9.29$, $p < 0.01$. The indoor relative humidity likely accounts for the effect of water vapour pressure on evaporation indirectly. A similar effect is not apparent in the regression lines for hot-humid climate (Fig. 14.4a). Indoor relative humidity is high (≥60%) more than 75% of the time in hot-humid climate. This result indicates that humidity influences the predicted neutral temperature in hot-dry climate but not in hot-humid climate.

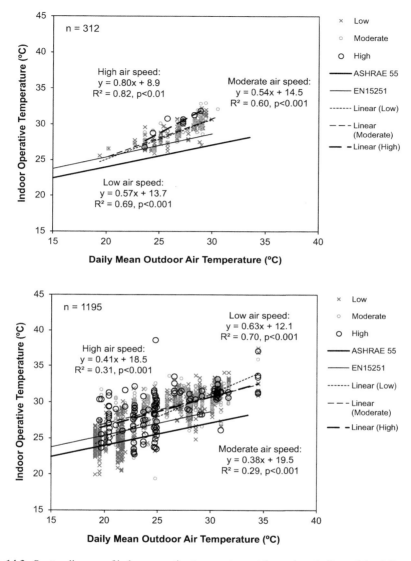

Fig. 14.3 Scatter diagram of indoor operative temperatures at thermal neutrality and the daily mean outdoor air temperatures at different indoor air speeds. (**a**) Hot-humid climate; (**b**) hot-dry climate (Source from [9])

14.4   Conclusion: Future Direction

This study highlights several key differences in the thermal adaptation of occupants in naturally ventilated buildings among climates and the existing standards [1, 2]. A basic set of adaptive thermal comfort criteria for naturally ventilated buildings in

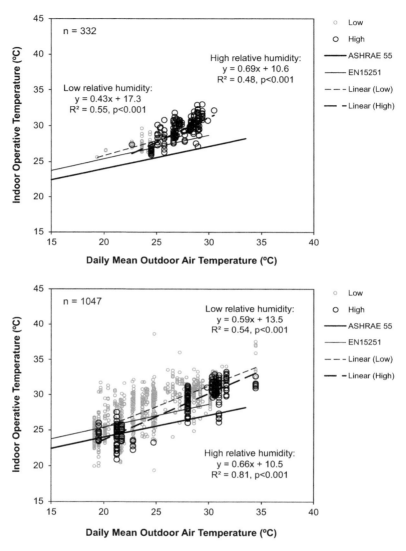

Fig. 14.4 Scatter diagram of indoor operative temperatures at thermal neutrality and the daily mean outdoor air temperatures at different indoor relative humidity. (**a**) Hot-humid climate; (**b**) hot-dry climate (Source from [9])

hot-humid climate is thus proposed in Table 14.2 based on the findings. It is anticipated that the new criteria can be incorporated as thermal comfort standards in hot-humid regions for better applicability (and saving energy).

One of the critical areas of future studies would be to develop an increased air speed allowance for hot-humid climate. Occupants in this climate likely adapt to

Table 14.2 Proposed adaptive thermal comfort equation and related criteria for naturally ventilated buildings in hot-humid climate

No.	Aspect	Criterion	Note
(i)	Climate type	All A climate types and summer season of Cfa climate type	Climate type refers to the Köppen-Geiger climate classification system
(ii)	Neutral operative temperature, T_{neutop} (°C)	$T_{neutop} = 0.57\,T_{outdm} + 13.8$	T_{outdm} is daily mean outdoor air temperature (°C), i.e. the 24-hour arithmetic mean for the day in question
(iii)	Daily mean outdoor air temperature, T_{outdm} (°C)	Range from 19.4 to 30.5	Recommended applicable range for criterion no. (ii)
(iv)	Lower comfort operative temperature limit, T_{lower} (°C)	No required limit	–
(v)	Upper comfort operative temperature limit, T_{upper} (°C)	$T_{upper} = T_{neutop} - 0.7$ for 80% comfortable thermal sensation votes	Graphical representation can be referred in Fig. 14.2a (continuous line in the right figure) for a different percentage of comfortable thermal sensation votes
(vi)	Indoor air speed, v (m/s)	<0.65 at and below neutral operative temperature; ≥0.65 above neutral operative temperature	Recommended to provide non-still air and occupants' control to adjust the indoor air speeds according to their preferences
(vii)	Indoor humidity, RH (%)	No required limit	–

Source from [9]

neutral temperatures by making use of air movement to aid evaporative heat loss in indoor high-humidity conditions. This can be seen in how cooling techniques have been implemented traditionally since the past, for example, in the Malay house (see Chap. 3) and the Chinese shophouse (see Chap. 37). It would be interesting to pay attention to a possible trade-off in thermal adaptation between securing lower indoor temperature by reducing ventilation rate with the outdoors (when the outdoor temperature is higher than the indoors) and increasing indoor air speed for sweat evaporation by improving natural ventilation while allowing temperature increase.

Acknowledgements We thank the support from the Ministry of Education, Malaysia, and Universiti Teknologi Malaysia for the Fundamental Research Grant Scheme (FRGS) Grant (No. R.J130000.7821.4F837). Our sincerest gratitude is given to Richard J. de Dear, Gail S. Brager, Donna Cooper, and all of the field study contributors to the ASHRAE RP-884 database. We greatly appreciate the advice offered by the editors of this book and Dr. Nakaya of Gifu National College of Technology.

References

1. ASHRAE (2013) ANSI/ASHRAE Standard 55-2013: thermal environmental conditions for human occupancy. American Society of Heating, Refrigerating and Air-Conditioning Engineers, Inc., Atlanta
2. BSI (2008) BS EN 15251:2007, Indoor environmental input parameters for design and assessment of energy performance of buildings addressing indoor air quality, thermal environment, lighting and acoustics. British Standards Institute, London
3. Humphreys MA, Nicol JF (1998) Understanding the adaptive approach to thermal comfort. ASHRAE T 104(1b):991–1004
4. de Dear RJ, Brager GS (2002) Thermal comfort in naturally ventilated buildings: revisions to ASHRAE Standard 55. Energ Buildings 34(6):549–561
5. Nicol F, Humphreys M (2010) Derivation of the adaptive equations for thermal comfort in free-running buildings in European Standard EN15251. Build Environ 45(1):11–17
6. Humphreys MA, Nicol JF, Raja IA (2007) Field studies of indoor thermal comfort and the progress of the adaptive approach. Adv Build Energ Res 1:55–88
7. Nicol F (2004) Adaptive thermal comfort standards in the hot-humid tropics. Energ Buildings 36(7):628–637
8. Toe DHC, Kubota T (2011) A review of thermal comfort criteria for naturally ventilated buildings in hot-humid climate with reference to the adaptive model. In: Bodart M, Evrard A (eds) Conference proceedings of the 27th International conference on passive and low energy architecture, Louvain-la-Neuve
9. Toe DHC, Kubota T (2013) Development of an adaptive thermal comfort equation for naturally ventilated buildings in hot-humid climates using ASHRAE RP-884 database. Front Archit Res 2:278–291
10. The University of Sydney (2010) ASHRAE RP-884 adaptive model project, data downloader. http://sydney.edu.au/architecture/staff/homepage/richard_de_dear/ashrae_rp-884.shtml. Accessed 3 Feb 2010
11. de Dear RJ, Brager G, Cooper D (1997) Developing an adaptive model of thermal comfort and preference. Final report on ASHRAE RP-884. Macquarie University, Sydney
12. NCDC (2012) Climate data online, global surface summary of day data version 7. http://gis.ncdc.noaa.gov/map/viewer/#app=cdo&cfg=cdo&theme=daily&layer=111&node=gis. Accessed 20 Apr 2012

Chapter 15
Comfort Temperature and Preferred Temperature in Taiwan

Ruey-Lung Hwang

Abstract Field experiments, using survey questionnaires and physical measurements simultaneously, were conducted in residences in Taiwan to investigate Taiwanese subjective thermal responses and comfort perception. Responses from those subjects suggest a thermal preference temperature, 25.0 °C lower than the thermal neutral temperature, 26.0 °C, by 1.0 °C. A new predicted formula (PD-TSV) of percentage of dissatisfied relating to mean thermal sensation votes is suggested. In comparison with the PMV-PPD model, the new formula reveals that besides an increase in minimum rate of dissatisfied from 5% to 11% and a shift of the TSV with minimum PD to the cool side of sensation scale is found. The limits of sensation votes corresponding to 80% acceptability are −1.50 and +0.70, and a suitable comfort zone of 80% acceptability for Taiwan range from 20.9 to 28.9 °C.

Another field study also conducted to investigate the thermal sensation of elderly people in Taiwan with an age greater than 60 years old. In the summer season, the thermal neutral temperature, 25.2 °C, for elderly is only slightly higher than the thermal preferred temperature, 25.0 °C. This indicates the optimal TSV for elderly is close to thermal neutrality. The PD-TSV model for elderly revealed that the sensation votes corresponding to 80% acceptability were ±0.75 for elders, about ±0.10 less than the levels projected by ISO 7730 model. The range of operation temperature for 80% thermal acceptability for elders in the summer was 23.2–27.1 °C, narrower than the range of 20.9–28.9 °C reported for non-elders.

Keywords Neutral temperature · Preferred temperature · Comfort range · Taiwan residences

R.-L. Hwang (✉)
Department of Industrial Technology Education, National Kaohsiung Normal University, Kaohsiung, Taiwan
e-mail: rueylung@nknu.edu.tw

© Springer Nature Singapore Pte Ltd. 2018
T. Kubota et al. (eds.), *Sustainable Houses and Living in the Hot-Humid Climates of Asia*, https://doi.org/10.1007/978-981-10-8465-2_15

15.1   Introduction

The geography of Taiwan spans between 22 and 25 °C in latitude and is of hot and humid climate characteristics. Almost all commercial buildings and over 80% residential houses use air-conditioning systems to maintain indoor thermal comfort during hot seasons. Although the air conditioning is prevailing in Taiwanese buildings, residential buildings, school classrooms and dormitories, and some small-scale office spaces are usually operated in hybrid ventilation mode, except for commercial buildings. The hybrid ventilation mode is a cooling operation scheme that the space is typically operated in natural ventilation mode; the air conditioning is operated as an auxiliary equipment only when the natural ventilation is insufficient to maintain indoor thermal comfort. Therefore, it is a worthwhile issue to identify the adaptive comfort model of Taiwan's residents and investigate their thermal adaptive behavior. Several studies conducted a series of in situ research and experiments of this issue. This chapter will discuss and introduce the adaptive comfort model for residential spaces of Taiwanese people and their corresponding adaptive behavior in Part III.

15.2   For Hybrid Ventilated Residences

Hwang et al. [1] carried out field experiments using thermal environment measurements with a questionnaire-based survey, respectively; in residences, 253 interviewees in 137 homes across Taiwan were visited during April to December. The examiner arrived at the visited sites upon previously arranged time. After the measurement instruments are set up and a brief introduction to the experiment procedures is given, the interviewees resume their routine activities and were asked to fill in a questionnaire after 20 min time. Contained in the survey sheet are the interviewee's demographic information, most preferred method of adaptation when they sensed thermal discomfort, and votes for thermal sensation, thermal acceptability, and thermal preference with regard to the current condition. The thermal sensation vote (TSV) is based on the ASHRAE seven-point sensation scale (-3, cold; -2, cool; -1, slightly cool; 0, neutral; $+1$, slightly warm; $+2$, warm; $+3$, hot). Thermal acceptability vote aims to understand whether the interviewee considers the current environment condition as acceptable or not. If the thermal condition is unacceptable, a further question is asked to see whether the discomfort is due to coolness or warmness. Thermal preference vote (TPV) employs McIntyre's scale of preference, namely, "I wish for a warmer or cooler thermal condition or no change."

A total of 648 residences were collected. Based on the collected data, a linear regression analysis is applied to understand the occupant's thermal sensitivity and neutral temperature. The best-fitted linear equations of mean thermal sensation votes (MTSVs) regressed against operative temperature (T_{op}) are shown as Eq. (15.1). The regression slopes, which stand for the correlation between the occupants' thermal sensitivity and indoor climate, are 0.28 °C in residences. The neutral temperature is defined as the temperature in which the occupants feel neither warm nor cool and can be obtained by applying a neutral thermal sensation vote (i.e., MTSV = 0) to the regression equation. The neutral temperature (T_n) is 26.0 °C for residences.

$$MTSV = 0.28\,T_{op} - 7.36 \qquad (15.1)$$

By applying a TSV value of ±0.85, the recommended PMV rang for thermal comfort by ASHRAE Standard 55 [2], into the corresponding linear regression equation, i.e., Eq. (15.1), the measured comfortable conditions of 80% acceptability can be determined. The comfort range of 80% acceptability is 23.0–29.0 °C for residences.

Although many studies have emphasized on climatic adaptation, before the following two issues have been clarified, one could not conclude directly that residents in Taiwan have well accommodated to hot and humid climate and thus are less critical to thermal conditions of indoor environments.

1. Does the thermal neutrality of occupant in Taiwan agree with the optimal thermal sensation?
2. Is the Predicted Mean Vote and Predicted Percentage of Dissatisfied (PMV-PPD) model proposed by Fanger suitable to predict the relationship between thermal sensation and the percentages of dissatisfaction of occupants in hot and humid area?

The term "thermal sensation" can be viewed as the interviewee's passive response or judgment of stimuli from the thermal environment to a certain extent. In comparison, thermal preference relates directly to the interviewees' subjective expectation to eliminate thermal discomfort stimuli via some measures of adjustment. To a certain degree, the preferred temperature, as determined by the mean TPV of interviewees, rather than the neutral temperature, should be treated as the optimal temperature. Results of regression analysis are as shown in Eqs. (15.2) and (15.3), as well as Fig. 15.1, which are obtained by applying the probit regression model of logistic analysis to TPV against T_{op}. Probit models are appropriate when attempting to model a binary outcome variable, e.g., yes/no, like/dislike, etc. The probit mode estimates the probability a value will fall into one of the two possible binary outcomes. The probit predictors can be written as probit $= ax + b$. Hence, whatever $ax + b$ equals, it can be transformed by the function to yield a predicted probability. The probit model uses the cumulative distribution function of the standard normal distribution.

$$\text{Requests for warmer}: \quad \text{probit} = -3.12\,T_{op} + 22.52 \qquad (15.2)$$
$$\text{Requests for cooler}: \quad \text{probit} = 3.76\,T_{op} + 28.15 \qquad (15.3)$$

The results of probit analysis indicate that the preferred temperature, which corresponds to the intersection of two fitted curves in Fig. 15.1, is 25.0 °C in residences. The preferred temperature is 1.0 °C lower than the neutral temperature.

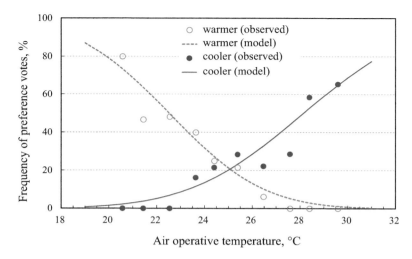

Fig. 15.1 Probit model for preferred temperature in hybrid ventilated residences [1]

It is commonly observed in studies on thermal comfort of the hot and humid climate zones that the occupants tend to have a preferred temperature lower than neutral temperature.

By applying the preferred temperature into the linear regressive model of MTSV and T_{op}, i.e., Eq. (15.1), we can identify the corresponding preferred MTSV of -0.27 for residences. This implies that in hot and humid Taiwan, the occupants' preferred thermal sensation does not equal to thermal neutrality but fell into the range between neutrality (0) and a slightly cool condition (-1).

This finding encouraged us to develop a new correlation between percentages of dissatisfied (PD) and mean sensation votes. When surveying thermal sensation, the TSV cast outside of the three central categories (-1; 0; $+1$) is generally considered an indication that the surveyed subjects are dissatisfied with the indoor microclimate. Meanwhile, votes for the two categories of warm ($+2$) and hot ($+3$) are regarded as indication of subject's hot discomfort; the two categories of cool (-2) and cold (-3) are regarded as cold discomfort. The probit method is used again to construct the relationship between the measured dissatisfaction percentage due to coolness or warmness and the mean sensation votes. Equations (15.4) and (15.5) give the best-fitting probit models for the percentages of dissatisfaction due to warmness and coolness, respectively.

$$\text{Hot discomfort}: \quad \text{probit} = 0.885\,\text{TSV} - 1.475 \tag{15.4}$$
$$\text{Cold discomfort}: \quad \text{probit} = -0.694\,\text{TSV} - 1.867 \tag{15.5}$$

Figure 15.2 demonstrates the results of analysis and the comparisons with the original correlation, as proposed by PMV-PPD curve (see Eq. (15.6) in ISO 7730 [3]). Thus a new best-fitting formula as Eq. (15.7), named TSV-PD, is established in a form similar to PMV-PPD equation. The comparison between PMV-PPD formula

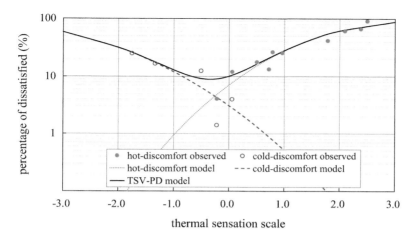

Fig. 15.2 Percentages of dissatisfied in relation to mean thermal sensation votes from the surveyed residents and its probit models

and TSV-PD formula reveals that the proposed formula features an increase in the minimum rate of dissatisfied from 5 to 9% and a shift of the TSV with minimum PD to the cool side of sensation scale. The lowest point of the TSV-PD curve corresponds to a TSV of -0.27.

$$PPD = 100 - 95.0\exp\left[-0.03353\,PMV^4 - 0.2179\,PMV^2\right] \tag{15.6}$$

$$PD = 100 - 89.4\exp\left[-0.00865(TSV + 0.27)^4 - 0.08494(TSV + 0.27)^2\right]$$
$$\tag{15.7}$$

Substituting the value of optimal TSV into .Equation (15.1) resulted a corresponding temperature of 25.3 °C, which is almost identical to the preferred temperature mentioned above. Furthermore, according to Eq. (15.1), the limits of sensation votes corresponding to 80% acceptability are -1.50 and $+0.70$ rather than on the values claimed in ASHRAE Standard 55 (-0.85 and $+0.85$). Therefore, a more suitable comfort zone of 80% acceptability for Taiwan should range from 20.9 to 28.9 °C.

15.3   For the Elders in Dwelling

With the population of over 60-year-old people exceeding 7% of the total population, Taiwan has entered an era of aging society since the year 2000. To provide a living environment which is appropriate to the increasingly growing group of the elders, the quality of the indoor environment in which the elders reside, particularly its thermal characteristics, needs to be better evaluated and realized, as the elders

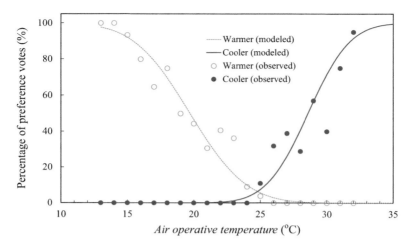

Fig. 15.3 Probit model for preferred temperature for elderly people [3]

typically spend more of their time and activities at home every day as compared to other age groups. Eighty-seven volunteers with an age of 60 years or greater (mean = 71 years), including 37 males and 50 females, were recruited in the study conducted by Hwang and Chen [4].

For the elders surveyed in this study, the best-fitted linear equations of mean thermal sensation votes (MTSVs) regressed against operative temperature (T_{op}) for the summer and the winter are expressed, respectively, as Eqs. (15.8) and (15.9):

$$\text{Summer} : \text{MTSV} = 0.39\,T_{op} - 9.84 \tag{15.8}$$
$$\text{Winter} : \text{MTSV} = 0.28\,T_{op} - 6.50 \tag{15.9}$$

The neutral temperatures obtained from Eqs. (15.8) and (15.9) for the elders were found to be 25.2 and 23.2 °C for summer and winter, respectively. The neutral temperature in the summer was 2.0 °C higher than its counterpart in the winter. The elders were thermally more sensitive when they are staying in a warm environment (summer; thermal sensitivity = 0.39 °C) than that in a cold environment (winter; 0.28 °C) by observing the slopes of Eqs. (15.8) and (15.9). That is, in summer the elders opt to raise their thermal sensation for one grade on the scale for every 2.6 °C increase in T_{op}, whereas in winter they lowered the grade only when the temperature drop was more than 3.6 °C.

As shown in Fig. 15.3, the best-fitted probit models of TPVs regressed against T_{op} for "want to be warmer" and "want to be cooler" are expressed, respectively, as Eqs. (15.10) and (15.11). The optimal temperature of thermal preference for the

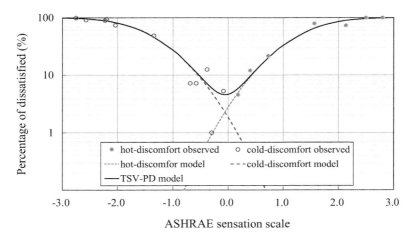

Fig. 15.4 Percentages of dissatisfied in relation to mean thermal sensation votes from the surveyed elders and its probit models [3]

elders occurred at the intersection of the two preference curves, corresponding to a T_{op} of 25.0 °C. While this value was only 0.2 °C lower than the neutral temperature for the elders.

$$\text{Warmer : probit} = -3.344\,T_{op} + 19.72 \tag{15.10}$$
$$\text{Cooler : probit} = 2.485\,T_{op} + 28.59 \tag{15.11}$$

Figure 15.4 shows the distribution of PD against MTSV from the surveyed elders and the probit models of logistic regression for estimating the overall PD, including the PD due to cold discomfort and the PPD due to hot discomfort. The probit models for estimating the cold-discomfort-originated PPD from the votes for the two categories of cool (-2) and cold (-3), hot-discomfort-originated PPD from votes for the two categories of warm ($+2$) and hot ($+3$), and overall PPD are mathematically expressed as:

$$\text{Cold discomfort :} \quad \text{probit} = -1.47\,\text{TSV} + 2.08 \tag{15.12}$$
$$\text{Hot discomfort :} \quad \text{probit} = 1.48\,\text{TSV} + 1.92 \tag{15.13}$$
$$\text{Elders :} \quad \text{PPD} = 100 - 97\exp\left(-0.3338\,\text{TSV}^2 - 0.01972\,\text{TSV}^4\right) \tag{15.14}$$

As determined from Eq. (15.14), the minimum PPD at any time for the elders in this study was 3%. The minimum PPD for the elders corresponds to a thermally neutral sensation (MTSV = 0). In Eq. (15.14), the MTSV equals ±0.85 when the PD reaches 20%. When the range of TSV = ±0.85 is used in Eqs. (15.8) and (15.9), the ranges of temperature corresponding to 80% thermal acceptability were found to be 20.2–27.1 °C in the summer and 20.5–25.9 °C in the winter. If the upper limit from the summer range is used to represent the upper bound of a yearly range, and the lower limit for the winter range is considered the lower bound, a temperature interval of 20.5–27.1 °C can be found to correspond to the range of temperature acceptability to 80% of the surveyed elders on a yearly basis.

15.4   Conclusions

This chapter depicted the requirements of Taiwan residents for thermal comfort in the indoor environment at home in Taiwan. In addition, the requirements of elderly people for thermal comfort were depicted to help realizing the variations between different age groups. The following conclusions were derived.

When the MTSVs were linearly regressed to the T_{op}, the slopes of best-fitted line, 0.28 for residents and 0.39 for elders during the summer, depict that the elders are more sensitive to the change in indoor microclimate. The thermally neutral temperature for Taiwan resident corresponded to a T_{op} of 26.0 °C, while the neutral temperature for elders is 25.2 °C in the summer season.

The preferred T_{op} for Taiwan residents occurred at 25.0 °C, which is lower than the neutral temperature by 1.0 °C. According to the proposed TSV-PD formula, the desired optimal sensation on the ASHRAE scale is not "neutral" but shifts to −0.27, a position between "slightly cool" and "neutral." This change will cause the comfort zone in the hot and humid regions to shift to cooler conditions. The limits of TSV that corresponds to 80% acceptability for hot and humid regions are −1.50 and +0.70 rather than −0.85 and +0.85 suggested by the ISO 7730, as well as an increase in the minimum rate of dissatisfaction from 5 to 9%. In light of this, the suitable comfort range of air operative temperature for Taiwan is 20.4–28.4 °C.

In the elderly group, that a 25.2 °C of preferred temperature is close to the corresponding neutral temperature, 25.0 °C, indicates that the optimal TSV for elderly occurred at thermal neutrality. The proposed PD-TSV model for the elders shows the minimum PPD of the elders was 3%, a level close to the minimum PPD of 5% projected by the ISO 7730 model; at a PPD of 20%, the corresponding TSVs were ±0.75 for the elders, about ±0.10 less than their counterparts projected by the ISO model, indicating that the elders were of a less flexible sensation toward acceptable indoor temperature than the ISO model would predict. The ranges of temperature corresponding to 80% thermal acceptability for the elders in the summer are found to be 23.2–27.1 °C.

References

1. Hwang RL, Lin TP, Cheng MJ, Ho MC (2009) Thermal perceptions, general adaptation methods and occupant's idea on trade-off among thermal comfort and energy saving in hot-humid regions. Build Environ 44:1128–1134
2. ASHRAE (2004) Thermal environmental conditions for human occupancy, Atlanta GA, American Society of Heating, refrigerating and air-conditioning engineers (ASHRAE standard 55-2004)
3. ISO (2005) Ergonomics of the Thermal Environment– Analytical Determination and Interpretation of Thermal Comfort Using Calculation of the PMV and PPD Indices and Local Thermal Comfort Criteria, Geneva, International Organization for Standardization (International Standard 7730)
4. Hwang RL, Chen CP (2010) Field study on behaviors and adaptation of elderly people and their thermal comfort requirements in residential environments. Indoor Air 20:235–245

Chapter 16
Thermal Comfort in Indian Apartments

Madhavi Indraganti and Kavita Daryani Rao

Abstract Socio-political and economic drivers can be historically traced behind the *apartment*'s evolutionary trajectory. Absence and poor adherence to norms appear to have caused thermal discomfort. It pushed occupants toward energy-intensive solutions. India needs to embrace the adaptive thermal comfort model to unburden her import-dependent energy balances. Occupant responses from real buildings underpin this model. This chapter focuses on the field studies in apartments. People in naturally ventilated apartments expressed comfort at 30.3 °C during the hot and warm-humid seasons. Thermal condition indoors varied adaptively with the outdoors. However, discomfort was high in summer. The subjects accustomed to air conditioners had lower comfort temperature. This cyclic path dependency works against India's sustainability agenda.

Keywords Indian apartments · Thermal comfort · Adaptive model · Warm-humid climate · Field survey · Comfort temperature

16.1 History of Apartments

Historically, the idea of providing an affordable living space close to work led to the *apartment* building. The oldest apartments *insulae* (circa 1 BC) exemplify this. Rich Roman landlords built these five-storied buildings inside the walled and overpopulated cities with scarce and expensive lands (e.g., Ostia). The apartments in the nineteenth-century Europe, UK, and USA also had a similar need [1]. Spread over 85% of the plot area, the deplorable *railroad and double dumbbell* plan types

M. Indraganti (✉)
Department of Architecture and Urban Planning, College of Engineering, Qatar University, Doha, Qatar
e-mail: madhavi@qu.edu.qa

K. D. Rao
Department of Architecture, Jawaharlal Nehru Architecture and Fine Arts University, Hyderabad, India

© Springer Nature Singapore Pte Ltd. 2018
T. Kubota et al. (eds.), *Sustainable Houses and Living in the Hot-Humid Climates of Asia*, https://doi.org/10.1007/978-981-10-8465-2_16

used for these apartment boxes exhibited little or no concern for thermal comfort or building standards.

16.1.1   Apartments in Pre- and Post-Independent India

In the nineteenth-century India, an upsurge in textile mills and a much similar need drove the apartment building. As a result, the millowners built *chawls* (multistoried buildings with multi-tenant rooms in each floor). Thermal comfort was never on their design agenda. However, these *chawls* with courtyards and singly loaded corridors all around behaved like transformed rural communities, transposed morphologically along a vertical axis (Fig. 16.1). They provided thermal relief through cross ventilation and mutual shading, necessary in warm-humid climates.

Postindependence, new socialist governments built large apartment complexes, in major cities for the working class. Private developers soon followed the suit. These complexes had large open spaces and semi-enclosed spaces, within and outside the private realm of the occupant. Until the 1990s, the bylaws included these areas in the salable carpet area. As a result, developers provided large semi-open areas for profit maximization [2].

Opening up of the economy in the nineties and twenty-first century led to rapid urbanization and an enormous need gap. Lax enforcement of changed bylaws encouraged many developers to flout the norms. Thermal comfort was soon

Fig. 16.1   A typical *chawl* of Mumbai with shops on the street front (Source: Marco Zanferrari)

compromised, with the open spaces and heavy thermal mass vanishing quickly [2]. Superstitious beliefs in popular *vaastu* (traditional Hindu science of architecture) and growing corruption in building sector also adversely affected the design and built form of the apartment [3].

16.1.2 Current Energy Demands of Residential Sector

India heavily relies of imported energy totaling to 62% of 528.3 million tons of oil equivalent of total final consumption, as of 2013. Residential electricity demand in India increased by 116% in 10 years reaching 207.27 TW.h in 2013. Among the Asia-Pacific Partnership countries, India ranks third in energy use in residential buildings, ranking highest in percentage terms at 34.6% of the total final consumption [4]. This meteoric rise is in part due to the rise in appliance usage, especially those used for thermal comfort. Energy transition from biomass to electricity and rapid urbanization also fuelled this rise.

About 73% of the energy in residential buildings is used for providing thermal comfort. India is yet to have adaptive thermal comfort standards to limit the energy use for comfort [5]. Following Western standards/practices verbatim leads to overcooling/heating and is counterproductive. For a country that majorly relies on imported energy, the need to reduce the energy use for comfort air conditioning cannot be overemphasized. The development of adaptive comfort standards, custom made to Indian climates and conditions, thus assumes great significance. And the thermal comfort field studies form its backbone.

16.2 Thermal Comfort in Apartments

Indraganti [6] conducted thermal comfort field studies in apartments in Hyderabad, India. She interviewed 113 subjects living in 45 mixed mode apartments for 3 months in hot-dry summer and warm-humid monsoon seasons of discomfort. All the four environmental variables affecting thermal comfort were concurrently measured in both longitudinal and transverse surveys between 7:00 and 23:00.

ASHRAE's Class II protocols were used in these paper-based surveys that yielded 3962 data sets (91.5% from naturally ventilated mode) (Table 16.1). The respondents in the age group 18–65 years participated voluntarily, of which 65% were women. Local meteorological station provided the outdoor environmental data. The subjects ticked their clothing and activity from the lists provided.

Table 16.1 Scales used in the survey

Scale	Thermal sensation (TS)	Thermal preference (TP)	Thermal acceptability (TA)
3	Hot		
2	Warm	Much cooler	
1	Slightly warm	A bit cooler	Unacceptable
0	Neutral	No change	Acceptable
−1	Slightly cool	A bit warmer	
−2	Cool	Much warmer	
−3	Cold		

Table 16.2 Outdoor and indoor environmental measurements and thermal comfort perceptions recorded in mixed mode apartments in Hyderabad, India [7]

Variable	May (n = 1405)		June (n = 1334)		July (n = 1223)	
	Mean	S.D.	Mean	S.D.	Mean	S.D.
T_o (°C)	33.8	0.6	29.7	1.2	28.8	0.6
RH_o (%)	31.5	6.0	54.4	3.6	58.9	4.3
T_a (°C)	34.7	1.6	30.9	1.2	30.3	1.1
Rh (%)	27.2	8.6	53.4	6.0	55.1	6.1
AH (g_a/kg$_{da}$)	9.2	2.2	14.9	1.4	14.8	1.1
T_g (°C)	34.5	1.8	31.2	1.2	30.7	1.1
V_a (m/s)	0.48	0.51	0.48	0.41	0.39	0.42
TS	1.8	1.0	0.5	0.8	0.42	0.7
TP	1.3	0.6	0.7	0.6	0.56	0.56
TA (%)	69		95		97	

T_o Outdoor daily mean temperature, RH_o Outdoor daily mean relative humidity, T_a Indoor air temperature, RH Indoor relative humidity, T_g Indoor globe temperature, V_a Indoor air velocity, AH Indoor absolute humidity, *S.D.* Standard deviation

16.2.1 Outdoor and Indoor Conditions and Thermal Responses

Both outdoors and indoors were very hot and dry during the summer (May) as seen in Table 16.2. Conditions improved substantially during the monsoon period (June and July). Figure 16.2 shows the distribution of thermal sensation and thermal preference votes in all the months. As it can be seen, most of the subjects had either warm or hot sensation in May and preferred to be cooler. As a consequence, only 31% accepted the environments. This is in part due to the poor design as discussed above. Most of the women wore traditional Indian ensembles like saris or *salwar-kameez*. Clothing insulation varied from 0.19 to 0.84 clo, averaging at 0.59 clo (standard deviation (SD), 0.12 clo) during the entire survey. Metabolic rates varied from 0.7 to 2.0 Met (resting to heavy work).

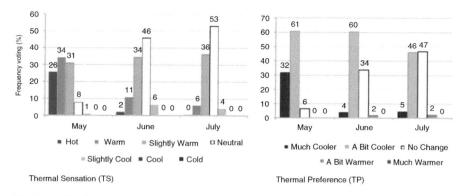

Fig. 16.2 Distribution of thermal sensation and thermal preference votes. People mostly voted on the warmer side of the sensation scale in summer (May), preferring cooler environments [6]

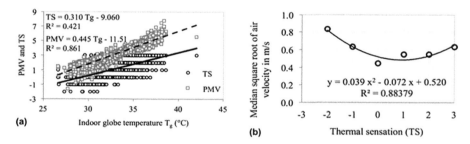

Fig. 16.3 (**a**) Regression of indoor globe temperature with thermal sensation and PMV. PMV was always more than the actual sensation [6]; (**b**) relationship between thermal sensation and indoor air velocity (all data)

16.2.2   Neutral Temperature

A regression of thermal sensation with indoor globe temperature returned a neutral temperature of 29.2 °C as shown below and Fig. 16.3. On the other hand, Fanger's predicted mean vote (PMV), a metric used in the industry to measure subjective sensation always exaggerated the subjective warmth of the residential environments. Humphreys and Nicol [8] also raised doubts about the validity of PMV in predicting the sensation of people in real-life situations. S.E. is standard error of the regression coefficient.

$$TS = 0.31\, T_g - 9.06 \qquad (16.1)$$

$(N = 3962, R^2 = 0.42, p < 0.001, S.E. = 0.006)$

Being a summer study, Indraganti [6] noted high indoor air velocities throughout the survey. In addition, humidity played a dominant role on comfort especially in warm-humid environments. Similar to other researchers [9], this study also noted highly a significant relationship between thermal sensation and indoor temperature and absolute humidity on multiple linear regression. It resulted in the following equation.

$$TS = 0.259\,T_g - 0.04\,AH - 6.808 \qquad (16.2)$$

($N = 3962$, $R^2 = 0.429$, $p < 0.001$, S.E.$_{Tg} = 0.009$, S.E.$_{AH} = 0.006$)

16.2.3 Comfort Temperature

Using 0.5 as the Griffiths' coefficient, comfort temperature (T_c) is estimated for all the responses recorded when air coolers and ACs were not in use. It averaged at 30.3 °C (S.D. = 1.8 °C), matching closely with the regression neutral temperature presented earlier. This is close to the comfort temperature reported in Japan [10] and Pakistan [11] studies. It also matched with other Indian apartment research in warm-humid and composite climates [12, 13]. Figure 16.4 reveals the adaptive relationship between T_c and outdoor daily running mean temperature (T_{rm}) as shown below.

$$T_c = 0.22\,T_{rm} + 23.33 \qquad (16.3)$$

($N = 3624$, $p < 0.001$, S.E. = 0.014).

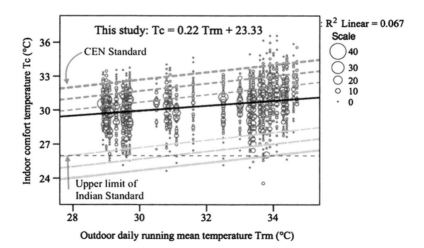

Fig. 16.4 Adaptive model of thermal comfort in NV mode: Regression of indoor comfort temperature (T_c) with outdoor daily running mean temperature (T_{rm}) on raw data (N = 3624) [6]. Green dashed lines indicate 95% confidence interval of slope

This relationship matches closely with the adaptive model reported for offices in India and is within the Commite European de Normalisation (CEN) standard limits [14, 15]. However, the equation lies way above the upper limit specified in the Indian standard [5].

16.3 Effect of Solar Exposure on Roof on Comfort Temperature

Indraganti [16] collected 31.8% data from the top floors with roofs exposed (RE) to direct sun. She noted higher indoor temperatures in them compared to the lower floors. Owing to continuous exposure to higher temperatures, subjects in RE flats expressed comfort at higher temperatures in all the months (Fig. 16.5a). The differences are more evident in warmer months.

16.4 Effect of Age, Gender, and Economic Group on Comfort Temperature

A higher percentage of women expressed comfort in home environments. While 74% female subjects voted in the central three categories of the sensation scale, only 69% men felt similarly. Mean comfort temperature of women was also found to be slightly higher than men in the monsoon season. Conversely, younger subjects

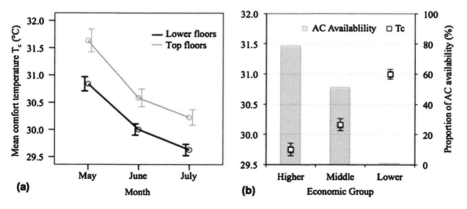

Fig. 16.5 Effect of (**a**) solar exposure (**b**) economic group on comfort temperature [16]. Roof exposed top floors recorded slightly higher comfort temperature in all the months. Subjects in lower economic groups had 1.2 K higher comfort temperature. Also shown is the proportion of AC availability. Error bars indicate 95% confidence interval (N = 3624)

(age < 40 years) expressed higher dissatisfaction (5–9%, depending on the gender). Thermal acceptance was highest in middle-aged women [17]. Karyono also noted similar gender based differences [18].

The data were divided into higher, middle, and lower economic groups. Subjects in higher economic groups had lower comfort temperature than their counterparts in lower economic groups. The difference (1.2 K) is statistically significant at 95% confidence interval as shown in Fig. 16.5b. Further to this, we also noted subjects in the higher economic group flats having greater and frequent access to the ACs and air coolers. For example, in upper two economic groups, ACs were available in 78.8% and 65% cases on all data and were in use in 9.7% and 2.8% cases, respectively [7]. Air coolers increase the humidity and cool by evaporation. They are very effective in dry summers.

Frequent exposure to lower temperatures in part could have lowered the comfort temperature in the upper economic groups. On the other hand, subjects in the lower economic group had a little/no access to ACs. They depended on cross ventilation, behavioral adaptation (e.g., hot midday siestas) to restore comfort during the overheated period. Possibly as a result, they also displayed higher tolerance to increased indoor temperatures.

Shove [19] commented on the sociological implications of uniform temperature regimes overriding the necessity of behavioral adaptation. Notable, a higher percentage of people in the upper economic group voted beyond the middle three categories of the sensation scale than that of the lower economic group. She rues about the air-conditioned way of life getting inscribed – hard wired –into the very fabric of buildings.

16.5 Concluding Remarks

This chapter explained various aspects of thermal comfort in Indian apartments as observed through field studies in warm climates. Outdoor and indoor temperatures in apartments during peak summer were very high leading to intense discomfort. This could partially be attributed to poor quality construction/design resulting from flouted bylaws. Thermal comfort was restored in the monsoon season.

Mean thermal sensation in summer in apartments was beyond the comfort band most of the time. This was in part due to (1) overheating resulting from solar exposure on roofs, (2) inadequate adaptive opportunities, and (3) high indoor temperatures (around 34 °C). Subjects could not opt for very high air speed during the overheated period as it exacerbated the condition through convective heat gain. On the contrary, air movement and humidity played crucial roles in monsoon months when the humidity was also high alongside moderately warm temperatures.

Indoor comfort temperature varied adaptively with the outdoor running mean temperature averaging at 30.3 °C. People living in upper floors had slightly higher comfort temperature. It was noted that subjects exposed to frequent AC usage such as those in upper economic group found comfort at lower temperatures than their counterparts in lower economic groups, without much access to ACs. Shove and Humphreys also noted a cyclic-loop of dependency on ACs, which pushed comfort to further lower temperatures [19, 20].

Acknowledgments Indraganti VS Prasad funded the Hyderabad thermal comfort surveys. Late Allakky Sreenivas, Rajyalakshmi, Sarma AVRJ, Ratnavati, C. Rajeswari, Lahari, Dhanya, Lohita, and Yashwanth assisted us. We thank all of them. Over a hundred residents responded, even foregoing their post-meal siestas. We are forever grateful to them.

References

1. Gallion BA, Eisner S (1963) The urban Pattern City planning and design. D. Van Nostrand, Princeton
2. Indraganti M (2009) Thermal comfort and adaptive use of controls in summer: An investigation of apartments in Hyderabad. PhD thesis. JNAFA University, Hyderabad
3. Bhattacharjee KP (2004) House desgin with local climate and materials. J Inidan Inst Archit 11:29–34
4. IEA (2013) Energy balances of non-OECD countries. International Energy Agency
5. BIS (2005) National building code 2005. Bureau of Indian Standards, New Delhi
6. Indraganti M (2010) Thermal comfort in naturally ventilated apartments in summer: findings from a field study in Hyderabad, India. Appl Energy 87(3):866–883
7. Indraganti M (2010) Adaptive use of natural ventilation for thermal comfort in Indian apartments. Build Environ 45:1490–1507
8. Humphreys MA, Nicol JF (2002) The validity of ISO-PMV for predicting comfort votes in every-day thermal environments. Energ Buildings 34(6):667–684
9. Sharma MR, Ali S (1986) Tropical summer index – a study of thermal comfort in Indian subjects. Build Environ 21(1):11–24
10. Rijal HB, Honjo M, Kobayashi R, Nakaya T (2013) Investigation of comfort temperature, adaptive model and the window-opening behaviour in Japanese houses. Archit Sci Rev 56 (1):54–69
11. Nicol JF, Roaf S (1996) Pioneering new indoor temperature standards: the Pakistan project. Energ Buildings 23:169–174
12. Singh S (2016) Seasonal evaluation of adaptive use of controls in multi-storied apartments: a field study in composite climate of north India. Int J Sustain Built Environ 5:83–98
13. Rajasekar E, Ramachandraiah A (2011) A study on thermal parameters in residential buildings associated with hot humid environments. Archit Sci Rev 54(1):23–38
14. Indraganti M, Ooka R, Rijal HB, Brager GS (2014) Adaptive model of thermal comfort for offices in hot and humid climates of India. Build Environ 74(4):39–53
15. CEN:15251 (2007) Indoor environmental input parameters for design and assessment of energy performance of buildings: addressing indoor air quality, thermal environment, lighting and acoustics. Commite European de Normalisation, Brussels
16. Indraganti M (2010) Using the adaptive model of thermal comfort for obtaining indoor neutral temperature: findings from a field study in Hyderabad, India. Build Environ 45(3):519–536
17. Indraganti M, Rao KD (2010) Effect of age, gender, economic group and tenure on thermal comfort: a field study in residential buildings in hot and dry climate with seasonal variations. Energ Buildings 42(3):273–281

18. Karyono TH (2000) Report on thermal comfort and building energy studies in Jakarta, Indonesia. Build Environ 35:77–90
19. Shove E (2002) Converging-conventions. Department of Sociology, Lancaster University, Lancaster. http://www.comp.lancs.ac.uk/sociology/papers/Shove-Converging-Conventions.pdf
20. Humphreys MA (1995) Thermal comfort temperatures and the habits of hobbits. In: Nicol J, Humphreys MA, Sykes O, Roaf S (eds) Standards for thermal comfort: indoor air temperature standards. Taylor & Francis, London, pp 3–13

Chapter 17
Comfort Temperature and Adaptive Model in Traditional Houses of Nepal

Hom Bahadur Rijal

Abstract Two surveys of the thermal environment and thermal sensations were conducted in the indoor and the semi-open spaces of traditional houses, during both summer and winter, in five districts of Nepal: Banke, Bhaktapur, Dhading, Kaski, and Solukhumbu. The surveys were carried out for 40 days, gathering a total of 7116 thermal sensations from 103 subjects. The results show that residents are highly satisfied with the thermal condition of their houses. The residents have higher comfort temperatures in semi-open spaces such as verandas than in indoor spaces. The findings reveal that people in the regions studied adapt well to the natural environment, as a result of which comfort temperatures are different in different climates. They are lowest in the cool climate, medium in the temperate climate, and highest in the subtropical climate. By using the relationship between indoors and outdoors, the adaptive model for dwellings was proposed to predict the comfort temperature.

Keywords Nepal · Traditional house · Semi-open · Field investigation · Neutral temperature

17.1 Introduction

Many field investigations related to comfort temperature (the temperature at which people feel neither cool nor warm) in houses have been conducted in different parts of world. However, the climate and living conditions of Nepal are unique, and in order to evaluate and improve the thermal environment of the houses in Nepal, a thermal comfort study was needed. Nepal is in the process of modernization, and in order to establish a standard for the indoor air temperature, first of all it is necessary to clarify the comfort temperature experienced by the Nepalese in the present

H. B. Rijal (✉)
Department of Restoration Ecology and Built Environment, Tokyo City University,
3-3-1 Ushikubo-nishi, Tsuzuki-ku, Yokohama, 224-8551 Japan
e-mail: rijal@tcu.ac.jp

© Springer Nature Singapore Pte Ltd. 2018
T. Kubota et al. (eds.), *Sustainable Houses and Living in the Hot-Humid Climates of Asia*, https://doi.org/10.1007/978-981-10-8465-2_17

conditions of traditional living. Moreover, it is thought that the comfort temperature experienced by the Nepalese, who live in traditional houses, may yield suggestions to building designers in considering sustainable lifestyles well-adapted to the local climate. If the comfort temperature is found to be different according to region or season, it is possible to increase the thermal comfort of the houses, with less energy used and consequently a lower impact on the environment.

This survey of the thermal environment and thermal sensations was conducted in real-life situations in summer and winter in five areas of Nepal in both indoor and semi-open spaces. Since no research has been conducted on indoor thermal comfort in Nepal, the objectives of this research were [1–3]:

- To evaluate thermal comfort of the residents
- To estimate the comfort temperature of the residents in indoor and semi-open spaces in the different climatic zones
- To develop the adaptive model for indoor and semi-open spaces

17.2 Method

17.2.1 Investigated Space

Investigated houses are shown in Chap. 6. The evaluated space was divided into indoor and semi-open space. The indoor-evaluated space was used for living and sleeping in the Banke and Bhaktapur areas; for cooking, dining, living, and sleeping in the Dhading and Kaski areas; and for prayer, living, and sleeping in the Solukhumbu area. In the Banke, Dhading, and Kaski areas, surveys were carried out in semi-open spaces as well because they are also used for living, working, and sleeping.

17.2.2 Thermal Comfort Survey

Surveys of both thermal comfort and the thermal environment were conducted. The scale used in the comfort survey assigned nine points for thermal sensation, four points for thermal comfort, and five points for thermal preference (Table 17.1). The questionnaires were translated into Nepali, the official language of Nepal, so that people could be interviewed [3]. From the relation between the two surveys, the comfort temperature of the residents was calculated. The survey was carried out for 40 days (5 areas × 4 days × 2 seasons). A total of 3552 thermal sensations were gathered in summer, and 3564 thermal sensations were gathered in winter from 103 subjects. The sensations were gathered at intervals of 1 h.

Table 17.1 Questionnaire form for thermal comfort survey [3]

(1) Thermal sensation	(2) Thermal comfort
−4. very cold	0. comfortable
−3. cold	1. slightly uncomfortable
−2. cool	2. uncomfortable
−1. slightly cool	3. very uncomfortable
0. neutral	(3) Thermal preference
1. slightly warm	−2. much warmer
2. warm	−1. slightly warmer
3. hot	0. no change
4. very hot	1. slightly cooler
	2. much cooler
(4) Skin moisture	(5) Activity
0. none	1. lying down
1. slightly	2. sitting resting
2. moderate	3. sitting working
3. profuse	4. standing
	5. moving around
(6) Do you have any hot/cold part in the body?	(7) Can you accept the present hot/cold environment or not?

17.3 Results

17.3.1 Thermal Satisfaction

In summer, the likelihood of (a) "neutral" on the thermal sensation scale, (b) "comfortable" on the thermal comfort scale, and (c) "no change" on the thermal preference scale was higher in semi-open spaces than in the indoor space of the subtropical climate. This was not so in the temperate climate. In the subtropical climate, the houses are open-type and incoming air makes the room hot, whereas in the temperate climate, the effect of the thermal mass of the stone houses keeps the room cool in the daytime. In winter, the likelihood of (a), (b), and (c) is higher in indoor spaces than in the semi-open spaces in both climates. Indoor spaces in the winter are warmer than the semi-open spaces because they are less influenced by the outdoor climate. However, the thermal sensations in the indoor and the semi-open spaces are similar, which agrees with the fact that the semi-open spaces are being used as a living space. Apart from the summer votes from the subtropical climate (Banke), the likelihood of (a), (b), and (c) is high [3], and it can be said that the thermal satisfaction of the evaluated space is high in summer and winter.

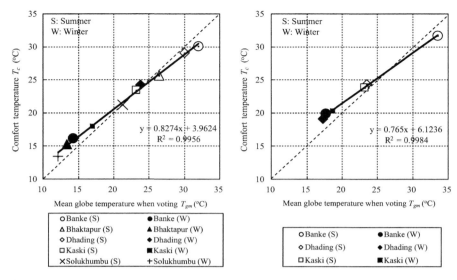

Fig. 17.1 Relation between the mean globe temperature and comfort temperature (left, indoor; right, semi-open) [3]

17.3.2 Regional Differences in Comfort Temperatures

In summer and winter, the indoor comfort temperatures are highest in the subtropical climate, medium in the temperate climate, and lowest in the cool climate areas, except for Dhading and Kaski in winter (Fig. 17.1, Table 17.2). The winter comfort temperature of Mustang district is lower than these districts (Fig. 17.2). In Dhading, indoor comfort temperature is higher both in summer and winter because of the higher room temperature caused by burning firewood [4]. Moreover, the difference in the indoor comfort temperature among areas is 8.9 K in summer and 4.6 K in winter, except for Dhading in winter. It is interesting that the difference among areas in summer is larger than in winter. Similar results are found in semi-open spaces. It can be deduced that the residents are fully adapted to the climate of each region.

Residents were adapting to the various thermal environments in various ways. Firstly, residents live in a natural environment and are therefore accustomed to coping with a dynamic changing environment. Even though the environment may change to a certain extent, they still feel comfortable. Secondly, the residents are used to living for long periods of time in high summer temperatures (e.g., 35 °C) in a subtropical climate and low winter temperatures (e.g., 0 °C) in a cold climate, which might produce a certain tolerance of heat and cold. Thirdly, their adaptation might be related to the different types of houses (open-type houses in subtropical climate, semi-open-type houses in temperate climates, and closed-type houses in cold climate) that achieve suitable indoor and outdoor temperature differences. Residents are used to living in hot and cold environments, and the indoor–outdoor temperature differences are small. In the investigated areas, even though the indoor and outdoor

Table 17.2 Regional and seasonal difference in the comfort globe temperature [3]

Evaluated space	Description	Banke	Bhaktapur	Dhading	Kaski	Solukhumbu	Regional difference (K)
Indoor	Summer (°C)	30.0	25.6	29.1	23.4	21.1	8.9
	Winter (°C)	16.2	15.2	24.2	18.0	13.4	4.5 (10.8[a])
	Summer–winter (K)	13.8	10.4	4.9	5.4	7.7	
	Summer–winter (clo)	0.36	0.58	0.27	0.12	0.72	
	Temp. equivalent to clo (K)	2.5~2.9	4.1~4.6	1.9~2.2	0.8~1.0	5.0~5.8	
	Summer–winter (m/s)	−0.01	−0.02	−0.04	0.30	0.00	
Semi-open	Summer (°C)	31.7		24.3	23.8		7.9
	Winter (°C)	19.9		19.1	20.3		1.2
	Summer–winter (K)	11.8		5.2	3.5		
	Summer–winter (clo)	0.56		0.26	0.21		
	Temp. equivalent to clo (K)	3.9~4.5		1.8~2.1	1.4~1.6		
Semi-open–indoor	Summer (K)	1.7		−4.8	0.4		
	Winter (K)	3.7		−5.1	2.3		

[a]Including Dhading district (effect of the firewood combustion)

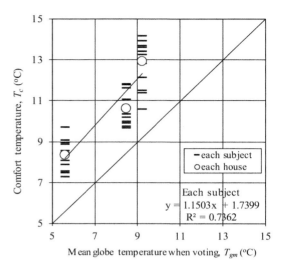

Fig. 17.2 Relation between comfort temperature and mean globe temperature when voting in winter in Mustang [6]

temperature differences are small [5], because of these differences, they feel comfortable. Fourthly, residents feel comfortable in both summer and winter using various methods such as clothing changes, transferring location indoors and outdoors, controlling apertures, burning firewood, sunbathing, taking cold showers, etc. According to the experiences of the author of this chapter, even though the author, coming from a temperate climate, felt hot in a subtropical climate and cold in a cold climate, the residents were, by voting "comfortable," saying that they were adapting to the heat or cold. Residents in the temperate climate also voted "comfortable," stating summer and winter comes gradually, and the human body adapts naturally to the environment.

17.3.3 Seasonal Difference in Comfort Temperature

The seasonal difference of indoor and semi-open spaces comfort temperature is high (Table 17.2). In the Dhading and Kaski districts, the indoor seasonal difference is smaller than in other areas, probably because of firewood use both in summer and winter. Nakamura and Okamura [7] pointed out, by analyzing the literatures of Iseki et al. [8] and Nicol et al. [9], that the seasonal difference in comfort temperature inside houses, taking account of clothing insulation, exceeds 6.0 K in Osaka (Japan) and Peshawar (Pakistan). The difference is confirmed by the survey by Kato et al. [10] and Sawachi et al. [11, 12] in Japan. We also obtain similar results to these studies. The seasonal difference may reflect the difference of clothing insulation, air movement, and adaptation level of the residents. According to Nicol et al. [9], the seasonal difference of 0.50 clo corresponds to the difference of 3.5–4.0 K in the comfort temperature. We too use this value to convert the seasonal difference in clothing insulation (Table 17.2). The seasonal difference in comfort temperature both for indoor and semi-open spaces by clothing insulation is approximately 1–6 K: the residents use clothing to adjust to the thermal environment.

Moreover, in Kaski, the seasonal difference in indoor air movement is 0.30 m/s. If the difference in air movement is considered to correspond to the difference of approximately 2–4 K in comfort temperature [9], the difference becomes approximately 4 K together with clothing insulation. In other areas, the seasonal difference in indoor air movement is very small. However, the seasonal difference in comfort temperature still remains: 1.4–10.9 K, even though the effects of clothing insulation and air movement were excluded. This remaining difference may be explained by the physiological or psychological adaptation of the residents to the seasons.

The seasonal difference in comfort temperature is smaller in the semi-open than in the indoor spaces except in Dhading district. There the seasonal difference in comfort temperature is greater, and a correspondingly greater seasonal difference in clothing insulation is slightly larger. The other explanation could be that the thermal conditions in the semi-open spaces are more severely influenced by the outdoor conditions.

17.3.4 Comparison of Comfort Temperature in Indoor and Semi-Open Space

The comfort temperature is higher in the semi-open spaces in the subtropical and the temperate climate (Kaski) both in summer and winter than those in the indoor spaces (Table 17.2). In the temperate climate of Dhading, the comfort temperature is higher in the indoor space, differing by 4.8 K in summer and 5.1 K in winter. It is similar in both seasons because of the quantities of firewood burning during the survey period (4 days), which are similar in summer (77 kg) and in winter (76 kg). Even though firewood is burnt in the Kaski district, the indoor comfort temperature is lower than in the semi-open spaces because the quantity of firewood burnt during the survey period (4 days) is lower in summer (35 kg) and winter (52 kg) than in the Dhading district. It may also be due to the difference in air movement. In winter, the average air movement is 1.4 (Banke), 1.5 (Dhading), and 2.2 (Kaski) times greater in the semi-open spaces than in the indoor spaces. However, the absolute air movement in semi-open spaces is small, and even in the Kaski district, the maximum value is 0.35 m/s. Furthermore, the physiological or psychological adaptation of the subjects could also be a factor. The difference in comfort temperature between indoor and semi-open spaces is larger in winter than in summer for all areas because the difference in air temperature itself is larger in winter. If the globe temperature is high, the comfort temperature tends also to be high, and so it is different in the indoor and semi-open spaces in a given area.

Fig. 17.3 Relation between comfort temperature and outdoor temperature

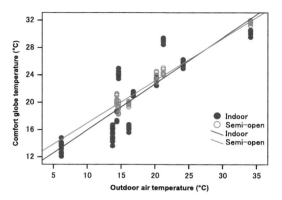

17.3.5 Adaptive Model

An adaptive model relates the indoor comfort temperature to the outdoor air temperature [13]. Figure 17.3 shows the relation between the comfort temperature and the mean outdoor temperature. Each data point represents the mean value for each subject. Most of data is from free running mode, and some data include the heating mode (firewood burning). The regression equations are given below.

Indoor
$$T_{cg} = 0.674T_o + 9.2 \left(n = 85, R^2 = 0.73, \text{S.E.} = 0.045, p < 0.001 \right) \tag{17.1}$$
Semi-open
$$T_{cg} = 0.617T_o + 10.8 \left(n = 62, R^2 = 0.96, \text{S.E.} = 0.015, p < 0.001 \right) \tag{17.2}$$

T_{cg} is comfort globe temperature by Griffiths' method (°C), T_o is the mean outdoor temperature (°C), n is the number of subject, R^2 is coefficient of determination, S.E. is the standard error for the regression coefficient, and p is the level of significance for the regression coefficient.

The regression coefficients are notably higher than that in the CEN standard (=0.33) [14]. The CEN standard is based on the field investigation in the office buildings and therefore may not apply to dwellings, where residents have more freedom to adapt.

17.4 Conclusions

Survey of the thermal environment and thermal sensations was conducted in summer and winter for residents in the traditional houses of the Banke, Bhaktapur, Dhading, Kaski, and Solukhumbu districts in Nepal. The results are:

1. The frequency of votes falling within the "thermal comfort zone" of thermal sensation, thermal comfort, and thermal preference is very high (86 to 100%)

indoors and in the semi-open spaces for both summer and winter. The residents are highly satisfied with the daytime thermal conditions of the houses in the spaces evaluated.

2. The indoor comfort temperature is highest in subtropical climate, medium in the temperate climate, and lowest in the cool climate areas both in summer (21.1–30.0 °C) and winter (13.4–24.2 °C), except in Dhading and Kaski districts in winter. The difference in the indoor comfort temperature among areas is higher in summer than in winter. Similar results are found in the semi-open spaces.

3. The comfort temperature has large seasonal differences (4.9 to 13.8 K). These differences might be due to the seasonal difference in clothing insulation, wind velocity, and the physiological or psychological adaptation of the residents to the seasons.

4. The comfort temperatures in semi-open spaces are 0.4 to 3.7 K higher in Banke and Kaski than in the indoor spaces both in summer and winter. The residents preferred higher temperatures in the semi-open spaces or rooms where firewood was burnt, which suggested that the higher ambient temperature raises the comfort temperature.

5. By using the relationship between indoors and outdoors, the adaptive model for dwellings was proposed to predict the comfort temperature. The regression coefficient of the adaptive model is higher than that in the CEN standard.

Acknowledgments We would like to give thanks to Prof. Harunori Yoshida, Prof. Noriko Umemiya, Prof. Michael Humphreys, and Prof. Fergus Nicol for their research guidance, to the investigated households for their cooperation, and to our families and friends for their support. This research is supported by the Japan Society for the Promotion of Science.

References

1. Rijal HB, Yoshida H, Umemiya N (2002) Investigation of the thermal comfort in Nepal. In: Proceedings of international symposium on building research and the sustainability of the built environment in the tropics, pp 243–262
2. Rijal HB, Yoshida H, Umemiya N (2003) Summer and winter thermal comfort of Nepalese in houses. J Archit Plann Environ Eng AIJ 68(565):17–24 (in Japanese with English abstract)
3. Rijal HB, Yoshida H, Umemiya N (2010) Seasonal and regional differences in neutral temperatures in Nepalese traditional vernacular houses. Build Environ 45(12):2743–2753
4. Rijal HB, Yoshida H (2002) Investigation and evaluation of firewood consumption in traditional houses in Nepal. In: Proceedings of the 9th international conference on indoor air quality and climate, vol 4. pp 1000–1005
5. Rijal HB, Yoshida H (2002) Comparison of summer and winter thermal environment in traditional vernacular houses in several areas of Nepal. Adv Build Technol 2(2):1359–1366
6. Rijal HB, Yoshida H (2006) Winter thermal comfort of residents in the Himalaya region of Nepal. In: Proceedings of international conference on comfort and energy use in buildings – getting them right (Windsor), Organised by the Network for Comfort and Energy Use in Buildings, Number of Pages: 15
7. Nakamura Y, Okamura K (1997) Analysis for appearance of seasonal acclimatization in optimum temperature based on some field investigations. J Archit Plann Environ Eng AIJ 495:85–91 (in Japanese with English abstract)

8. Iseki K, Isoda N, Yanase T, Hanaoka T (1988) A survey of residential thermal environments (Part 2): effect of thermal conditions on the residents. J Home Econ Jpn 19(8):879–884 (in Japanese with English abstract)

9. Nicol F, Jamy GN, Sykes O, Humphreys M, Roaf S, Hancock M (1994) A survey of thermal comfort in Pakistan toward new indoor temperature standards. Oxford Brookes University, School of Architecture, Oxford

10. Kato T, Yamagishi A, Yamashita Y (1996) Difference between winter and summer of the indoor thermal environment and residents' thinking of detached houses in Nagano city. J Archit Plann Environ Eng AI J 481:23–31 (in Japanese with English abstract)

11. Sawachi T, Matsuo Y (1989) Daily cycles of activities in dwellings, in the case of housewives: study on residents' behavior contributing to formation of indoor climate, Part 2. J Archit Plann Environ Eng AIJ 398:35–46 (in Japanese with English abstract)

12. Sawachi T, Matsuo Y, Hatano K, Fukushima H (1987) Determinants of heating and air conditioning behavior and acceptable ranges of temperature based on behavior: Study on residents' behavior contributing to formation of indoor climate, Part 1. J Archit Plann Environ Eng AIJ 382:48–59 (in Japanese with English abstract)

13. Humphreys MA, Rijal HB, Nicol JF (2013) Updating the adaptive relation between climate and comfort indoors; new insights and an extended database. Build Environ 63(5):40–55

14. Comité Européen de Normalisation (CEN) (2007) EN 15251: indoor environmental input parameters for design and assessment of energy performance of buildings addressing indoor air quality, thermal environment, lighting and acoustics. CEN, Brussels

Chapter 18
Comfort Temperature and Adaptive Model in Japanese Dwellings

Hom Bahadur Rijal

Abstract In order to quantify the seasonal differences in the comfort temperature and to develop a domestic adaptive model for Japanese dwellings, thermal measurements, a thermal comfort survey and an occupant behaviour survey were conducted for 4 years in the living rooms and bedrooms of dwellings in the Kanto region of Japan. We have collected about 36,114 thermal comfort votes from 244 residents of 120 dwellings. The results show that the residents are highly satisfied with the thermal environment of their dwellings. People are highly adapted in the thermal condition of the dwellings, and thus the comfort temperature has large seasonal differences. An adaptive model for housing was derived from the data to relate the indoor comfort temperature to the prevailing outdoor temperature. Such models are useful for the control of indoor temperatures.

Keywords Japanese dwellings · Field survey · Comfort temperature · Adaptive model

18.1 Introduction

Indoor temperatures are an important factor in creating comfortable homes. An understanding of the locally required comfort temperature can be useful in the design of dwellings and their heating and cooling systems to avoid excessive energy use. Comfort temperatures in dwellings have been widely investigated, with key studies in Japan, China, Singapore, Malaysia, Indonesia, Nepal, India, Pakistan, Iran and UK [1, 2]. However there are limitations in the research to date with some studies conducted over short time periods and some based on small samples. Comfort temperatures vary according to the month and season, requiring long-term data to fully understand perceptions and behavioural responses to temperatures in the home.

H. B. Rijal (✉)
Department of Restoration Ecology and Built Environment, Tokyo City University,
3-3-1 Ushikubo-nishi, Tsuzuki-ku, Yokohama, 224-8551 Japan
e-mail: rijal@tcu.ac.jp

© Springer Nature Singapore Pte Ltd. 2018
T. Kubota et al. (eds.), *Sustainable Houses and Living in the Hot-Humid Climates of Asia*, https://doi.org/10.1007/978-981-10-8465-2_18

In 2004 ASHRAE [3] introduced an adaptive standard for naturally ventilated buildings, and CEN [4] proposed an adaptive model for free-running naturally ventilated buildings. No Japanese data was included in these adaptive models, and little of the data was from dwellings. Occupant behaviour is different in the office and at home, and thus the existing adaptive models may not apply to dwellings.

In order to record seasonal differences in the comfort temperature and to develop a domestic adaptive model for Japanese dwellings, thermal measurements and a thermal comfort survey were conducted for 4 years in the living rooms and bedrooms of dwellings in the Kanto region of Japan [5].

18.2   Field Investigation

A thermal comfort survey and the thermal measurement were conducted in 120 dwellings in Kanto region (Kanagawa, Tokyo, Saitama and Chiba) of Japan from 2010 to 2014 [5–7]. It has hot humid summer, cold winter and mild spring and autumn. The monthly mean outdoor air temperature (relative humidity) of Tokyo is 5.7 °C (42%) in January and 28.6 °C (72%) in August.

The indoor air temperature and the relative humidity were measured in the living rooms and bedrooms, away from direct sunlight, at 10-min intervals using a data logger. The globe temperature was also measured in the living room in some surveys. The number of subjects was 119 males and 125 females. Respondents completed a questionnaire several times a day in the living rooms and twice in the bedroom ("before go to bed" and "after wake-up from the bed").

The ASHRAE scale is frequently used to evaluate the thermal sensation, but the words "warm" or "cool" imply comfort in Japanese, and thus the modified ASHRAE scale is also used to evaluate the thermal sensation (Table 18.1). To avoid a possible misunderstanding of "neutral", it is explained as "neutral (neither cold nor hot)". It is also said that the optimum temperature occurs on the cooler side in summer and on the warmer side in winter [8]. We have collected 36,114 thermal comfort votes. Outdoor air temperature and relative humidity were obtained from the nearest meteorological station.

18.3   Results and Discussion

The data were divided into three groups: the FR mode (free running), CL mode (cooling by air conditioning) and HT mode (heating). First we have determined the CL and HT modes based on actual cooling and heating used. Some in these

Table 18.1 Questionnaires for thermal comfort survey

No.	ASHRAE scale	Modified ASHRAE scale
1	Cold	Very cold
2	Cool	Cold
3	Slightly cool	Slightly cold
4	Neutral	Neutral (neither cold nor hot)
5	Slightly warm	Slightly hot
6	Warm	Hot
7	Hot	Very hot

Table 18.2 Distribution of outdoor and indoor air temperature during voting

	FR			CL			HT		
Variables	N	Mean	S.D.	N	Mean	S.D.	N	Mean	S.D.
T_i (°C)	25,195	23.7	5.3	6532	27.3	1.9	3582	18.9	2.9
T_g (°C)	11,012	23.5	4.5	2951	27.6	1.7	2256	19.6	2.8
T_o (°C)	25,339	18.9	8.0	6802	27.6	2.7	3604	7.2	4.2
RH_i (%)	25,195	59	11	6532	57	9	3582	48	11
RH_o (%)	24,495	68	18	6789	76	11	3603	56	19

T_i indoor air temperature, T_g globe temperature, T_o outdoor air temperature, RH_i indoor relative humidity, RH_o outdoor relative humidity, N number of sample, $S.D.$ standard deviation

categories used window opening to provide ventilation. All data that were in neither the CL nor the HT mode were classified as being in the FR mode.

18.3.1 Distribution of Outdoor and Indoor Temperature

The mean indoor air temperature and globe temperature are very similar (Table 18.2). The Japanese government recommends the indoor temperature of 20 °C in winter and 28 °C in summer. The results showed that the mean indoor temperatures during heating and cooling were close to the recommendation.

18.3.2 Distribution of Thermal Sensation

Table 18.3 shows the percentage of modified thermal sensation (mTSV) in each mode. Even though residents used the heating or cooling, they sometimes felt "cold" or "hot". As there are many "neutral" votes, it can be said that residents were generally satisfied in the thermal environment of the dwellings. This may be due to the adaptation of the residents to the local climate and culture.

Table 18.3 Percentage of thermal sensation in each mode

Mode	Items	Modified thermal sensation (mTSV)							
		1	2	3	4	5	6	7	Total
FR	N	115	1292	4200	14,248	4011	1364	281	25,511
	Percentage (%)	0.5	5.1	16.5	55.9	15.7	5.3	1.1	100
CL	N	7	39	504	4751	1265	257	58	6881
	Percentage (%)	0.1	0.6	7.3	69.0	18.4	3.7	0.8	100
HT	N	62	372	854	2292	87	5	–	3672
	Percentage (%)	1.7	10.1	23.3	62.4	2.4	0.1	–	100

N number of sample

18.3.3 Thermal Comfort Zone

To locate the thermal comfort zone, probit regression analysis was conducted for the modified thermal sensation vote (mTSV) categories and the temperature for FR mode. The analysis method is ordinal regression using probit as the link function and the temperature as the covariate.

The results of the probit analysis are shown in Table 18.4. The temperature corresponding to the median response (probit = 0) is calculated by dividing the constant by regression coefficient. For example, the mean temperature of the first equation will be 1.0/0.211 = 4.7 °C (Table 18.4). The inverse of the probit regression coefficient is the standard deviation of the cumulative normal distribution. For example, the standard deviation of air temperature of the FR mode will be 1/0.211 = 4.739 °C (Table 18.4). These calculations are fully given in Table 18.4. Transforming the probits using the following function into proportions gives the curve of Fig. 18.1a–b. The vertical axis is the proportion of votes.

$$\text{Probability} = \text{CDF.NORMAL (quant, mean, S.D.)} \tag{18.1}$$

where "CDF.NORMAL" is the cumulative distribution function for the normal distribution and "quant" is the indoor air temperature (°C) or globe temperature (°C); the "mean" and "S.D." are given in Table 18.4.

The highest line is for category 1 (very cold) and so on successively. Thus, it can be seen that the temperatures for thermal neutrality (a probability of 0.5) are around 24 °C (Fig. 18.1a, b).

Reckoning the three central categories as representing thermal comfort and transforming the probits into proportions give the bell curve of Fig. 18.1c. The result is remarkable in two respects. The proportion of people comfortable at the optimum is very high, only just less than 100%, and the range over which 80% are comfortable is wide—from around 17 to 30 °C. This is presumably because people in their own dwellings are free to clothe themselves according to the room temperature, without the constraints that are apt to apply at the office.

Table 18.4 Results of the probit analysis

Variable	Equation	Mean (°C)	S.D. (°C)	N	R²	S.E.
T_i	$P(\leq 1) = 0.211T_i\text{-}1.0$	4.7	4.739	25,177	0.48	0.002
	$P(\leq 2) = 0.211T_i\text{-}2.5$	11.8				
	$P(\leq 3) = 0.211T_i\text{-}3.8$	18.0				
	$P(\leq 4) = 0.211T_i\text{-}6.2$	29.4				
	$P(\leq 5) = 0.211T_i\text{-}7.2$	34.1				
	$P(\leq 6) = 0.211T_i\text{-}8.2$	38.9				
T_g	$P(\leq 1) = 0.223T_g\text{-}1.5$	6.7	4.484	11,008	0.43	0.003
	$P(\leq 2) = 0.223T_g\text{-}2.8$	12.6				
	$P(\leq 3) = 0.223T_g\text{-}4.1$	18.4				
	$P(\leq 4) = 0.223T_g\text{-}6.6$	29.6				
	$P(\leq 5) = 0.223T_g\text{-}7.7$	34.5				
	$P(\leq 6) = 0.223T_g\text{-}8.7$	39.0				

T_i indoor air temperature (°C), T_g globe temperature (°C), $P_{(\leq 1)}$ is the probit of proportion of the votes that are 1 and less, $P_{(\leq 2)}$ is the probit of the proportion that are 2 and less and so on, *S.D.* standard deviation, *N* number of sample, R^2 Cox and Snell R^2, *S.E.* standard error of the regression coefficient

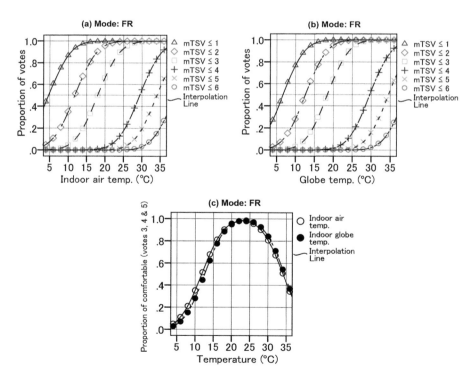

Fig. 18.1 Proportion of modified thermal sensation vote (mTSV) or comfortable (mTSV 3, 4 or 5) for indoor air or globe temperature

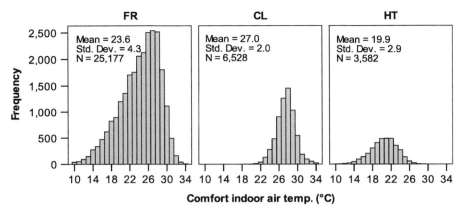

Fig. 18.2 Prediction of comfort temperatures from each observation by Griffiths' method [5]

18.3.4 Comfort Temperature

The comfort temperature is predicted by the Griffiths' method [2, 9, 10].

$$T_c = T + (4 - C)/a \qquad (18.2)$$

T_c is comfort temperature by Griffiths' method (°C), T is indoor or globe temperature, C is modified thermal sensation vote and a is rate of change of thermal sensation with room temperature, and 0.50 is used for this study [11].

The mean comfort globe temperatures by the Griffiths' method are 23.5 °C in FR mode, 27.2 °C in CL mode and 20.4 °C in HT mode which are similar to the comfort air temperatures (Fig. 18.2). We chose to use the Griffiths' method because the presence of adaptation ordinary regression analysis and probit analysis, although commonly used, can give misleading values for the comfort temperatures. In our data powerful adaptation to the seasonal variation of indoor temperature necessitates the use of the Griffiths' method.

18.3.5 Seasonal Difference in Comfort Temperature

In this section, to clarify the seasonal difference, the comfort temperature for each month and season is investigated (Figs. 18.3 and 18.4). The comfort temperature does not vary much within the winter or summer seasons. However, it is quite changeable in the spring and autumn. The results showed that the comfort temperature changes according to the season, and thus it is related to the changes in indoor and outdoor air temperature which occur in spring and autumn. The comfort air temperature by the Griffiths' method is 17.6 °C in winter, 21.6 °C in spring, 27.0 °C in summer and 23.9 °C in autumn in FR mode. Thus, the seasonal difference of the

Fig. 18.3 Profiles of the monthly mean comfort temperature, indoor temperature and outdoor temperature with 95% confidence intervals [5]

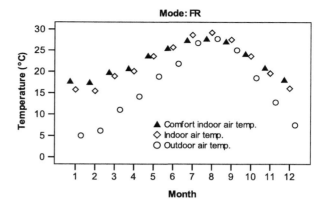

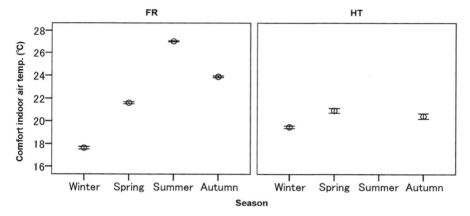

Fig. 18.4 Seasonal variation of comfort temperature with 95% confidence intervals [5]

mean comfort temperature is 9.4 K which is similar to the value found in previous research in Japan and Nepal [2]. The comfort temperature of the heating HT mode also changes significantly from season to season (Fig. 18.4). The comfort temperature found in previous research in Japan, Nepal, Pakistan and UK ranges from 8.4 to 30.0 °C [2]. The wider range may suggest that the comfort temperature has regional differences.

18.3.6 The Adaptive Model

An adaptive model relates the indoor comfort temperature to the outdoor air temperature [3, 4]. Figure 18.5 shows the relation between the comfort air temperature calculated by the Griffiths' method and the running mean outdoor temperature. The regression equations are given below.

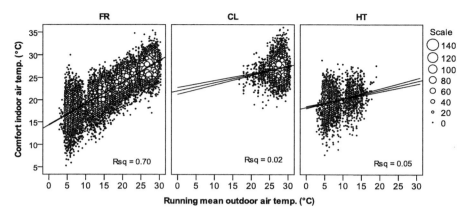

Fig. 18.5 Relation between the comfort indoor air temperature and the running mean outdoor temperature [5]

FR mode $T_{ci} = 0.480T_{rm} + 14.4$

$$\left(n = 25,177, R^2 = 0.70, \text{S.E.} = 0.002, p < 0.001\right) \quad (18.3)$$

FR mode $T_{cg} = 0.417T_{rm} + 15.9$

$$\left(n = 11,008, R^2 = 0.65, \text{S.E.} = 0.003, p < 0.001\right) \quad (18.4)$$

CT mode $T_{ci} = 0.180T_{rm} + 22.1$

$$\left(n = 6,528, R^2 = 0.02, \text{S.E.} = 0.014, p < 0.001\right) \quad (18.5)$$

CL mode $T_{cg} = 0.200T_{rm} + 21.7$

$$\left(n = 2,951, R^2 = 0.04, \text{S.E.} = 0.018, p < 0.001\right) \quad (18.6)$$

HT mode $T_{ci} = 0.193T_{rm} + 18.3$

$$\left(n = 3,582, R^2 = 0.05, \text{S.E.} = 0.014, p < 0.001\right) \quad (18.7)$$

HT mode $T_{cg} = 0.149T_{rm} + 19.2$

$$\left(n = 2,256, R^2 = 0.03, \text{S.E.} = 0.017, p < 0.001\right) \quad (18.8)$$

T_{ci} is comfort indoor air temperature by Griffiths' method (°C), T_{cg} is comfort globe temperature by Griffiths' method (°C) and T_{rm} is the exponentially weighted running mean outdoor temperature for the day (°C). (S.E. is the standard error of the regression coefficient.)

The regression coefficient and the correlation coefficient in the FR mode are higher than in the CL and HT modes. The regression coefficient in the FR mode is notably higher than that in the CEN standard (=0.33) [4]. The CEN standard is based on the field investigation in the office buildings and therefore may not apply to dwellings, where residents have more freedom to adapt.

18.4 Conclusions

A thermal comfort survey of the residents of the Kanto region of Japan was conducted for 4 years. The thermal environment in living rooms and bedrooms was investigated. The following results were found:

1. The residents proved to be highly satisfied with the thermal environment of their dwellings, as indicated by the high proportion of "neutral" responses.
2. The seasonal difference (summer and winter) in comfort indoor temperature was very high at 9.4 K.
3. An adaptive relation between the comfort temperature indoors and the outdoor air temperature could be an effective tool for predicting comfort temperature and for informing control strategies.

Acknowledgements We would like to give thanks to Prof. Michael Humphreys and Prof. Fergus Nicol for their research guidance, the households who participated in the survey and the students for data entry. This research was supported by Grant-in-Aid for Scientific Research (B) Number 25289200.

References

1. Nicol F, Roaf S (1996) Pioneering new indoor temperature standards: the Pakistan project. Energ Buildings 23:169–174
2. Rijal HB, Honjo M, Kobayashi R, Nakaya T (2013) Investigation of comfort temperature, adaptive model and the window-opening behaviour in Japanese houses. Archit Sci Rev 56 (1):54–69
3. ASHRAE Standard 55 (2004) Thermal environment conditions for human occupancy. American Society of Heating Refrigeration and Air-conditioning Engineers, Atlanta
4. Comité Européen de Normalisation (CEN) (2007) EN 15251: indoor environmental input parameters for design and assessment of energy performance of buildings addressing indoor air quality, thermal environment, lighting and acoustics. CEN, Brussels
5. Rijal HB, Humphreys M, Nicol F (2015) Adaptive thermal comfort in Japanese houses during the summer season: behavioral adaptation and the effect of humidity. Buildings 5(3):1037–1054
6. Rijal HB, Humphreys MA, Nicol JF (2014) Development of the adaptive model for thermal comfort in Japanese houses. In: Proceedings of 8th Windsor Conference: Counting the Cost of Comfort in a changing world, Cumberland Lodge, Windsor, UK, 10–13 April 2014. Network for Comfort and Energy Use in Buildings, London. http://nceub.org.uk
7. Rijal HB, Humphreys M, Nicol F (2015) Adaptive thermal comfort in Japanese houses during the summer season: Behavioral adaptation and the effect of humidity. Buildings 5 (3):1037–1054
8. McIntyre DA (1980) Indoor climate. Applied Science Publishers, Ltd, London
9. Griffiths ID (1990) Thermal comfort in buildings with passive solar features: field studies. Report to the Commission of the European Communities. EN3S-090 UK: University of Surrey, Guildford
10. Nicol F, Jamy GN, Sykes O, Humphreys M, Roaf S, Hancock M (1994) A survey of thermal comfort in Pakistan toward new indoor temperature standards. Oxford Brookes University, Oxford
11. Humphreys MA, Rijal HB, Nicol JF (2013) Updating the adaptive relation between climate and comfort indoors: new insights and an extended database. Build Environ 63:40–55

Chapter 19
Thermal Comfort Survey in Japan

Takashi Nakaya

Abstract In this chapter, thermal comfort survey data is used to evaluate the thermal environment selected by the occupants, considering the thermal adaptation and limitations. Field surveys were conducted in occupied houses in Hyogo and Osaka, Japan, from August to September 2003.

The occupants lived in a summer hot side area of the thermal comfort zone, with around 80% of the occupants accepting the hot environment. But there was a temperature upper limit for the hot environment. Within the scope of the investigation conducted herein, wet-bulb globe temperature (WBGT) 29 °C is thought to be the upper limit. The thermal environment selected by the occupants is not in the thermal comfort zone but is within the safety range and is thermally acceptable.

Keywords Field surveys · House · Mixed mode · Occupants behaviour · Comfort zone · Adaptation and limitation

19.1 Introduction

To design for thermal comfort in hot environments, understanding occupants' behaviours in buildings is necessary. Reducing the use of air conditioning (AC) in summer is an essential part of reducing the energy consumption of buildings.

Health risks, such as heatstroke, become prevalent in hot environments. Reducing the use of air conditioners will increase the temperature in indoor environments and consequently increase the thermal stress and risk of heatstroke. By understanding the cooling, saving and the hot limit of adaptation, one can safely reduce energy consumption. In this chapter, the following items are considered, obtained from the survey of Japanese occupants taken during the summer [1, 2].

T. Nakaya (✉)
Department of Architecture, Shinshu University, Nagano, Nagano, Japan
e-mail: t-nakaya@shinshu-u.ac.jp

© Springer Nature Singapore Pte Ltd. 2018
T. Kubota et al. (eds.), *Sustainable Houses and Living in the Hot-Humid Climates of Asia*, https://doi.org/10.1007/978-981-10-8465-2_19

What thermal environment conditions are chosen by the occupants?
Do the occupants accept the indoor thermal environment?
What is the limit of the thermal environment in which occupants live?

19.2   Method

Field surveys [1, 2] were conducted in Hyogo and Osaka, Japan. The maximum
temperature during the measurement period was 34.5 °C. The surveys were
conducted over a 2-month period, from August to September 2003, by researchers
conducting thermal measurements and questionnaires at individual houses. The
survey was conducted at 31 detached houses in Hyogo Prefecture and 31 apartments
in Osaka Prefecture (Fig. 19.1). Seventy people (6 men and 64 women) took part in
the survey. Their ages ranged from 24 to 73 years (average, 46 years). Most of the
subjects were women (housewives).

Environmental and human factors were measured in the living room. The indoor
air temperature was measured using a thermistor sensor. The globe temperature was
measured using the globe thermometer with a 150 mm diameter copper sphere,
which was painted black. The relative humidity was measured using the humidity
sensor. The air velocity was measured using an omnidirectional sensor. These
instruments were placed at the centre of the living room. The measurements were
performed at a height of 0.6 m above the floor level, as this height was considered to
be appropriate for someone sitting on the floor. Measurements were taken every
5 mins. The outdoor temperature data were taken from the weather station nearest to
the survey areas. The metabolic rate and insulation of clothing were estimated based
on the answers to the questionnaire [3]. The operative temperature was determined
using the average indoor air temperature and the globe temperature [3]. The wet-bulb
temperature was calculated using measured air temperature and relative humidity

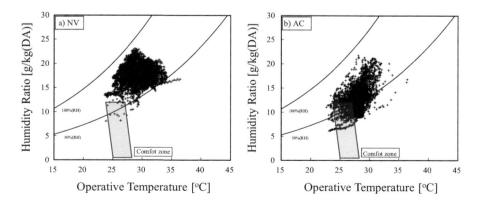

Fig. 19.1 Psychrometric chart during the ventilation (**a**) Natural ventilation and (**b**) AC [2])

measurement data. WBGT was calculated from globe temperature and wet-bulb temperature. The PMV index predicts the mean response of a large group of people according to the ASHRAE thermal sensation scale [6].

19.3 Results and Discussion

19.3.1 Occupant Conditions

The average indoor air temperature was 28.9 °C. The relative humidity was 66%, the mean radiant temperature was 28.9 °C, and the air velocity was 0.11 m/s. The mean metabolic rate was 1.3 met, and the clothing insulation was 0.29 clo. The representative example of clothing was the combination of a thin, short-sleeved shirt, short trousers and underwear. In an environment with high temperature and humidity, there was a tendency for people to live wearing relatively thin clothes. When evaluated by PMV, the PMV is 1.26 and the PPD is 38%. When evaluating data with PMV, the discomfort rate for the indoor thermal environment was high.

The subjective scale used in the analysis was based on thermal acceptability, which was measured on a 6-point scale (Table 19.1). For the analysis, data was recorded as binary data containing "acceptable" and "unacceptable" ratings. A total of 459 votes were obtained. However, the proportion of "unable to accept the indoor thermal environment" was 20.5% of all subjective votes, 5.4% when using air conditioning, and 22.0% when using natural ventilation. Furthermore, 80% of the occupants had accepted the indoor thermal environment.

19.3.2 Psychrometric Chart

19.3.2.1 Thermal Comfort Zone

The temperature and humidity data were plotted on the psychrometric chart. Further, the thermal comfort zone was added to the psychrometric chart [3]. The range of the comfort zone is defined in PMV.

Figure 19.1a shows the indoor data during natural ventilation (NV). Temperature and humidity in the room were plotted on the psychrometric chart. The temperature in the room was widely distributed, with a maximum temperature of approximately

Table 19.1 Subjective vote on thermal acceptability

How are you feeling in the moment?	
1: very unacceptable	4: slightly unacceptable
2: acceptable	5: unacceptable
3: slightly acceptable	6: very unacceptable

35 °C. Figure 19.1b shows the indoor data when air conditioning (AC) was used for cooling. When using AC, the temperature and humidity were lower than during natural ventilation. Data within the comfort zone was 0.2% for natural ventilation and 17.5% for AC. The use of AC was temporary, and the temperature and humidity could not be controlled within the comfort zone. However, the acceptable rate of natural ventilation is 78.0% and of AC is 94.6%. Most of the occupants had accepted the thermal environment.

19.3.2.2   The Heatstroke Danger Threshold

Extreme thermal environments affect human health, especially hot environments. In hot environments, it is not possible to dissipate heat from the human body, and the risk of heatstroke increases. Figure 19.2 shows the graph obtained by adding WBGT line and heatstroke-related information to Fig. 19.1. WBGT is the index used for examining the temperature limits of the hot environment [4]. The open square in the figure corresponds to heatstroke deaths of healthy male US soldiers assigned to sedentary duties in Midwestern army camp offices [5]. Figure 19.3 shows the frequency of WBGT. There is little data exceeding WBGT 28 °C, and there is no data above WBGT 29 °C. Therefore, the thermal environment during natural ventilation and air conditioning was both below the heatstroke danger threshold. In both conditions, the occupants stayed in the hot area more than in the thermal comfort zone. They accepted the hot environment. However, the occupants did not stay in the environment with temperatures higher than the heatstroke danger threshold.

The use of natural ventilation and electric fans has the effect of increasing the transmission rate and promoting heat dissipation from the human body. However, when the difference in temperature and humidity between the human body and the environment becomes small, transmission cannot be expected. In other words, in the hot environment, it is necessary to lower the temperature and humidity by air conditioning, and there is a hot limit of adaptation using air flow.

It is considered that the ET* line of 35 °C is the heatstroke danger threshold (ASHRAE fundamental 2013). The ET* line of 35 °C corresponds to approximately WBGT 29 °C. Further, under the condition of light work, the hot limit of adaptation was WBGT 29 °C (acclimatization) and WBGT 30 °C (no acclimatization). Judging from survey data, literature and standards, the hot limit of natural ventilation is considered to be approximately WBGT 29 °C. In addition, the hot limit should consider the human's acclimatization ability. Acclimatized humans can tolerate hot temperature environments. Elderly people and children should be set lower critical temperature than healthy adults.

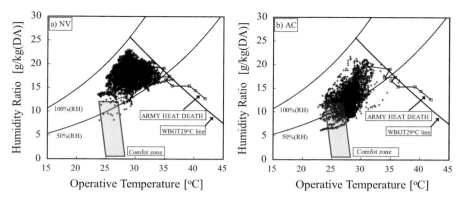

Fig. 19.2 Psychometric chart of indoor thermal environment in summer, (**a**) natural ventilation (NV) and (**b**) air conditioning (AC) [2]

Fig. 19.3 Frequency of the indoor WBGT [2]

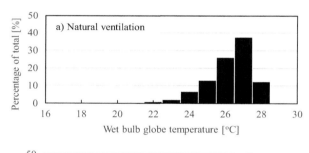

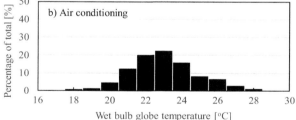

19.4 Conclusions

In this chapter, the following conclusions were reached, based on the summer field survey of Japan's occupants:

Occupants continued to live in a thermal environment higher than the thermal comfort zone. However, approximately 80% of the occupants accepted the thermal environment.

There was a hot limit for hot environment. Judging from survey data, literature and standards, the hot limit of natural ventilation is considered to be approximately WBGT 29 °C.

Occupants were not comfortable with thermal environment but were able to accept thermal discomfort, living in the safe environment.

References

1. Nakaya N, Matsubara N, Kurazumi Y (2005) A field study of thermal environment and thermal comfort in Kansai region. J Environ Eng AIJ 597:51–56
2. Nakaya N, Matsubara N, Kurazumi Y (2008) Use of occupant behaviour to control the indoor climate in Japanese. In: Proceedings of conference: air conditioning and the low carbon cooling challenge network for comfort and energy use in buildings residences
3. ASHRAE Standard 55 (2013) Thermal environmental conditions for human occupancy. Atlanta American Society of Heating, Refrigerating, and Air-conditioning Engineers Inc
4. ISO 7243 (1989) Hot environments – Estimation of the heat stress on working man, based on the WBGT index (wet bulb globe temperature)
5. ASHRAE (2013) ASHRAE fundamental chapter 9 health. American Society of Heating, Refrigerating, and Air-conditioning Engineers Inc.

Box A: Thermal Comfort in Japanese and Indian Offices

Madhavi Indraganti, Ryozo Ooka, and Hom Bahadur Rijal
Department of Architecture and Urban Planning, College of Engineering, Qatar University, Doha, Qatar
Institute of Industrial Science, The University of Tokyo, Tokyo, Japan
Department of Restoration Ecology and Built Environment, Tokyo City University, Tsuzuki-ku, Yokohama, Japan
Email: madhavi.indragaganti@fulbrightmail.org; madhavi@qu.edu.qa; ooka@iis.u-tokyo.ac.jp; rijal@tcu.ac.jp

Abstract Japan and India face complex energy challenges. They rely heavily on imported fuels. Energy use in buildings is on the rise, majorly fuelled by air conditioners. In this light, the application of the adaptive model in reducing the building energy demand assumes significance. This box gives an overview of the adaptive comfort work and comfort temperatures in Japan and India.

Keywords Office buildings · Thermal comfort · Adaptive model · Warm-humid climate · Field survey · Comfort temperature

Introduction

Japan and India face serious energy challenges. Both rank poorly on the level of energy self-sufficiency (132 and 83 in the world). Japan produced 6% of its total primary energy supplied, and India produced 67% in 2013. Buildings in Japan and India consume 36–38 % of total final energy consumed, and air conditioning (AC) majorly contributes to this [1]. To depart from nuclear power, Japan administered *setsuden* (energy saving) guidelines. These lack scientific basis. Indian indoor temperature norms ignore the diversity in building stock and climates [2]. Reducing AC use seems vital, and the application of adaptive thermal comfort standard is a significant step in that direction. It is yet to be built for Japan and India. This process relies on field studies. The following sections briefly explain the thermal comfort field studies in Japan and India undertaken in this context.

Thermal Comfort in Japanese Offices

Indraganti et al. [3] surveyed 83 office spaces for 3 months in summer (July 4 to September 11). From 435 respondents, they collected 2042 responses. Women were 33%. The four offices investigated were run in naturally ventilated (NV) and

© Springer Nature Singapore Pte Ltd. 2018
T. Kubota et al. (eds.), *Sustainable Houses and Living in the Hot-Humid Climates of Asia*, https://doi.org/10.1007/978-981-10-8465-2

Table 1 Scales used in the survey

Scale	Thermal sensation (TS)	Thermal preference (TP)	Thermal acceptability (TA)
3	Hot		
2	Warm	Much cooler	
1	Slightly warm	A bit cooler	Unacceptable
0	Neutral	No change	Acceptable
−1	Slightly cool	A bit warmer	
−2	Cool	Much warmer	
-3	Cold		

air-conditioned (AC) modes. Table 1 shows the scales used. We used standard Japanese translations for these. The clothing insulation varied from 0.38 to 0.97 clo during the study. Indoor temperatures above 28 °C (summer *setsuden* limit) were noted in 80% of the NV and 30% in AC environments. However, both NV and AC environments had higher air speed: in 60–70% cases, it was more than 0.2 m/s. About 84% were comfortable (voted −1 to +1 on TS) with ACs on, and 14% lesser without. Fanger's PMV differed substantially (with 95% confidence interval) from TS in all the cases. Preference for cooler environments was higher when the ACs were turned off. TP$_{mean}$ was 0.32 with ACs on and 0.77 without them. TA (enquired from a direct question) was 92% with ACs and 76% without.

Comfort Temperature and Thermal Adaptation

Policy decisions on indoor climate management require knowledge of the comfort temperature (T$_C$). T$_C$ estimated with a Griffiths' coefficient of 0.33 K^{-1} was 25.8 °C (N = 423, S.D. = 3.7 °C) in NV mode and 27.2 °C (N = 1979, S.D. = 3.3 °C) in AC mode [3]. In a yearlong study in Tokyo and Yokohama, Rijal et al. [4] obtained 25.0 °C and 25.9 °C, respectively [4]. On the other hand, Tanabe et al. [5] reported 26.2 °C in AC mode in a Tokyo study. In Sendai, Tsukuba and Yokohama Goto et al. [6] found the preferred SET* to be 26.0 °C. Figure 1 presents the Indraganti et al. [3] data superimposed over the CEN standard [7]. Most of the data is within the classes I and II of the CEN standard. They also noted that in 32% and 42% of the cases, T$_C$ was beyond the Japan's *setsuden* limits in NV and AC modes, respectively [3].

Thermal adaptation appears to be the main reason for this wide range in comfort temperature. The adaptation was chiefly through clothing and increased air speeds in all the environments. For example, air speeds were more than 0.2 m/s in 30% and 40% of the NV and AC environments, respectively.

Thermal Comfort in Indian Offices

In a yearlong thermal comfort field study in Indian offices, Indraganti et al. [8] recorded 6048 comfort responses involving 2787 people. These were from

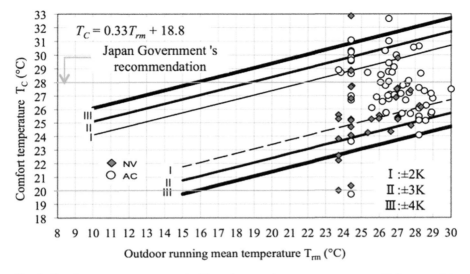

Fig. 1 Comfort temperature regressed with outdoor running mean temperature with Japanese data superimposed [3]. (Each point represents mean value for a room)

13 mixed mode (MM) buildings, 14 fully AC buildings, and 1 NV building [9]. The data were collected in Hyderabad and Chennai in NV and AC modes. They measured TS, TP, and TA using the scales mentioned above, besides many indoor environmental variables. The outdoor daily mean temperature varied from 21.5–35.5 °C averaging at 27.8 °C. Indian summers were very hot, monsoon warm-humid, and winters very mild. The indoor globe temperature (T_g) oscillated between 21.8 and 37.7 °C with 26.9 °C as its mean. Mean clothing insulation was 0.7 clo (S.D. = 0.08 clo).

In India, majority of the subjects voted on the warmer side of the sensation scale ($TS_{mean} = 0.4$, S.D. = 1.3) and preferred cooler indoors ($TP_{mean} = 0.4$, S.D. = 0.8) in NV environments. On the other hand, AC environments were sensed on the cooler side of the TS scale ($TS_{mean} = -0.1$, S.D. = 1.3). Actual sensation vote was always away from Fanger's PMV. Mode of operation seems to have little affected TA. It averaged at 72 % and 71% in NV and AC, respectively. Japan recorded much higher TA, partly due to their deep-rooted cultural ethos [10].

Adaptive Model and the Comfort Temperature

Using 0.5 K^{-1} as the Griffith's coefficient, T_{C_mean} was estimated as 28.0 and 26.4 °C in NV and AC modes, respectively. These values are well above the limits set in the Indian codes [2] and the applicability ranges others suggested [11]. Regression of T_C with T_{rm} reflects thermal adaptation of the people. Figure 2 features the adaptive relationships Indraganti et al. [8] obtained for India in both NV and AC modes. The relating T_{rm} ranged from 23.6 to 33.0 °C (NV mode) and 23.6 to 35 °C (AC mode).

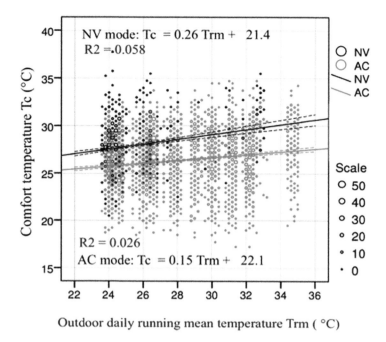

Fig. 2 Adaptive model of thermal comfort for India [8]. T_C regressed with outdoor running mean temperature for NV and AC modes on raw data

Higher slope in NV mode indicates faster adaptation (3.8 K for unit change in T_{rm}) much similar to the Europeans (Fig. 1) than in AC mode.

Another study [12] from Indian offices reported neutral temperatures ranging from 19.6 to 28.5 °C (NV mode) and 21.5 to 28.7 °C (AC mode). The corresponding T_{rm} ranged between 12.5 and 31.0 °C (NV mode) and 13.0 and 38.5 °C (AC mode). They presumed the buildings to be switching into the AC mode, when T_{rm} crosses 31.0 °C. On the other hand, Sharma and Ali [13] from Roorkee NV buildings reported tropical summer index (TSI, synonymous to comfort temperature) to be varying between 25 and 30 °C.

Concluding Remarks

This box highlighted the need to adopt the adaptive comfort model for Japan and India. Simulation/design of indoor environment requires knowledge on comfort temperature. In Japan, it varied between 25.8 °C (NV mode) and 27.2 °C (AC mode) in summer. The comfort temperatures in India were found to be 28.0 and 26.4 °C in NV and AC modes, respectively. These differed significantly from the respective national standards.

References

1. IEA (2015) IEA Energy Atlas, International Energy Agency
2. BIS (2005), National Building Code, Bereau of Indian Standards
3. Indraganti M, Ooka R, Rijal HB (2013) Thermal comfort in offices in summer: Findings from a field study under the 'setsuden' conditions in Tokyo, Japan. Build Environ 61:114–132
4. Rijal HB, Humphreys MA, Nicol JF (2017) Towards an adaptive model for thermal comfort in Japanese offices, Building Research & Information https://doi.org/10.1080/09613218.2017.1288450
5. Tanabe S, Iwahashi Y, Tsushima S, Nishihara N (2013) Thermal comfort and productivity in offices under mandatory electricity savings after the Great East Japan earthquake. Archit Sci Rev 56(1):4–13
6. Goto T, Mitamura T, Yoshino H, Tamura A, Inomuta E (2007) Long-term field survey on thermal adaptation in office buildings in Japan. Build Environ 42:3944–3954
7. CEN:15251 (2007) Indoor environmental input parameters for design and assessment of energy performance of buildings: addressing indoor air quality, thermal environment, lighting and acoustics. Brussels: Commite European de Normalisation
8. Indraganti M, Ooka R, Rijal HB, Brager GS (2014) Adaptive model of thermal comfort for offices in hot and humid climates of India. Build Environ 74 (4):39–53
9. Indraganti M, Ooka R, Rijal HB, Brager GS (2015) Thermal comfort in offices in India: Behavioral adaptation and the effect of age and gender. Energy Build 103:284–295
10. Indraganti M, Ooka R, Rijal HB (2014) Thermal Comfort and Acceptability in Offices in Japan and India: A Comparative Analysis, In Annual Conference of the Society of Heating Air-conditioning and Sanitary Engineers of Japan (SHASE) Sep 2014, Akita, Japan
11. ASHRAE (2010) ANSI/ASHRAE Standard 55-2010, Thermal environmental conditions for human occupancy, American Society of Heating, Refrigerating and Air-Conditioning Engineers, Inc, Atlanta
12. Manu S, Shukla Y, Rewal R, Thomas L, de Dear RJ (2016), Field studies of thermal comfort across multiple climate zones for the subcontinent: India Model for Adaptive Comfort (IMAC). Build Environ 98:55–70
13. Sharma MR, Ali S (1986) Tropical Summer Index—a study of thermal comfort in Indian subjects. Build Environ 21(1):11–24

Part III
Adaptive Behavior

Chapter 20
Principles of Adaptive Behaviours

J. Fergus Nicol and Michael A. Humphreys

Abstract This short chapter introduces the causal relationship between the behaviour of building occupants, their comfort and the energy used by the buildings. It suggests that much of the behaviour is motivated by the desire of the occupants to make themselves comfortable and to optimise the environment. The provision of comfortable conditions in domestic buildings in hot-humid climates is highlighted. An annex also introduces ways in which the comfort-related behaviour can be understood and allowed for in predictive simulations of indoor temperature and energy use.

Keywords Occupant behaviour · Adaptive comfort · Energy used · Hot-humid climates · Cooling strategies

20.1 Basic Principle of Adaptive Behaviour

The behaviour of building occupants has often been seen as a problem by environmental engineers and building simulators. Their actions can often appear to be random in both their motivation and their effects. But they can make a big difference to the success of any strategy for ensuring comfort in a building. In particular behaviour can radically change the amount of energy used by mechanical systems. Buildings which are physically identical can vary in the amount of energy they use because the building occupants differ in their use of the building. This can lead to an increase in the energy

J. Fergus Nicol (✉)
Low Energy Architecture Research Unit, Sir John Cass Faculty of Art, Architecture and Design, London Metropolitan University, London, UK
e-mail: f.nicol@londonmet.ac.uk

M. A. Humphreys
School of Architecture, Faculty of Technology, Design and Environment, Headington Campus, Oxford Brookes University, Oxford, UK
e-mail: mahumphreys@brookes.ac.uk

© Springer Nature Singapore Pte Ltd. 2018
T. Kubota et al. (eds.), *Sustainable Houses and Living in the Hot-Humid Climates of Asia*, https://doi.org/10.1007/978-981-10-8465-2_20

Fig. 20.1 basic model of
ways in which building
occupants achieve comfort
in a building (Nicol)

This has to be done within the existing climatic, social,
economic, **architectural** and cultural context. **Buildings
should be designed to provide acceptable conditions**

use, but properly targeted behaviour can also help to cut the use of energy and allow
the building to remain comfortable without using any energy for much of the time.

The adaptive principle 'If a change occurs such as to produce discomfort, people
react in ways which tend to restore their comfort' is the basis of the adaptive
approach to thermal comfort [1]. It introduces the idea that in a changing environment behavioural, responses may be needed to ensure human comfort. These
responses will be directed towards restoring comfort and can therefore be seen as
arising from the thermal environment through the mediation of the human subject. In
their paper *Understanding the Adaptive Approach*, Humphreys and Nicol list over
30 common actions a person might take to make themselves comfortable [2].

Over time it is essential to balance the metabolic heat produced by the body with
the heat lost to the environment in order to keep the body temperature constant. They
identify five types of behavioural actions which change the physical/physiological
heat balance in five basic ways:

1. Regulating the rate of internal heat generation (e.g. change activity)
2. Regulating the rate of body heat loss (e.g. change clothing)
3. Regulating the thermal environment (e.g. open a window, turn down the heating)
4. Selecting a different thermal environment (e.g. move to another room, go out)
5. Modifying the body's physiological condition (e.g. vasoregulation, sweating,
   shivering and changes of posture)

These changes mean that the actions people take either change their own response
to achieve comfort in an existing environment or change the environment to suit
them (Fig. 20.1). The means they can use to enable them to keep the thermal balance
(the windows, shades, blinds as well as any heating or cooling system) are often
referred to as 'adaptive opportunities' [3].

Some actions show aspects of more than one type of behaviour. Opening a
window, for instance, can increase the air movement inside a room which will
help cool occupants by convection, encourage evaporative cooling and this will
increase the temperature at which they can be comfortable. At the same time, it will
encourage mingling of the inside and the outside air which will cool or heat indoor
air depending on the temperature difference between the two. It is important

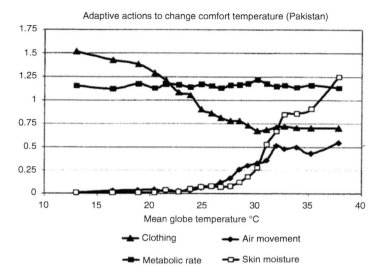

Fig. 20.2 The responses of Pakistani office workers to indoor temperature (Nicol)

to remember that there are other reasons for opening a window – maybe to improve the view out, to reduce indoor pollution or to hear the birds singing in the garden. The window may be closed to reduce the glare or keep out the smell and noise of traffic.

The interaction between the environment and behaviour is illustrated in Fig. 20.2 based on surveys in Pakistan. As the indoor temperature increases, the mean values for clothing insulation (in clo units) falls and the air speed (ms^{-1}) rises as windows are opened or fans turned up. The metabolic rate (wm^{-2}) is not noticeably effected as these Pakistani office workers go about their business, but mean skin moisture [on a scale suggested by Webb [4] 0 (none), 1 (slight), 2 (moderate) and 3 (profuse)] rises as the physiological response of the body seeks to increase heat loss through evaporation.

20.2 Comfort and Climate: Choosing the Right Behavioural Control System

An adaptive approach to design will as far as possible allow the building occupants to remain comfortable for as much of the time as possible and with the minimum use of energy. The accepted vision of the built environment is that the building and its services should provide comfort for the occupants. In contrast the adaptive approach suggests that the building (and its services) should provide the occupants with means to make themselves comfortable.

Any particular building has to deal with the specific climate in which it is built. In addition to a heating system, a building designed for a site in a cool climate might thus concentrate on providing ways the occupant can capture and store 'natural' heat in the warmer times of the day or the year. There are many methods used for the capture and storage of solar heat. These are ways that have been developed to allow building occupants to access solar or other sources of natural energy when it is needed. In the built environment, the most common site for energy storage is in the thermal mass of the building. Examples of energy storage methods are Trombe walls, which absorb solar heat and pass it to the building's interior, heated cores which store heat in hot weather and pass it on as needed and earth-coupled heat transfer which can store heat in the ground during hot weather and use it to heat the building in colder seasons. In addition to this there are other 'passive' ways to achieve heating or cooling; examples are window shades, opening windows which let in outside air and encourage air movement and so on.

In climates where overheating is more likely to be the problem than getting too cold, then systems will concentrate on providing 'coolth' rather than warmth. These may include shading to keep out the sun, air movement to encourage heat loss by convection and evaporation, night ventilation to cool down the thermal mass using low night-time air temperatures and so on.

One assumption of the adaptive approach is that the building occupants will understand the role of the different adaptive opportunities provided and be sufficiently familiar with their purpose and the way they work to use them in the way the designer intended. If they are to use the opportunities to make themselves comfortable, they must know what they do, how they do it and when they will be most effective. The designer also should ensure that the controls do not conflict with one another – or indeed with other aspects of the building such as security.

20.3   Seasonal Changes

Most areas of the world have a climate which varies seasonally from one part of the year to another. In the equatorial tropics, the seasonal temperature change is relatively small as is the daily temperature range. This is the classic situation in what is referred to as a hot-humid climate. Because of the high level of evaporative heat in the air, any change in the temperature involves relatively large amounts of heat exchange. The climatic variation from one part of the year to another is kept small by the generally high level of humidity. In these areas the seasonal change of the climate will change not from hot to cold but from a relatively humid or 'rainy' season to a relatively dry season. It is also possible to have a 'hot humid' season where the summer and the winter seasons are quite different but the hotter months behave like the equatorial climate. The way in which the adaptive opportunities are used in the hot-humid season will be similar to those in the equatorial climate, but the designer has to keep in mind that there may be contrasting requirements in the cooler times of year.

20.4 Behaviour in a Hot-Humid Climate

Understanding of the human thermal response to the hot-humid climate is complicated by the supposed effect of the high humidity on the thermal sensation. The high humidity will affect the ability of the body to keep cool by the use of the evaporation of sweat from the skin surface. A number of estimates of this effect have been made and are surveyed by Nicol [5]. He found that the effect on comfort of humidity, though real, is actually smaller than is often imagined. A high humidity is equivalent to a rise in the temperature of about 1 K, compared to a low humidity. The high humidity will also reduce the range of temperatures which are comfortable. The effect of high humidity may lead to other types of discomfort, such as increased sweating, but measured in terms of the ASHRAE scale, the effect is small.

An important cooling strategy in the hot-humid climate is the use of air movement. This can cool by convection, if the air is cool, and by evaporation. The evaporative effect will be most important to understand in the hot-humid climate because most people will have some sweat, but the rate of evaporation may be reduced by the humidity of the air. Movement of the air over the skin is therefore necessary to ensure heat is lost by evaporation. Heat loss by the use of night cooling of the building can be important, though the temperature of the mass of the building may not be as different from the daytime as could be expected in a dry climate. Nicol [5] estimated from a survey in Pakistan that the comfort temperature in a room with fans running is about 2 K higher than when the fans are not running over an outdoor temperature range of 20 °C to 32 °C. Above 32 °C very few offices did not have fans running.

The aim of the designer in the hot-humid climate must be to use low-energy technologies to make occupant comfortable. The solar heat must be kept out, so roofs and ceilings should be well insulated and where possible extended by the use of shading especially over openings. Air movement by cross ventilation should be readily available and controllable by the occupants. Low-energy fans are essential. Buildings in equatorial sites should also where possible be arranged with their major axis is an east-west direction to reduce the solar heating of the major walls which can be most easily shaded on the south and north sides. There is a possible conflict if the hot-humid weather is in one particular season and the building needs to encourage solar heating in another season. Careful and thoughtful design can overcome some of the problem by using well-designed moveable shading or the clever use of deciduous vegetation.

20.5   Mechanical Cooling and Heating and Adaptive Comfort

The control of indoor conditions using mechanical cooling or heating is always a powerful option if the system is provided by the building and may be essential in more extreme weather. Nicol [6] has found that the availability of mechanical control can result in a wide range of indoor temperatures in domestic buildings and must be controlled by occupants. Indeed the indoor temperature range in heated and cooled buildings is found in most cases to be greater than that in free-running buildings. In the mechanically conditioned building, the choice of indoor temperature is left to the occupants who are free to decide on the basis of their own preference (and their ability to afford to run the buildings at high or low temperature). In the free-running building, it is the layout and materials of the building itself which decide indoor temperature according to the physics of the situation, and the occupants will need to adjust themselves to this.

20.6   Annex: Adaptive Behaviour as a Stochastic Phenomenon

The scientific method suggests that we should propose a model of the underlying process of, say, the opening of a window or windows in a building motivated by indoor temperature and then check this proposition against measured data from the field. In Fig. 20.3 a we are considering a single occupant with two possibilities: window open or window closed.

If we start from the bottom left of the diagram with a low temperature and the window closed, as we move right we will get to a point where the temperature is too hot with the window closed and the temperature reaches a 'trigger temperature' (c/o) at which the occupant will open the window to cool the room. At temperatures above this trigger temperature, we can assume that the window will be open.

If we start from the top right of the diagram and the room cools, we will reach a second 'trigger temperature' (o/c) at which the occupant will close the window to prevent the temperature from falling any further. At temperatures below this trigger temperature, we can assume the window will be closed.

Between o/c and c/o, we cannot be sure whether the window is open or closed as this would depend on the way the temperature has changed. It could, for instance, have increased to c/o. The window is opened and the indoor temperature falls, but the window will remain open unless it drops below o/c.

In Fig. 20.3b we suggest what the effect will be if there are a number of occupants. Each occupant will have a slightly different value for o/c and c/o which

Fig. 20.3 Developing a
model of window opening
behaviour. (**a**) Single
occupant window opening.
(**b**) The effect of numerous
different occupants. (From
Rijal et al [8])

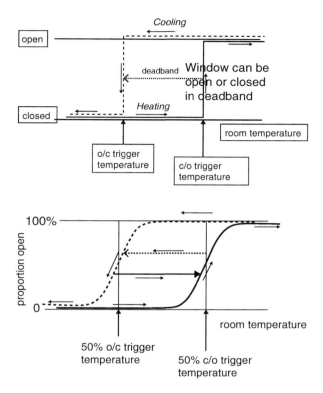

means that instead of a sharp trigger temperature there will a sigmoid one where the
average value of the c/o and the o/c will be the temperature at which there is a 50%
chance of the window being open or closed. Between the two sigmoid curves, the
window can be either open or closed.

Figure 20.4 shows how the likelihood of the window being open in occupied
rooms actually changes with temperature. Each dot on the graph is the average
number of windows open in 25 rooms at the temperature shown on the x-axis. As the
temperature rises, the number of windows that are open increases. The middle of the
three lines on the graph is the 50% line, and 83% of points lie between the two outer
lines. Such a graph can be used to estimate the likelihood a window will be open.
The method is given in Rijal et al. [7]. It can be used for other similar behavioural
responses to environmental stimuli such as the running of fans [9] or use of lights
and allows the simulator to estimate the indoor temperature from the usual input
variables for a simulation.

Fig. 20.4 Measured values
of the proportion of
windows open with indoor
globe temperature. Each
point is the mean proportion
of windows open at the
given temperature. Data
from surveys in the UK
reported in Rijal et al [7].

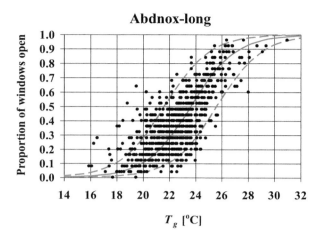

20.7   Conclusions

This chapter has looked at the behaviour of building occupants in relation to their
thermal comfort and buildings' energy use. Adaptive comfort suggests that much
occupant behaviour is directed towards avoiding discomfort as suggested by the
adaptive principle. The palette of behaviours which will help ensure indoor comfort
will depend on the climate.

In the hot-humid climate, the key adaptive behaviours are to exclude solar
radiation and to ensure a source of variable air movement. Minimising the heating
effect of solar radiation is essential in an environment where even a small increase in
temperature can give rise to overheating. The roof of the building is the most
vulnerable surface especially in equatorial regions where the altitude of the sun in
the sky normally is high, but it is also important to shade walls and openings.

Air movement will help the body to lose heat by evaporation but needs to be
controllable. The possibility of controllable air movement is essential to fit the
different comfort preferences of occupants. In many cases cross ventilation can be
used and it can be augmented by fans.

Night cooling of the building structure is possible but is less effective because the
diurnal variation of outdoor air temperature is relatively small and in addition a lightweight structure is often preferred in a hot-humid context.

Mechanical cooling and air conditioning is increasingly used and is relatively
controllable, but there is a considerable energy cost involved with both the cooling
load and the need for air circulation.

References

1. Nicol F, Humphreys M, Roaf S (2013) Adaptive thermal comfort, principals and practice. Routledge, London, p 8
2. Humphreys M, Nicol F (1998) Understanding the adaptive approach to thermal comfort. ASHRAE Trans 104(1):991–1004
3. Baker N, Standeven M (1995) A behavioural approach to thermal comfort assessment in naturally ventilated buildings. In: Proceedings of the CIBSE national conference, Eastbourne, pp 76–84
4. Webb C (1959) An analysis of some observations of thermal comfort in an equatorial climate. BJIM 16(3):297–310
5. Nicol F (2004) Adaptive thermal comfort standards in the hot-humid tropics. Energy Build 36 (7):628–637
6. Nicol F (2017) Temperature and adaptive comfort in heated, cooled and free-running dwellings Building Research and Information. doi: https://doi.org/10.10180/09613218.2017.1283922
7. Rijal HB, Tuohy P, Humphreys M, Nicol F, Samuel A, Clarke J (2007) Using results from field surveys to predict the effect of open windows on thermal comfort and energy use in buildings. Energy Build 39(7):823–836
8. Rijal HB, Tuohy P, Humphreys MA, Nicol JF, Samuel A (2012) Considering the impact of situation-specific motivations and constraints in the design of naturally ventilated and hybrid buildings. Archit Sci Rev 55(1):35–48
9. Rijal HB, Tuohy P, Humphreys MA, Nicol F, Samuel A, Raja IA, Clarke J (2008) Development of adaptive algorithms for the operation of windows, fans and doors to predict thermal comfort and energy use in Pakistani buildings. ASHRAE Trans 114(2):555–573

Chapter 21
Behavioural Changes May Affect Changes in Comfort Temperature of Indonesian People

Tri Harso Karyono

Abstract The development of the Indonesian economy has changed the capability of its people to use more modern technology and equipments. The use of the new technology has changed people's behaviour. More people tend to abandon using bicycle and foot to go to work and school; they started to use motorized vehicles. More buildings and houses are now air-conditioned, neglecting the natural ventilation cooling. The behaviour of wearing clothes has also changed. Traditional and conventional local clothes have been changed by the modern, particularly the western clothes. Changing on clothings and also changing of the building cooling system have changed the comfort temperatures of people who have changed them. A number of thermal comfort studies show that the more people being exposed to the lower temperatures would reduce their comfort temperature. Changing a certain people's behaviour may affect to the changing in their comfort temperatures.

Keywords Behavioural changes · Comfort temperature · Economic development · Indonesia · Modern technology · Vernacular society

21.1 Introduction

As a country with thousands of islands, inhabited by multiethnic groups of people, Indonesia has various cultures which lead to the various behaviours. Gained her independence in 1945 from the Dutch, this country is still in the process of development. Having many islands spread out throughout the area of 1,904,569 km² [1], there is a problem in the equality of the nation development. Some regions have been rapidly developed so that people can enjoy a better living, practicing modern life with the support of modern technology. In this kind of development stage, people live and behave in a modern way equal to their fellows in the developed world. They live in the air-conditioned (AC) homes or apartments, working in the AC offices, going to school in the AC classrooms, enjoying shopping in the AC shopping malls,

T. H. Karyono (deceased)
School of Architecture, Tanri Abeng University, Jakarta, Indonesia

© Springer Nature Singapore Pte Ltd. 2018
T. Kubota et al. (eds.), *Sustainable Houses and Living in the Hot-Humid Climates of Asia*, https://doi.org/10.1007/978-981-10-8465-2_21

etc., and develop their experience to be living in the low indoor temperature since the outdoor is quite warm or even hot. On the other hand, many people have been living in the small islands or small towns, far away from the capital city of Jakarta. They experienced living in the lower stage of development. A number of these people are still living in the forest, practicing life in the Stone-Age way of living. Some of them are hunters and even build their shelters in the trees. They behave naturally to adjust to the nature, using a very simple technology to support their life. They try to keep their ancestors' way of life to blend with the nature. Sometimes, they even don't know about life outside their territory. They don't know that their fellow Indonesians have already enjoyed a modern civilization, facilitated by modern technology and thus behave differently. However, most of the Indonesians are now living in between, neither too modern nor too primitive. They have enjoyed some modern way of life, although keeping the traditional way is still becoming part of things that must be preserved by them. In this sense, Indonesians' way of life, which is very much affected by the economical status, is various, ranging from a Stone Age, such as some people in Papua, to a very modern way of life, such as those living in the big cities as Jakarta.

Since Indonesia is in the equator belt, this country possesses a warm and humid tropical climate. This climate provides ambient temperatures higher than in the cold climate with an average of 28 °C in the coastal areas and less than 24 °C in the highland or mountainous regions. This kind of climate is, in fact, quite suitable for human beings to be lived in naturally, without any mechanical support. In a number of lowland and coastal cities, for instance, the range of prevailing ambient temperatures is between 18 and 32 °C, which is relatively close to the comfort temperatures of the average people that live in this climate. Based on some comfort studies, people in this climate were comfortable at about 27 °C [2–4].

Moreover, in this case people may not need energy for heating nor for cooling if they wish as the prevailing ambient temperature is close to the required comfort temperature. People require less energy to achieve the required comfort environment. In the previous time, people have lived comfortably without air conditioners (AC) in their homes and working places. A number of homeless people could sleep well on the open spaces, such as under the bridges or under the shop's terraces, wearing normal tropical clothes.

From time to time, the nation's economy has been growing quite well, improving people capabilities to afford more things and facilities needed. Their living standard has been improved. Since then, a great number of Indonesians living in the big cities have changed their lifestyles from traditional to modern ways. People changed their behaviours to be more modern. In terms of travelling, people tend to abandon using the bicycle and using foot to reach a short and medium distances. They started to use motorized vehicles as they can afford them. In terms of building, people are no longer opening the windows or turning on the fans when they feel warm; they turn on the AC instead.

21.2 Economic Development and the Changes of People's Behaviour

If we look back to the so-called primitive world or vernacular society in the tropical countries, such as Indonesia, people had been living in a sustainable manner, consuming fewer natural resources, using less energy and producing less negative impact to the environment. The members of the vernacular community obey the wisdom from their ancestors they believed to be correct. The wisdom might be saying that any harmful act to the natural environment would create a natural disaster. Vernacular people tend to believe that all things (plants, trees, rivers, soil, etc.) in their surrounding environments have spirits, just like human beings, who can take revenge if they are attacked. People had to look after the environment, creating a harmonious living and conserving the natural environment. There were no law enforcements for anybody breaking the wisdom; however, people obeyed the wisdom because of the social punishment for those against it.

In Indonesia, they are a number of people who belong to particular vernacular communities who have already known about modern development and modern civilization, but they don't want to follow this kind of modernism. They keep their own cultures and, thus, their behaviours as they are. They are keeping their distinctive way of life, rejecting the use of modern technologies, such as electricity, vehicles, etc. One of them is the vernacular society of Baduy in Banten, West Java. This community with a total population of around 7200 people inhabits a particular reserve area of 5200 hectares. They consist of two groups: the Inner and the Outer Baduy. The Inner Baduy consists of around 350 people, while the remaining population is the Outer Baduy [5]. The Baduys applied a traditional way of life; they forbid to use vehicles when travelling. These vernacular people tend to apply less modern technologies and depend more to their own muscle power. They walked barefoot to go to anywhere. They don't use umbrellas to protect themselves from the rain but instead use banana leaves as a means of rain protections.

Looking at the way this kind of people respond to the modern civilization, we can see that this has a relationship with their daily behaviours. Walking for a medium and long distance will raise people's metabolic rate, and thus more heat would be generated by their bodies. Consequently, people will wear a minimum and thinner clothes, such as a short pant or a mini trouser, and even wear no shirt, in order to dissipate the heat easily from their bodies to their immediate environments.

21.3 Behavioural Changes Would Affect to the Changing of Comfort Temperatures

A study on thermal performance of some vernacular dwellings in East Nusa Tenggara, Indonesia, shows inhabitants feeling comfortable in their modest dwellings, although the indoor and outdoor temperatures were relatively high [6]. Again, people were wearing thin clothes to adjust their clothing to respond to their surrounding climate.

Behaviour affects the person's lifestyle which leads to the amount of energy they may consume. People who keep their behaviour to use suits in their office in a tropical climate would require lower indoor temperatures than those who wear standard tropical clothes.

The invention of new technology, such as an air conditioner, has helped people to have more choice to cool their immediate space easily. The new technology, which helps people to be living conveniently and easily, would change people's habit to be more dependent on this technology. Technology would change people's behaviour. People tend to be adapted to the lower indoor temperature when they are always exposed to the cool air-conditioned rooms.

As the country's economy is developing and more people can afford to use more modern technology, such as air conditioners, there would be more people changing their behaviour to rely on AC rooms than opening their windows to cool the rooms. This kind of changes would consequently change people's thermal comfort requirements, in which people tend to be more comfortable in lower temperatures than before.

The more the economy is developed, the more advanced technologies are offered. The advanced technology tends to offer a more convenient way to support people's life, though it will cost more as it uses more energy to operate and also produce more negative impact to the environment. People tend to change their way of life by using more advanced technology and, consequently, would be changing the human behaviour towards life. The more the economy is developed, the more people in Indonesia are able to provide air conditioners in their rooms. More people would change their behaviour to be more dependent to air conditioners than with natural ventilation to make their rooms thermally comfortable.

A comfort study done by Karyono et al. [7] showed that subjects who had been exposed more frequently and within a longer time in AC rooms were comfortable at a lower temperature than those who were regularly being exposed in the naturally ventilated buildings. It was found from this study that comfort temperature of subjects of university students who were fully exposed in the AC rooms, both in their classrooms and accommodations, was comfortable at about 1 to 1.5 °C lower than those who were partly exposed in the naturally ventilated rooms.

Changes in the behaviour to use AC more frequently have decreased human comfort temperature. The human body tends to be adjusted to the lower temperature when they had been exposed to such a long period in the lower indoor temperatures of AC rooms, so that they would be thermally more comfortable in the lower temperatures. This has led to the increase of building energy consumption due to the increase of building cooling load. Behaviour has affected human thermal comfort and energy demand.

A thermal study by Kwon and Choi [8] in South Korea revealed that subjects felt comfortable wearing different clothes in different seasons with different air temperatures. Between January and February where the outdoor temperature was 5 °C, subjects felt comfortable with clothing insulation of 1.87–3.14 clo (very thick clothes). While between July and August when the outdoor temperature was 29 °C, subjects were comfortable with clothing insulation of 0.4–1.01 clo (thin clothes). This study shows that by wearing thin clothes, people would still be comfortable with the exposed temperature of 29 °C.

Changing behaviour towards clothing would change people's comfort temperature. In Indonesia, the national clothes, Batik, normally made from thinner fabrics and usually worn as a formal dress in the formal meeting, would help people to be more comfortable in a higher room temperature than the Western suits. In this case, the use of energy for cooling would be reduced, thus, minimizing the carbon emissions. A number of thick uniforms, worn by some Indonesian government officers, might be altered to be thinner with a lower clothing value, so that cooling energy in the rooms can be reduced [9].

21.4 Conclusions

The Indonesian economic development gives its people to have more access to the modern technology. The massive use of modern technology, such as air conditioner, has changed people's behaviour, both physically and psychologically, to adapt to the lower temperatures created by the air conditioning system. The more this country's economy is developed, the more likely the buildings would be air conditioning and the more people would change to be comfortable in the lower indoor temperatures. The comfort temperatures of the people tend to change to be lower due to the long exposure to the air conditioning spaces, and this would increase the building energy consumption, enhancing the global warming. Changes on the clothing thermal insulation might help people to be still comfortable in the higher indoor temperature, thus minimizing the cooling energy in the building.

References

1. CIA (2017) The world factbook: Indonesia. Available at https://www.cia.gov/library/publica tions/the-world-factbook/geos/id.html. Accessed 23 Aug
2. Karyono TH (2000) Report on thermal comfort and building energy studies in Jakarta. J Build Environ 35:77–90
3. Karyono TH, Sri E, Sulistiawan JG, Triswanti Y (2015) Thermal comfort studies in naturally ventilated buildings in Jakarta, Indonesia. Buildings 5:917–932
4. Karyono TH (1996) Thermal comfort in the Tropical South East Asia Region, Archit Sci Rev, v39(3): 135-139
5. World Rain Forest Movement (2017), The Baduy People of Western Java – living tradition. http://wrm.org.uy/oldsite/bulletin/87/Indonesia.html. Accessed 23 Aug
6. Karyono TH, Suwantara IK, Nugrahaeni R, Suprijanto I, Vale R (2012) Temperature performance and thermal comfort study in vernacular houses in East Nusa Tenggara, Indonesia. In: Proceedings of 7th Windsor Conference: The changing context of comfort in an unpredictable world Cumberland Lodge, Windsor, UK, 12–15 April 2010

7. Karyono TH, Heryanto S, Faridah I (2015) Air conditioning and the neutral temperature of the Indonesian university students. Archit Sci Rev 8:174–183
8. Kwon J, Choi J (2012) The relationship between environmental temperature and clothing insulation across a year. Int J Biometeorol 56(5):887–893
9. Karyono TH (2014) Green architecture, an introduction to sustainable building in Indonesia (in Bahasa Indonesia). Raja Grafindo Press, Jakarta

Chapter 22
Window-Opening Behaviour in Hot and Humid Climates of Southeast Asia

Hiroshi Mori, Tetsu Kubota, and Meita Tristida Arethusa

Abstract This chapter presents the results of a comparative analysis of the window-opening behaviour of residents in Southeast Asian cities. Using face-to-face interviews with a questionnaire, a total of 1,315 samples were obtained from typical households in 3 cities: Surabaya, Indonesia; Bandung, Indonesia; and Johor Bahru, Malaysia. Then, a hierarchical cluster analysis was conducted to analyse the window-opening behaviour in the households. The results showed that overall, daily window-opening behaviour can be classified into four patterns. Pattern A shows opening windows or doors during the morning and evening, Pattern B depicts opening windows or doors during the daytime, Pattern C shows opening windows or doors throughout the day except for sleep time and Pattern D describes opening windows or doors throughout the day. Almost all of the respondents in Surabaya and Johor Bahru were categorised as Patterns B, C or D, indicating that they usually keep windows or doors open during the daytime, but the opening patterns varied during the periods after coming home and during sleeping. In contrast, most of the respondents in Bandung were categorised into Pattern A, showing that they briefly open windows and doors during the morning and evening but close them during the daytime. Furthermore, the daily usage patterns of air conditioners were also analysed in this chapter.

Keywords Window-opening · Hot-humid climate · Adaptive behaviour · Passive cooling · Southeast Asia

H. Mori (✉)
YKK AP R&D Center, PT. YKK AP Indonesia, Tangerang, Banten, Indonesia
e-mail: h-mori@ykkap.co.id

T. Kubota
Graduate School for International Development and Cooperation (IDEC), Hiroshima University, Hiroshima, Japan
e-mail: tetsu@hiroshima-u.ac.jp

M. T. Arethusa
PT. Hadiprana Design Consultant, Jakarta, Indonesia

22.1   Introduction

There is concern that the spread of air conditioners (AC) among urban residential buildings in hot and humid Southeast Asia will contribute to further increases in primary energy consumption and will therefore increase CO_2 emissions in the near future [1]. Higher-income occupants are using AC more, and it is expected that the AC ownership rate will increase as incomes increase [2]. Natural ventilation is recommended to achieve indoor thermal comfort in a manner that reduces energy use. Window-opening behaviour is one of the key adaptive behaviours in hot-humid climates, but there have been relatively few studies conducted in this region [3, 4]. This study aims to reveal the typical daily patterns of occupants' window-opening behaviour in Southeast Asian cities through a comparative analysis. Most of the studies in this field have attempted to relate window-opening behaviour with indoor/outdoor thermal conditions, particularly air temperatures [5, 6]. This is because the window-opening behaviour is generally driven by attempts of occupants to avoid discomfort due to high temperatures. In fact, recent studies showed that seasonal changes of outdoor temperatures have a significant relationship with the duration of windows remaining open [7]. Nevertheless, seasonal changes are almost absent in most of the tropical cities, except for the precipitation and wind conditions. This study, therefore, examines occupants' daily average patterns of window-opening without considering the seasonal changes of outdoor weather conditions.

22.2   Profile of Respondents

Four sets of survey data were used for this study (a total of 1,315 samples). As shown in Table 22.1, they include 347 samples obtained in the apartments of Surabaya, Indonesia (Oct. 2013); 299 samples in the apartments of Bandung, Indonesia (Sept. 2014); 303 samples in the apartments of Johor Bahru, Malaysia (Apr.–Jun. 2006) and 366 samples in the terraced houses of Johor Bahru (Sept.–Oct. 2004). All of the surveys were conducted through face-to-face interviews with typical households using a questionnaire form. The profile of respondents and their houses, the daily average patterns of window-opening behaviour and the usage of cooling appliances (AC, standing/ceiling fans) were investigated in the surveys.

The average household size of the respondents was 1.9 and 3.4 people in Indonesia and 4.6 and 5.4 people in Malaysia (Table 22.1). The yearly household income was the highest among respondents in the terraced houses of Johor Bahru (income ratio, 9.4), followed by those in the apartments of Johor Bahru (8.8), the apartments of Bandung (8.3–9.0) and the apartments of Surabaya (7.6). The above income ratios were calculated by considering the yearly currency exchange rates.

The altitude of Bandung City is relatively high, approximately 700 m above sea level on average. Therefore, the average air temperature was not as high as those in the other cities. The ownership level of AC was approximately 62% in the case of

Table 22.1 Brief profile of respondents

Case study city	Bandung (Indonesia)	Surabaya (Indonesia)	Johor Bahru (Malaysia)	Johor Bahru (Malaysia)
House type	Apartment	Apartment	Apartment	Terraced house
Survey year	2014	2013	2006	2004
Sample size	299	347	303	366
Average household size (people)	1.9	3.4	4.6	5.4
Median of yearly household income (ratio)[a]	8.3–9.0	7.6	8.8	9.4
AC ownership (%)	34.8	10.7	35.3	62.0
Daily mean air temperature (°C)				
Maximum	26.1	30.7	31.1	31.1
Average	21.5	27.4	26.9	26.9
Minimum	17.6	23.9	23.5	23.5

[a]The values of income ratio were obtained with the following formula: income values on national currency were multiplied by the exchange rate into USD
(Source: The World Bank, Indicators) and their natural logarithm were taken

terraced houses in Johor Bahru and approximately 35% in the apartments of Johor Bahru and Bandung. The low AC ownership level of 11% in Surabaya was probably not due to its climatic conditions but due to the relatively lower income levels (7.6).

22.3 Daily Window-Opening Patterns

This section analyses the daily patterns of occupants' window-opening behaviour using all of the samples (N=1,315). We asked respondents whether they open one of their operable windows or doors in each hour during a typical day during the dry season. A binary variable was used to indicate whether a window or door was open or closed during each hour. The weighted average of weekday and weekend conditions was calculated. Then, a principal component analysis and a hierarchical cluster analysis (squared Euclid's distance, Ward's method [8]) were carried out to classify all the samples into several groups with similar window-opening patterns. As a result, four types of daily window-opening patterns were determined as shown in Table 22.2 and Fig. 22.1. In Fig. 22.1, the x-axis indicates the time (24 h), and the y-axis shows the percentage of respondents who open their windows/doors at the time.

Pattern A, 'morning and evening', depicts the daily behaviour in which the respondents usually open and close windows or doors two times a day. The windows and doors are closed during the daytime. This is the only pattern in

Table 22.2 Summary of daily window-opening patterns

	N	Share (%)	Average duration per day (hour)
Pattern A, morining and evening	260	19.8	3.6
Pattern B, daytime	544	41.4	9.4
Pattern C, all-day except for sleep Time	288	21.9	13.9
Pattern D, all-day	223	17.0	20.1
Total	1315	100.0	11.0

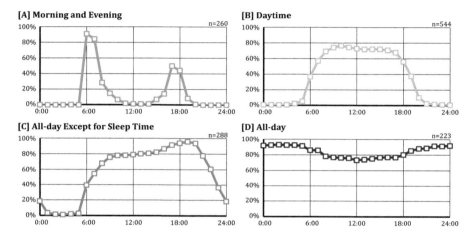

Fig. 22.1 Daily patterns of window-opening behaviour in various opening patterns

which windows and doors are closed during the day. In Pattern A, (20% of respondents), the average duration in which windows are open is 3.6 h per day, which is the shortest among the four patterns. Pattern B, 'daytime', illustrates the daily behaviour in which the respondents open windows or doors during the daytime between 5:00 and 20:00. This pattern accounts for the largest share of respondents (approximately 41%), with an average opening duration of 9.4 h per day. Pattern C, 'all-day except for sleep time', households also open the window during the daytime until the night (24:00). This pattern implies that respondents close windows/doors when they go to bed. The average opening duration is 13.9 h per day. Pattern D, 'all-day', depicts the behaviour in which the respondents keep the windows or doors open throughout the day. In pattern D, the windows/doors are opened even during sleeping time. Consequently, the average opening duration is the longest in this case, which is 20.1 h per day.

Figure 22.2 presents the proportions of these four window-opening patterns in each of the case studies. As shown, the proportions are clearly different between the sample from Bandung and the other sample. It can be seen that, in the three samples from Surabaya and Johor Bahru, most of the respondents open windows or doors during the daytime, i.e. Patterns B, C and D. In particular, Pattern B, 'daytime',

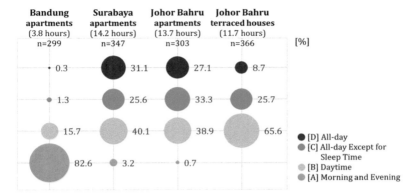

Fig. 22.2 Shares of respective window-opening patterns. Figures in brackets show average durations of window-opening per day

makes up the largest share in the three cases, namely, 40.1%, 38.9% and 65.6%, respectively. Meanwhile, the opening patterns vary during the periods after coming home and during sleeping, although they keep windows or doors open during the daytime. In contrast, more than 80% of respondents in Bandung are categorised as Pattern A, 'morning and evening'. Unlike the other cases, most of the respondents open neither windows nor doors during the daytime. Many previous studies in temperate regions [7] showed that outdoor air temperature affects occupants' window-opening behaviour. Hence, it can be concluded that the difference in window-opening behaviour is likely due to the relatively cool climate of Bandung.

22.4 Factors that Adversely Affect Window-Opening Behaviour

There are various factors that adversely affect window-opening behaviour other than temperature [9]. Figure 22.3 shows the reasons for the respondents not to open windows. As shown in the left figure, on average, insects, security, rain and privacy are the top four reasons among all samples. This section analyses the relationships between the window-opening patterns and the reasons for not opening windows. The respondents were divided into two groups, depending on the reasons chosen, i.e. 'insects', 'security', 'rain' and 'privacy', respectively. Then, the window-opening patterns were compared between the two groups (whether or not they chose it as the reason for not opening windows). As mentioned previously, the window-opening patterns in Bandung are clearly different from other cases. In this analysis, we dealt with only the major opening patterns in each of the respective cases, i.e. Patterns B to D are analysed in Surabaya and Johor Bahru (Fig. 22.4), while Patterns A and B are used in Bandung (Fig. 22.5).

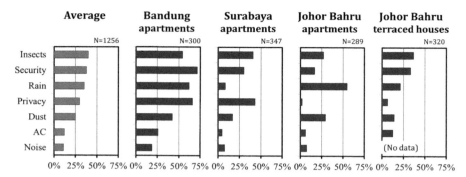

Fig. 22.3 Reasons for not opening windows

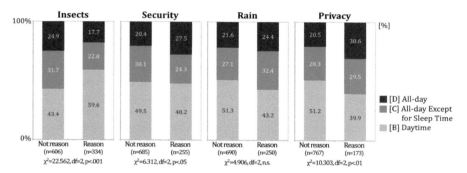

Fig. 22.4 Differences of daily window-opening patterns between the respondents of whether they chose each of the reasons in Surabaya and Johor Bahru

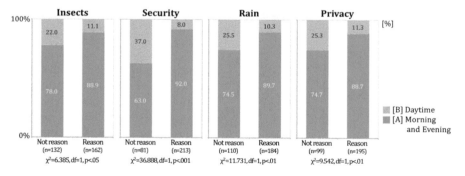

Fig. 22.5 Differences of daily window-opening patterns between the respondents of whether they chose each of the reasons in Bandung

In Surabaya and Johor Bahru, the window-opening patterns vary between the respondents depending on whether they chose 'insects' as a reason for not opening windows. The respondents who are less concerned about the insects tend to open windows more during the evening and night. As shown in Fig. 22.4, approximately

57% of respondents open windows or doors even during the evening if they did not choose 'insects' for not opening windows (i.e. Pattern C or D). In contrast, significant differences are not observed across window-opening patterns with regard to rain. However, the opening patterns differ whether they chose 'security' or 'privacy' as a reason for not opening windows. If the respondents are concerned about security or privacy, they tend to close windows more during the night. The interpretation for this result has yet to be confirmed, but the concern about security or privacy is probably not a factor influencing the occupants' window-opening behaviour at night.

As shown in Fig. 22.5, in Bandung, Pattern A always takes larger shares among the respondents who are concerned with each of the reasons than among those who are not concerned. The difference between Pattern A and B is whether windows are opened during the daytime. Therefore, it is considered that the concerns about insects, security, rain and privacy are factors that hinder window-opening during the daytime.

22.5 Usage Patterns of Air Conditioners and Their Relationships with Window-Opening Patterns

In this section, we analyse the daily usage patterns of air conditioners (AC) and then examine their relationships with their window-opening patterns. First, as before, a principal component analysis and a hierarchical cluster analysis were conducted to examine the daily usage patterns of AC. The results revealed that daily AC usage of its owners (N=467) can be categorised into three patterns, as shown in Table 22.3 and Fig. 22.6.

Pattern I, 'evening', depicts the usage pattern of which the respondents use their AC between 16:00 and 24:00. Pattern I accounts for approximately 11% of all AC owners. The average usage time is 4.8 h per day. Pattern II, 'night-time', shows the usage pattern of which the respondents operate AC between 18:00 and 7:00. Pattern II is the largest group, with 74% of the share. The average usage time is 7.0 h per day. Pattern III, 'mainly daytime', illustrates the usage pattern in which the respondents use their AC for the longest period, with an average usage time of 10.3 h per day. More than 80% of them usually use AC between 12:00 and 15:00.

Figure 22.7 presents the shares of the respective AC usage patterns in each of the case studies. The grey circles indicate the percentages of non-AC owners. Meanwhile, Fig. 22.8 illustrates the relationships of daily window-opening patterns with

Table 22.3 Summary of daily AC usage patterns

	N	Share (%)	Average duration per day (hour)
Pattern I, evening	50	10.7	4.8
Pattern II, night-time	344	73.7	7.0
Pattern III, mainly daytime	73	15.6	10.3
Total	467	100.0	7.3

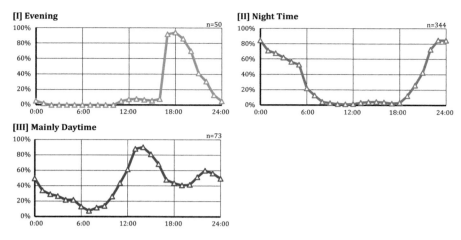

Fig. 22.6 Daily patterns of AC usage in various usage patterns

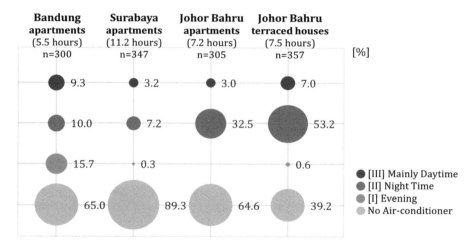

Fig. 22.7 Shares of respective AC usage patterns. Figures in brackets show average usage time of AC per day

daily AC usage patterns. As shown in Fig. 22.7, Pattern I, 'evening', accounts for the largest proportion (45% among the AC owners) in Bandung but is rarely observed in the other cities. Among the AC owners who follow Pattern I, 85.7% open windows or doors during the morning and evening (i.e. Pattern A) (Fig. 22.8). These rates indicate that both windows/doors and AC are used simultaneously (but probably in different rooms) during the evening in this case. On the other hand, approximately 68–92% of AC owners in Surabaya and Johor Bahru use them during the night (i.e. Pattern II) (Fig. 22.7). The window-opening patterns with AC usage Pattern II

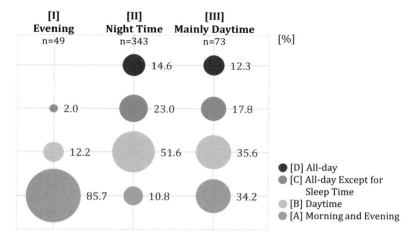

Fig. 22.8 Relationships of daily window-opening patterns with daily AC usage patterns

vary as indicated in Fig. 22.8. Most respondents (approximately 75%) open windows or doors during the daytime (i.e. Pattern B or C).

It can be concluded that the most typical tendency found in this survey is that occupants open windows/doors during the daytime from 5:00 to 20:00 for 9.4 h per day in the living room and use AC for approximately 7 h from 18:00 to 7:00 in the bedroom.

22.6 Conclusions

The main findings of this chapter are summarised as follows:

- Daily patterns of occupants' window-opening behaviour were categorised into four distinct patterns. It was found that, in Surabaya and Johor Bahru, almost all of the respondents open windows/doors at least during daytime. The results of Bandung showed an exception because of its relatively cool climate and most of the respondents close windows/doors during daytime.
- In particular, concern about insects was a factor that adversely affects window-opening behaviour after coming home and during sleep in Surabaya and Johor Bahru. Meanwhile, concerns about insects, security, rain and privacy adversely affected the opening behaviour during daytime in Bandung.
- The most typical tendency found in Surabaya and Johor Bahru was that occupants open windows/doors during daytime in the living room and use AC during night-time for about 7 h per day in the bedroom.

References

1. Asian Development Bank (2010) Key indicators for Asia and the Pacific special chapter: the rise of Asia's middle class
2. Uno T et al (2003) Survey on thermal environment in residences in Surabaya, Indonesia: use of air conditioner. J Asian Archit Build Eng 2(2):15–21
3. Kubota T et al (2009) The effects of night ventilation technique on indoor thermal environment for residential buildings in hot-humid climate of Malaysia. Energy Build 41:829–839
4. Ekasiwi SNN et al (2013) Field survey of air conditioner temperature settings in hot, humid climates, part 1: questionnaire results on use of air conditioners in houses during sleep. J Asian Archit Build Eng 12(1):141–148
5. Brager GS et al (2004) Operable windows, personal control, and occupant comfort. ASHRAE Trans 110(2):17–35
6. Yun GY et al (2009) Thermal performance of a naturally ventilated building using a combined algorithm of probabilistic occupant bebahviour and deterministic heat and mass balance models. Energ Build 41:489–499
7. Rijal HB et al (2013) Investigation of comfort temperature, adaptive model and the window opening behavior in Japanese houses. Architect Sci Rev 56(1):54–69
8. Architectural Institute of Japan (ed) (2012) Kenchiku toshi-keikaku no tameno chousa bunseki houhou (Investigation and analysis method for building and city planning). Inoue Shoin, Tokyo
9. Suzuki T et al (2002) Structure of the causes of window opening and closing behavior from summer to autumn. J Archit Plann Environ Eng AIJ 556:91–98

Chapter 23
Survey of Thermal Environment of Residences Using Air Conditioners in Surabaya, Indonesia

Tomoko Uno, Shuichi Hokoi, and Sri Nastiti N. Ekasiwi

Abstract This chapter reports the results of a survey on the attitudes of residents towards the use of air conditioners in Indonesia. A questionnaire survey and measurements of the thermal environment were carried out. With increasing income, the percentage of residents who felt that air conditioning was necessary also increased. Once residents start to use air conditioners, they continue their use. When an air conditioner is used, the lowest room temperature ranges from 23 to 29 °C, which is lower than that observed in Japan. Air conditioners are typically used around 14:00 and during sleeping time; this is similar to the situation in Naha, a subtropical area of Japan. Also, the duration of air conditioner use is longer than that in Japan. As a result of these findings, the consumption of energy used for cooling is expected to increase in this area.

Keywords Hot-humid climate · Field survey · Air conditioning on/off · Operating time of air conditioner · Temperature setting of air conditioner · Consciousness of use of air conditioner · Thermal environment

23.1 Introduction

In hot-humid climates such as Indonesia, the use of air conditioners has increased [1, 2]. They start to use air conditioners not only in offices but also residences, and this has led to the problem of increased energy consumption.

This chapter was revised from the paper [11] in references, based on the recent researches.

T. Uno (✉)
Department of Architecture, Mukogawa Women's University, Nishinomiya, Hyogo, Japan
e-mail: uno_tomo@mukogawa-u.ac.jp

S. Hokoi
Kyoto University, Kyoto, Japan

Southeast University, Nanjing, China

S. N. N. Ekasiwi
Department of Architecture, Institut Teknologi Sepuluh Nopember, Surabaya, Indonesia

© Springer Nature Singapore Pte Ltd. 2018
T. Kubota et al. (eds.), *Sustainable Houses and Living in the Hot-Humid Climates of Asia*, https://doi.org/10.1007/978-981-10-8465-2_23

235

Several researchers in Japan and other Asian countries have reported the indoor thermal conditions [3–5], operating times and temperature settings [6–8] for residential air conditioner use, as well as the attitudes of residents towards air conditioner use [7, 9–12]. In particular, operating times and patterns differ depending on region, e.g. the use of air conditioners at nighttime is more frequent in lower latitude areas [3, 6, 8]. Compared with daytime, a low temperature is required during nighttime hours [5]; moreover, a lower temperature setting is preferred in low-latitude areas [9, 10]. The desire of residents for air conditioners is closely related to energy consumption. In order to reduce energy use for cooling, it is necessary to understand thermal conditions and the use of air conditioners, together with the identification of suitable housing designs for cooling.

In this chapter, we report data on the measurement of thermal environments and the results of a questionnaire survey of residences where air conditioners were used. We focus on the attitudes of residents towards air conditioner use.

23.2 Survey of Thermal Environment

23.2.1 Area and Climate

The surveyed area is in Surabaya (7°S 133°E), which is located in the eastern region of Jawa Island, Indonesia. It has a hot, humid climate and has two seasons: dry and wet. Precipitation is low in the dry season (May to October). The monthly average temperature varies from 27.2 to 29 °C, and the hottest months are October and November (Fig. 23.1). The annual mean temperature is 28 °C. The monthly average of the relative humidity ranges from 67% to 80%. The results reported in this chapter are mainly those of a survey conducted in July, when the temperature is at its lowest.

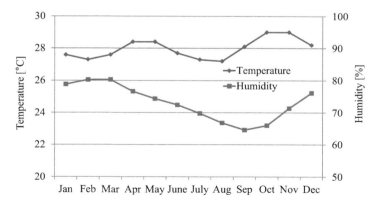

Fig. 23.1 Monthly average temperature and humidity (The monthly temperature and humidity are the data between 1991 and 2010 at Juanda Airport meteorological station, which are calculated by Meteonorm ver7)

23.2.2 Outline of Questionnaire Survey

The questionnaire and measurement surveys were carried out 1st–15th of July, 2002, during the dry season in Surabaya.

A questionnaire survey relating to indoor thermal conditions was completed by low-, middle- and high-income residents (Table 23.1). The questionnaire results are classified by income (Table 23.2). In total, 118 questionnaires were completed (Tables 23.3 and 23.4), with 34 residents (18 houses) in the low-income group,

Table 23.1 Contents of questionnaire [11]

1.	Address, number of family members, income
2.	House plan, floor area
3.	Electrical energy consumption
4.	Thermal comfort, thermal sensation, reasons for any discomfort, measures during discomfort (daytime and nighttime)
5.	Number of electric fans and air conditioners, how often and how long they are used, whether they are necessary or not

Table 23.2 Classification by income (1 $ ≒ 8300 IDR @2003) [11]

Income class	Low	Middle	High
Monthly income	< 1,500,000 IDR (< $180)	1,500,000 IDR–2,500,000 IDR ($180–300)	> 2,500,000 IDR (> $300)

Table 23.3 Relation between income and air conditioner (AC) possession [11]

Income	With AC	Without AC	Total
Low	0 persons	34 persons	34 persons
	(0 houses)	(18 houses)	(18 houses)
Middle	6 persons	33 persons	39 persons
	(4 houses)	(18 houses)	(22 houses)
High	28 persons	7 persons	35 persons
	(13 houses)	(3 houses)	(16 houses)
Student	2 persons	8 persons	10 persons
Total	36 persons	82 persons	118 persons
	(17 houses)	(39 houses)	(56 houses)

Table 23.4 Occupations by income [11]

	Income		
Occupation	Low	Middle	High
Housewife	9	4	9
Worker	18	18	13
Student	5	7	7
Servant	0	1	0
Unemployed	1	1	1
No answer	1	8	5

39 residents (22 houses) in the middle-income group and 35 residents (16 houses) in the high-income group. Most of the respondents were office workers or housewives (Table 23.3). Four of the 22 houses in the middle-income group, and 13 of the 16 houses in the high-income group, had air conditioners. In addition, 10 students lived alone in a rented room, with two having an air conditioner in their room. The number of males and females was approximately equal.

23.2.3 Outline of Measurements

Along with the questionnaire survey, the temperature and humidity in five houses were measured. Four of the five houses had one or two ACs in their bedroom or living room, whereas the remaining house did not have an AC. Three households with ACs were classified as high income, and the other houses were classified as middle income.

The outdoor temperature, relative humidity and global solar radiation were measured in an experimental house built on the campus of the Institut Teknologi Sepuluh Nopember (ITS).

23.3 Results of Survey

Based on the results of the questionnaire survey and the measured thermal conditions, resident attitudes on the use of ACs will be discussed in the following section.

23.3.1 Necessity of Air Conditioners

No low-income households had ACs. On the other hand, 18% of the middle-income houses and 81% of the high-income houses had ACs. It is clear that possession of ACs is closely related to income.

Figure 23.2 shows the attitudes of residents towards the use of ACs. It is evident that 26% of low-income residents and 23% of middle-income residents consider that air conditioning is necessary (Fig. 23.2a). This percentage rises to 71% among high-income residents. Therefore, the higher the income, the greater is the perceived need for an AC. The percentage of respondents answering 'unnecessary' was 14% in the low-income bracket, which was less than that for the middle-income bracket (41%). Because of the generally poor thermal conditions in low-income houses, ACs are regarded as essential for improving the thermal condition. Seventeen percent of the low-income residents regarded ACs as a waste of money; however, such an answer was not returned by the high-income residents. This indicates that the number and use of ACs will increase with increasing income.

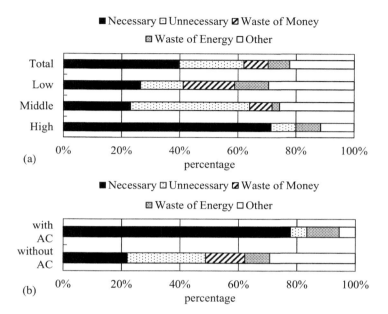

Fig. 23.2 Attitudes towards air conditioners (AC) in relation to (**a**) income group and (**b**) possession [Figure (**a**) shows the attitudes towards air conditioners in relation to low-, middle-, and high-income groups and their total. Figure (**b**) shows the attitudes towards air conditioners in relation to the possession of ACs.] [11]

The attitude of respondents towards the possession of ACs is shown in Fig. 23.2b. More than 77% of people who had an AC considered its possession essential. From this result, it is expected that once people start using an AC, they will continue its use.

23.3.2   Thermal Conditions in Houses with and Without Air Conditioners

Figure 23.3 shows temperatures measured over 2 days in selected houses. Figure 23.3a shows the measured results in the living room and bedroom of a middle-income house (House A), which does not have ACs. Figure 23.3b shows measured temperature results in two houses (House B and House C), where ACs are installed in the bedroom.

In the house without ACs, the temperature ranged from 27 to 31 °C in the living room and between 28 and 29 °C in the bedroom during daytime. At night, the temperature in the living room became lower than that in the bedroom, at around 27 °C.

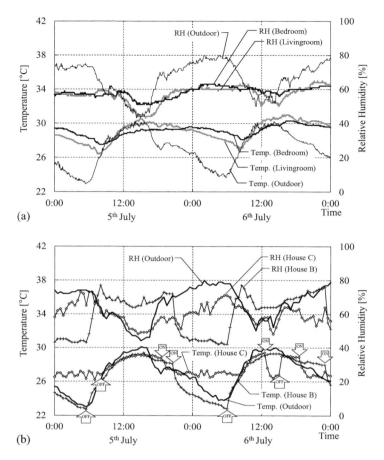

Fig. 23.3 Measured temperature and humidity [11]: (**a**) House A without an AC, and (**b**) Houses B and C with an AC. The on/off conditions of the ACs are shown by arrows

The measured air temperature both in the living room and bedroom fell rapidly before 7:00, suggesting that residents open doors and windows in the morning; this is in accordance with the survey responses. However, doors and windows were kept open, and the room temperatures increased relatively rapidly after 8:00 due to the influence of the outdoor temperature. This temperature increase in the bedroom was less than that in the living room. This is because the area open to the outside is less than that of the living room, which limits the amount of solar radiation coming into the room. In hot climates, it is very important to prevent solar radiation from entering a room.

At nighttime, doors and windows were closed because of mosquitoes and for security reasons. While the temperature in the living room continued to decrease until morning, the temperature in the bedroom remained at around 29 °C, which is an uncomfortable condition for sleeping.

In houses with ACs, the temperature fell rapidly (Fig. 23.3b) when residents turned on the AC. In House B, the AC was used at night, from 23:00 (5th of July) until 6:00 in the morning (6th of July). In this house, the AC was used every nighttime (Figs. 23.3b and 23.6). The temperature when the AC was turned on was around 28 °C. After this, the temperature decreased to 23 °C—the low-level temperature setting. Once the AC was turned on, the temperature continued to decrease until morning. Because of insufficient AC capacity and the large heat loss of the room, the temperature did not decrease to the desired value.

In House C, the AC was used at both nighttime and daytime. The operation was commenced at 20:30 and continued until 6:00 the following morning. In the daytime, the AC was used from 14:00 (July 6). In this house, the AC was used during almost every nighttime but only sometimes during the day. The AC was turned on at around 28.5 °C, and the room temperature was kept constant at around 27 °C. The AC in House C could control the room temperature according to the residents' desire. It should be noted that a temperature of around 27 °C was preferred by the residents of House C. This temperature is the same as that generally chosen by residents in Japan [13, 14].

After the AC was turned off at around 6:00 (House B) or around 8:00 (House C), the temperature and humidity increased to the level of the outdoor air because the residents opened the doors and windows at this same time. If they had kept the doors and windows closed, the room temperature would have remained lower during daytime.

In both Houses B and C, the ACs were continuously used at nighttime, probably to enable comfortable sleeping conditions.

23.3.3 Temperature When Air Conditioner Is Used

Figure 23.4 shows the cumulative frequency of the temperature at which the residents start to use an AC and the resulting lowest temperature achieved. Figure 23.4a shows the results for July (dry season), and Fig. 23.4b shows results for February (wet season). Here the term, 'lowest temperature' is used for the lowest temperature measured during the operation of the AC, and the term 'starting temperature' is used for the temperature at which AC use was initiated.

In July, the starting temperature was typically between 28 and 30 °C during daytime and between 26 and 29 °C during nighttime. The residents started to use ACs at slightly lower temperature during nighttime than during daytime. The lowest temperature ranged from 23 to 29 °C during daytime and from 23 to 27 °C during nighttime.

In February, the residents started to use ACs at between 28 and 32 °C, with the temperature during nighttime lower than that during daytime. The lowest temperature was from 25 to 30 °C during daytime and from 24 to 28 °C during nighttime.

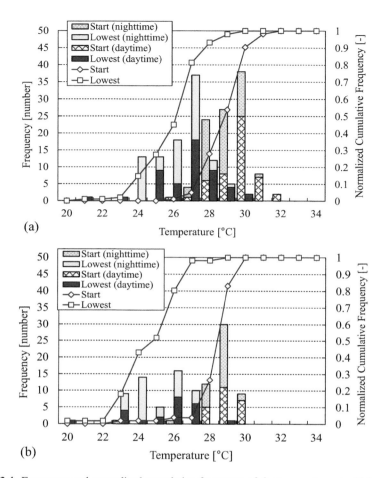

Fig. 23.4 Frequency and normalised cumulative frequency of the temperature at which resident starts to use ACs, together with the resulting lowest temperature [11]: (**a**) dry season (July), (**b**) wet season (February)

The starting temperatures in February and July were almost the same (Fig. 23.5). However, the lowest temperature in July was lower than that in February. Since the outdoor temperature and humidity ratio in July (dry season) was lower than that in February,[1] the ACs worked well to decrease room temperature to a greater extent in July than in February. The lowest temperature achieved is affected by AC efficiency.

---

[1]Generally, the monthly average temperature in July and in February is not so different, and the humidity in July is lower than that in February. However, the temperature in July 2002 was higher than that in February 2001. The average, minimum and maximum of the measured temperature during the survey were 27.5 °C, 24.4 °C, and 29.8 °C in July 2002, and 28.1 °C, 25.8 °C, and 31.2 °C in February 2001.

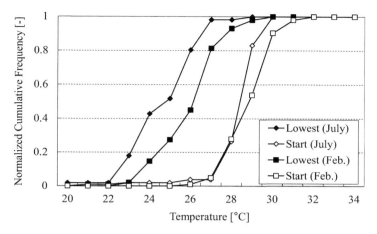

Fig. 23.5 Normalised cumulative frequencies of the temperature at which residents start to use ACs, together with the lowest temperature achieved (comparison between July and February) [11]

In both seasons, the starting temperature and lowest temperature during nighttime were 0.3–0.4 °C lower than those during daytime. The residents preferred lower temperature conditions during nighttime. This tendency for a preferred low temperature setting during nighttime is also seen in Japan [6]. One of the reasons that the air-conditioned temperature in the nighttime is lower than that in the daytime is that the AC works well, because the outdoor temperature when AC is operated is higher during daytime than that during nighttime. However, the measured temperature results in House B during nighttime (Fig. 23.3b) indicate the possibility that the room temperature did not reach the residents' desired level. This is likely due to insufficient AC capacity and/or poor thermal performance of the house. As such, the residents desire a much lower room temperature.

The lowest temperature achieved here for Surabaya ranged from 23 to 29 °C, which is slightly lower than that for Japan [13, 14]. This shows that some people in Surabaya prefer much lower temperature condition.

23.3.4   Temperature Settings of Air Conditioners

According to the questionnaire answers of residents having an AC in their house, the temperature settings of the AC ranged from 16 to 30 °C, with 93% of such respondents setting it under 25 °C (Fig. 23.6). The temperature settings are lower than the realised conditions in the house (Fig. 23.3b). The desired condition may not be achieved because of the insufficient capacity of the AC or the suboptimal thermal construction of the house. Since the houses had large openings and were not airtight,

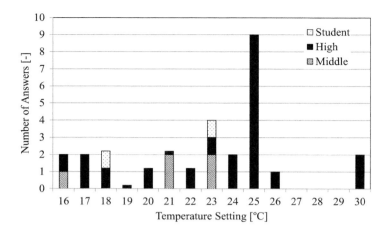

Fig. 23.6 Temperature settings from the questionnaire results

room temperature is greatly influenced by outdoor conditions. Therefore, even when an AC was used, it may not have maintained the desired room temperature.[2]

Some respondents also provided information about the temperature settings in their office. These temperatures ranged from 16 to 22 °C, much lower than those set in residences.

23.3.5 Times When Air Conditioners Are Used

In relation to the times when ACs are used, most people answered that they use ACs from 21:00 to 6:00 during nighttime, and 13:00–15:00 during daytime.

In Houses B and C, the residents used their AC almost every night and sometimes during daytime too, as can be seen from Fig. 23.7. This is consistent with the results observed during the wet season [11]. The pattern of AC use was similar throughout the year. In Japan (other than in Naha), residents use ACs at around 14:00 and in the evening, before turning it off before going to bed [3, 4]. On the other hand, people in Surabaya continue to use ACs until morning. The same situation during nighttime can be seen in Naha [3], which is located in a subtropical climate in Japan, and in Johor Bahru in Malaysia [8].

The duration of operation observed in the present study was from 8 to 14 h (measured results from four houses); this is longer than that in Japan.

---

[2]In Japan, the air conditioner is mainly used when the residents are active. On the other hand, the residents in Surabaya use it mainly in their sleeping time. The temperature setting of the air conditioner and the comfort temperature might be different between active and sleeping times, although it is not discussed here. It may need more consideration.

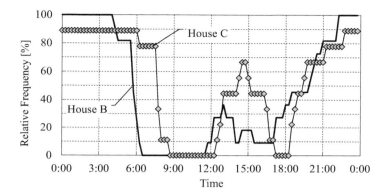

Fig. 23.7 Operation of ACs (House B and House C) [11]. This figure shows the percentage of using AC in each time during measurement period

Based on these results, we consider that an increase in the energy consumed for the purpose of residential cooling could become a serious problem in Indonesia.

23.4 Conclusions

This chapter reports the results of field research focusing on the attitudes of residents towards the use of ACs. A questionnaire survey and residence temperature measurements were carried out in Surabaya, Indonesia.

The percentage of households possessing ACs and the need to use ACs increase with an increase in household income. These findings suggest that once a resident starts using an AC, they will continue its use.

The lowest temperature achieved when residents use an AC is from 23 to 29 °C, which is lower than that observed in Japan. Most of the respondents reported an AC temperature setting of 16–25 °C, which was lower than the realised conditions. This might be due to the insufficient capacity of the ACs and/or the poor thermal performance of the houses.

The times when ACs are used are around 14:00 and during nighttime. This is similar to the findings for Naha and Malaysia in subtropical or tropical areas. The duration over which ACs are used is longer than that in Japan.

The operating hours and low temperature settings of ACs in low thermal performance houses will likely lead to an increase in the energy consumed for the purpose of cooling. This energy consumption could become a serious problem in Indonesia. Because factors such as (1) the areas in which ACs are used and (2) the operating hours of residential ACs are limited, thereby enhancing the thermal performance of rooms containing ACs is of high importance.

References

1. IEA Energy Balances of Non-OECD Countries (2000) International Energy Agency
2. http://www.bps.go.id/ Statistics Indonesia
3. Suzuki K, Matsubara N, Morita D, Sawachi T, Bohgaki K (1995) A survey on heated and cooled environment in the municipal apartment-houses at Sapporo, Kyoto and Naha, analysis of regional characteristics of the resident's living style and consciousness about heating and cooling, and consideration of energy conservation measures Part 1. J Archit Plann Environ Eng AIJ 60(475):17–24
4. Bougaki K, Sawachi T, Yoshino H, Suzuki K, Akabayashi S, Inoue T, Ohno H, Matsubara N, Hayashi T, Morita D (1998) Study on the living room temperature and it's regional differences in summer & winter season, study of energy consumption in residential buildings from the viewpoint of life style, on the basis of national scale survey Part 2. J Archit Plann Environ Eng AIJ 60(505):23–30
5. Habara H, Narumi D, Shimoda Y, Mizuno S (2005) A study on determinations of air conditioning on/off control in dwellings based on survey. J Archit Plann Environ Eng AIJ 70 (589):83–89
6. Bougak K, Sawachi T, Yoshino H, Suzuki K, Akabayashi S, Inoue T, Ohno H, Matsubara N, Hayashi T, Morita D (1998) Study on the heating & cooling pattern and heating & cooling period in residential buildings on the basis of national scale surveys. J Archit Plann Environ Eng AIJ 63(509):41–47
7. Lin X, Umemiya N (2016) Actual state and consciousness of households' air-conditioner use during 2001 post-disaster electricity shortages. J Archit Plann Environ Eng AIJ 81 (727):785–794
8. Kubota T (2007) A field survey on usage of air-conditioners and windows in apartment houses in Johor Bahru city. J Archit Plann Environ Eng AIJ 72(616):83–89
9. Ekasiwi SNN, Majid NHA, Hokoi S, Oka D, Takagi N, Uno T (2014) Field survey, of air conditioner temperature settings in hot, humid climates, Part 1: questionnaire results on use of air conditioner in houses during sleep. J Asian Archit Build Eng 12(1):141–148
10. Majid NHA, Takagi N, Hokoi S, Ekasiwi SNN, Uno T (2014) Field survey of air conditioner temperature setting in a hot dry climate (Oman). HVAC&R Res 20:751–759
11. Uno T, Hokoi S, Ekasiwi SNN (2003) Survey on thermal environment in residences in Surabaya, Indonesia: use of air conditioner. J Archit Plann Environ Eng 2(2):15–21
12. Uno T, Hokoi S, Ekasiwi SNN, Funo S (2003) A survey on thermal environment in residential houses in Surabaya, Indonesia. J Archit Plann Environ Eng AIJ 68(564):9–15
13. Tomioka S, Akabayashi S, Sakaguchi J, Yamagishi A, Tominaga Y, Sasaki Y (2001) Research about the indoor warm and temperature environment and the shelter performance in a residence of Niigata prefecture, Part 10: the operation method and the amount of energy consumption of the air conditioning in a residence. Summ Tech Pap Annu Meet AIJ D2:211–212
14. Tanimoto J, Hagishima A, Katayama T (2001) A fundamental study on characteristics of cooling loads applied Markov model to the air conditioning on/off control part 3: a survey on transition characteristic of on/off control for air conditioning system based on a field measurement. Trans Soc Heat AC Sanit Eng Jpn 82:59–66

Chapter 24
Occupants' Behavior in Taiwan

Ruey-Lung Hwang

Abstract The thermal adaptation behaviors of Taiwan residents and elders at homes when they sense thermal discomfort was depicted. Among the habitual methods to achieve thermal comfort, the most common strategy of thermal adaptation was, in a descending order, window-opening, use of electrical fan, and turning on the air conditioner for residents, while window-opening, adjustment in clothing, and use of electrical fan for the elders. The habitual adaptation method of interviews was influenced by the three factors: effectiveness, accessibility, and cost in relieving thermal discomfort.

Keywords Thermal adaptation behaviors · Taiwan residents · Elders

24.1 Introduction

The adaptive model of thermal comfort describes that the indoor occupants may seek to adapt to thermal conditions and subsequently to achieve better thermal comfort through active behavioral changes when such mechanisms are available. In addition to a variety of adaptation methods, occupants in residences have more opportunities to choose the methods of adaptation to achieve thermal comfort depending on their needs and their preferences. Does the difference in opportunities to choose from a variety of methods to achieve thermal comfort affects thermal perceptions of occupants? It is worth to discuss in detail. Additionally, understanding occupants' most preferred method of adaptation may help to understand the implementation result of some low-cost or zero-cost methods, as advocated by the energy-related department of Taiwan government in order to reduce energy consumption in the use of A/C systems.

R.-L. Hwang (✉)
Department of Industrial Technology Education, National Kaohsiung Normal University, Kaohsiung, Taiwan
e-mail: rueylung@nknu.edu.tw

© Springer Nature Singapore Pte Ltd. 2018
T. Kubota et al. (eds.), *Sustainable Houses and Living in the Hot-Humid Climates of Asia*, https://doi.org/10.1007/978-981-10-8465-2_24

The elderly people are physiologically and psychologically unique and, thus, have requirements toward an indoor microclimate different from those of other age groups, often more demanding. For instance, the elders in general are of a lesser capacity to adapt to the change in the ambient temperature, as human metabolic rate decreases with aging. Significant efforts have been made to investigate the thermal comfort in association with the indoor microclimate. However, few focused on studying the thermal adaptive behaviors used by the elders.

The behaviors of thermal adaptation are classified as personally based (e.g., clothing adjustment and water/soft drink drinking for hydration) and environmentally based (e.g., window-based ventilation, fan, curtains/blinds, and air conditioning). This chapter will discuss and introduce the predominant strategies of thermal adaptive behavior in residences and for elders in Taiwan.

24.2   General Methods of Adaptation for Hot and Humid Conditions

Hwang et al. [1] carried out a questionnaire-based survey with 253 interviewees in 137 homes across Taiwan during April to December to investigate the behaviors of thermal adaptation including personally based means (e.g., clothing adjustment and water/soft drinks for preventing dehydration) and environmentally based means (e.g., window-based ventilation, fan usage, curtains/blinds deployment, and air-conditioning usage). In a field survey, the thermal adaptive behaviors commonly adopted by the occupants can be evaluated by asking:

> What method do you use to make yourself more comfortable (checking from a list of options) when you feel the environment is too hot (or cold)?

According to the Hwang's study [1], Fig. 24.1 illustrates the relative frequencies of the most preferred method of adaptation suggested by Taiwan's residents at homes when they sense thermal discomfort. Among the habitual methods to achieve

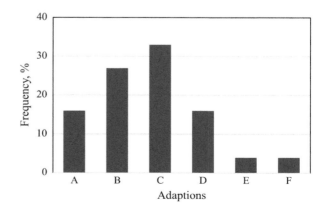

Fig. 24.1 The respondents' most preferred thermal adaptation methods when discomfort is caused by warmness (Note: A: Use A/C; B: Use fan; C: Open window; D: Change clothing: E: Take a shower or wash the face: F: Others)

thermal comfort, 33% of occupants who choose to open the windows is the highest, followed by using electrical fan (25%), turning on the air conditioner (16%), adjusting clothing level (15%), and other methods.

It can be inferred from Fig. 24.1 that the habitual adaptation method of interviews was influenced by the following three factors.

(1) Effectiveness in relieving thermal discomfort: Lowering the temperature and increasing air velocity are the two most effective methods to quickly relieve thermal discomfort. The top three preferred methods of adaptation by most interviewees are using A/C, using electrical fan, and opening window. All these means are aimed to meet either one of both requirements.
(2) Accessibility: The fact that over 90% of all the interviewees chose to turn on the air conditioner, open the window, use the electric fan, and adjust clothing level, which is the four easiest measures to adjust thermal condition, simply verifies the influence of convenience on thermal adaptation behaviors. Those methods are deemed more accessible to occupants.
(3) Cost: All the visited spaces were equipped with A/C system, yet the occupants' tendencies change from the costly method of using A/C systems to low-cost and even zero-cost methods of using electrical fan and opening window.

24.3   The Elders' Thermal Adaptive Behaviors in Dwelling

In order to understand the predominant strategy of thermal adaptation for elders in dwelling, Hwang and Chen [2] carried out a questionnaire survey on elderly people in Taiwan with an age greater than 60 years old. A total of 352 valid questionnaires were collected, 160 in the summer and 192 in the winter. Among the questionnaires collected in the summer and the winter, 47% and 43% were generated by males, respectively. Figure 24.2 summarizes the behaviors frequently adopted by the elders in summer and winter for adaptation to the indoor thermal environment when deemed necessary. In summer, window-opening (38%), adjustment in clothing,

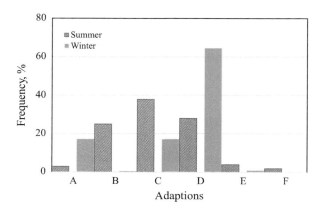

Fig. 24.2 Behaviors of thermal adaptation commonly adopted by elderly people to indoor environments as surveyed in summer and winter (Note: A: Use cooing/heating; B: Use fan; C: Open window; D: Change clothing; E: Take a shower or wash the face; F: Others)

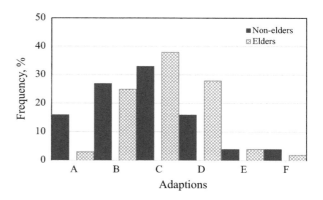

Fig. 24.3 Behaviors of thermal adaptation to indoor microclimate commonly adopted in summer by elderly people and by non-elders (Note: A: Use A/C; B: Use fan; C: Open window; D: Change clothing; E: Take a shower or wash the face; F: Others)

and use of electrical fan (25%) were the three most common strategies the elders used to adapt to the microclimate when they found the indoor environment is thermally uncomfortable. To a much lesser extent, the elderly would select mechanical cooling, bathing, and other means of adaptation. In winter (Fig. 24.2), the majority of the studied elders alleviated themselves from thermal discomfort by means of clothing adjustment (64%). The window-closing (17%) and mechanical heating (17%) were also favorable means for thermal adaptation. No other approaches were attempted by the elders for thermal adaptation in the winter.

Figure 24.3 compares the adaptive behaviors in summer preferred by elders ($\geqq$60 years) and non-elders (<60 years) at home from the study by Hwang et al. [1]. As shown in Fig. 24.3, both the elders and the non-elders opted for window-opening as the most selected strategy for thermal adaptation. The elders in favor of window-opening as their predominant approach (38%) are 4% more than the non-elders (34%). Approximately 25% of people from both groups used electrical fan in summer as an adaptive mechanism. Significant difference in thermal adaptive behaviors between both groups occurred in the use of mechanical cooling and clothing adjustment. Twenty-eight percent of the elders chose clothing adjustment as the main approach to adapt to the microclimate, whereas only 16% of the non-elders opted for it. Regarding the mechanical cooling, 3% of the elders consider it as a favorable approach which was much less than the non-elders (16%). The elderly participants in this study were much inclined toward using "mild" adjustments that lie in the domain of personal control.

24.4   Conclusions

This chapter depicts the variations of thermal adaption behaviors between different age groups in their requirements of indoor thermal comfort. The following conclusions were derived:

- Among the habitual methods to achieve thermal comfort, 33% of occupants who choose to open the windows is the highest, followed by using electrical fan (25%), turning on the air conditioner (16%), adjusting clothing level (15%), and other methods.
- The most common strategy of thermal adaptation in the summer that the elders adopted was, in a descending order, window-opening, adjustment in clothing, and use of electrical fan. In the winter, approximately 64% of the elders adapted to the indoor environment by adjusting clothing while 17% by using mechanical heating.
- The habitual adaptation method of interviews was influenced by the following three factors: effectiveness, accessibility, and cost in relieving thermal discomfort.

References

1. Hwang RL, Lin TP, Cheng MJ, Ho MC (2009) Thermal perceptions, general adaptation methods and occupant's idea on trade-off among thermal comfort and energy saving in hot-humid regions. Build Environ 44:1128–1134
2. Hwang RL, Chen CP (2010) Field study on behaviors and adaptation of elderly people and their thermal comfort requirements in residential environments. Indoor Air 20:235–245

Chapter 25
Occupant Behavior in Indian Apartments

Madhavi Indraganti

Abstract Occupant adaptation happens to be the key mechanism behind achieving thermal comfort in buildings. User's thermal adaptation is under researched in India, although it is important to understand and limit the energy use in buildings. Relying on filed study data, this chapter looks at various methods of adaptation available and in use in apartments in India. Operating windows, doors, fans, air coolers, and air conditioners (AC) was noted to be robustly correlating with outdoor and indoor temperatures and thermal sensation. Their adaptive operation was limited by many non-thermal factors. People used traditional Indian attires for thermal comfort. These allowed occupants to adapt through changing the drape (and the insulation) within the same ensemble, when the indoor temperature moved outside the comfort range.

Keywords Indian apartments · Thermal comfort · Adaptive model · Occupant adaptation · Field survey · Personal environmental controls

25.1 Occupant Behavior

Passive buildings presuppose active users. User adaptation to form, space, and micro-climate is critical to overall thermal comfort, especially when the active systems are not in charge. Further, it transforms the house into a dynamic breathing whole rather than a static configuration of masses and voids. Adaptation and migration are some of the oldest of thermal strategies used in finding the most favorable microclimate within and outside of the house [1]. This chapter looks at occupant behavior in apartments in India [2]. It examines the adaptive actions users undertake to effectively moderate indoor thermal discomfort, by using personal environmental controls (PEC). The common PECs are fans, air conditioners (AC), windows, doors, etc. The use of PECs expands the comfort regime in many dimensions.

M. Indraganti (✉)
Department of Architecture and Urban Planning, College of Engineering, Qatar University, Doha, Qatar
e-mail: madhavi@qu.edu.qa

© Springer Nature Singapore Pte Ltd. 2018
T. Kubota et al. (eds.), *Sustainable Houses and Living in the Hot-Humid Climates of Asia*, https://doi.org/10.1007/978-981-10-8465-2_25

25.2 Adaptive Use of Natural Ventilation: Windows

Operating windows is one of the most common methods of occupant adaptation in naturally ventilated (NV) buildings. Not many researched on the occupant behavior in residential apartments in India [3]. Indraganti [2] observed the thermal adaptation of residents in apartments in Hyderabad [4]. In summer and monsoon seasons, she recorded 3962 responses and simultaneous thermal environmental measurements constituting of air temperature (T_i), globe temperature (T_g), air speed (V_a), and relative humidity (RH) in five naturally ventilated apartments (KD, SA, RA, KA, and RS).

Indraganti [2] noted all the environments fitted with operable windows of varying sizes and materials (wood, steel, and aluminum). These windows were adaptively operated. The status of all the PECs were recorded as binary data (open/in use, 1; closed/unused, 0). The proportion of open windows (p_w) increased as the indoors became warmer. It peaked at about 70% when indoor globe temperature was between 32 and 34 °C (Fig. 25.1). When T_g was warmer than this, occupants adaptively closed the windows. This usually happened during the hot midday in summer.

Rijal et al. [5] also found the indoor and outdoor temperatures to be the major stimuli. Summer in composite climates is usually very hot and dry. Opening the windows during the overheated period usually brings convective heat gains, reflective glare, and discomfort. During such overheated periods (i.e., between 12:00 and 15:00), occupants express *hot* (+3) sensation. While keeping the windows and balcony doors closed, the occupants then resort to other means of environmental control, such as the use of fans, ACs, and air coolers, as elucidated in the following

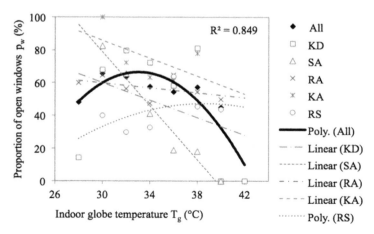

Fig. 25.1 Adaptive operation of windows varying with indoor globe temperature (T_g) (binned data, N = 3962). Lines represent variation in the proportion of open windows (p_w) with temperature, in individual buildings [2]

sections. Greatest changes in p_w can be noted in peak summer months. Occupants operate the balcony doors similar to the windows.

25.2.1   Challenges to Adaptive Operation of Windows

Mere existence of a control does not guarantee thermal comfort [6]. Privacy, convenience, safety, sun penetration, and occupant attitudes also affect the window operation as noted by Indraganti [2]. She observed these nonthermal factors seriously hindering the window operation. For example, windows opening into the public realm, (such as main passage ways), or the ones with inadequate shading devices, or malfunctioning hardware (hinges and the casement furniture), were seldom operated adaptively. Conversely, she noted a higher percentage of balcony doors open. This could possibly be because, their operation was not limited by the above constraints.

25.3   Adaptive Use of Fans

Behavioral use of controls relates to the physiology and psychology of the body and the building characteristics [7]. Users maintain a dynamic association with the available controls in the building. In addition to using windows, occupants in apartments use many other electrical controls adaptively. These could be fans, air coolers (evaporative coolers), and air conditioners, with fans being very common. Understandably, ceiling fans induce air movement distribution and consume a lot less energy than ACs.

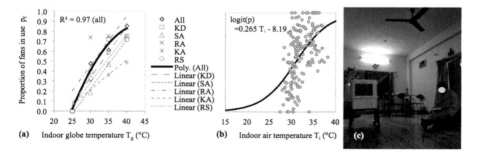

Fig. 25.2 (**a**) Adaptive operation of fans varying with indoor globe temperature (T_g) (binned data, N = 3962) [3]. Lines represent proportion of fans in use (p_f) in individual buildings. (**b**) Logistic regression of p_f with indoor air temperature with dots indicating the binned data from individual buildings with bin size varying from 10 to 25 sets of data (N = 3798, $p < 0.001$). (**c**) Subjects adaptively using the air cooler after switching off the fan and closing the windows

Table 25.1 Proportion of environmental controls available and in use across buildings [3]

| Building | N | Proportion of control (%) | | | | | | | |
| | | Fan | | Air cooler | | | AC | | |
		Off	On	Available	Off	On	Available	Off	On
KA	708	23.4	76.6	20.8	19.9	0.8	0.0		
KD	1295	39.2	60.8	21.5	11.6	9.9	78.8	69.2	9.7
RA	844	57.5	42.5	19.9	18.5	1.4	44.2	41	3.2
RS	689	62.1	37.9	16.8	15.8	1	1.7	1.7	0
SA	426	31.5	68.5	92.7	83.1	9.6	65	62.2	2.8

Indraganti [3] noted proportion of fans in use (p_f) correlating robustly with T_g, T_i, and outdoor daily mean temperature (T_o). She found about 84% fans in use when T_g was at 35–40 °C (Fig. 25.2). Logistic regression shown in Fig. 25.2b relates the proportion of fans in use with indoor air temperature ($p < 0.001$). The actual data (shown superimposed) matches closely with this. In top floor roof-exposed (RE) flats, using fan during the overheated period recirculates hot air. In these tenements, subjects often use pedestal fans, air coolers, and ACs if available, instead of ceiling fans (see KD in Table 25.1 and Fig. 25.2c).

25.3.1 Barriers to Adaptive Use of Fans

Oftentimes, many nonthermal constraints limit the ceiling fan usage. In top floor roof-exposed (RE) flats, using fan during the overheated period recirculated hot air accumulated under the ceiling. In these tenements, subjects often use pedestal fans, air coolers, and ACs if available, instead of ceiling fans. In addition, Indraganti [3] noted other constraints. Many fans had either dysfunctional air speed regulators or were totally nonexistent. Some of the fans were noisy, and their efficacy in air distribution was not uniform. These factors also limited the fan usage.

25.4 Adaptation Through Behavioral Control Actions

Behavioral adaptation immediately follows the physiological adaptation in naturally ventilated environments. In all the months, Indraganti [3] recorded the occupants' many adaptive behavioral control actions. These were noted down as binary data (0, not in use; 1, in use) as listed in Fig. 25.3. A higher proportion of subjects undertook a diverse range of actions when the environment was hot and harsh. On the contrary, behavioral adaptation was very limited during the milder months (July).

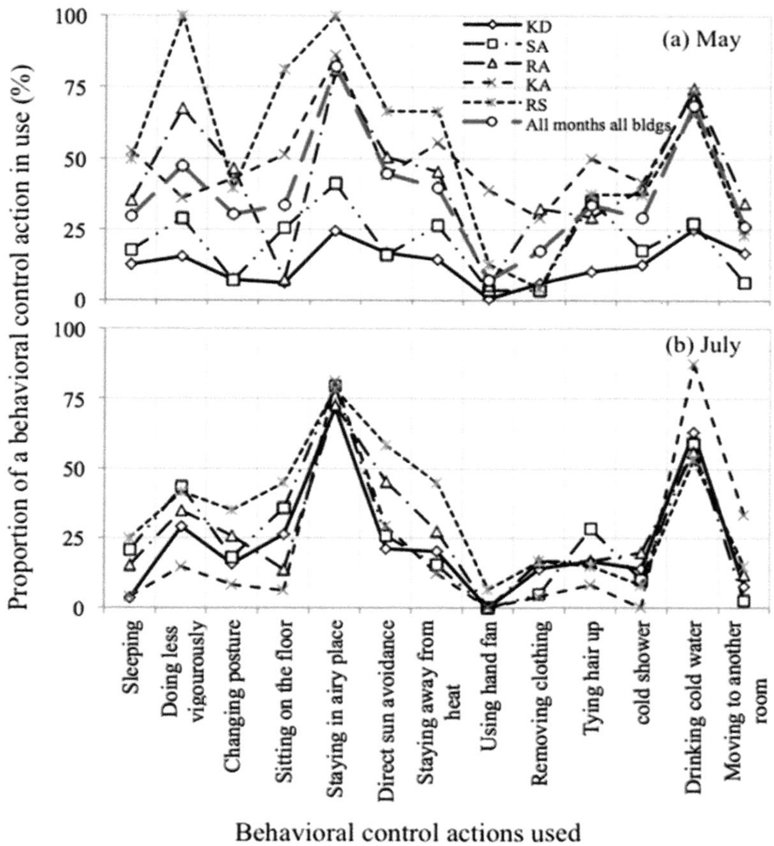

Fig. 25.3 Occupant adaptation through several behavioral control actions in various buildings (KD, KA, SA, RA, RS) in (**a**) May (n = 381) and (**b**) July (n = 327) (all months all buildings, N = 1042). Behavioral adaptation was higher in summer (May) [3]

While *doing things less vigorously* and *staying in airy place* were the actions most used in May, only the latter was much preferred in July, when the environment was warm and humid. Inter-building variations were very evident in May (Fig. 25.3a), and not much in July (Fig. 25.3b). This in part could be due to varying availability/use of other controls such as windows, balcony doors, fans, air coolers, and ACs in some buildings. She noted higher availability of electromechanical controls limiting the behavioral adaptation. For example, very little behavioral adaptation was noted in KD in contrast to RS in May. Incidentally, KD recorded highest availability and use of ACs, while RS had none at all and also recorded lowest use of fans.

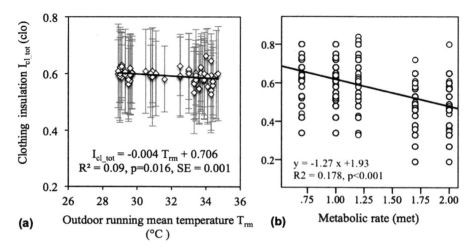

Fig. 25.4 Variation in clothing insulation with (**a**) outdoor running mean temperature (**b**) metabolic rate. (Each square dot represents the daily average in a building with the error bars representing the standard deviation; *SE* Standard error). Subjects adaptively chose lighter clothing as the metabolic activity became vigorous and as the temperature moved up (all data N = 3962) [4]

25.5 Adaptation Through Metabolism and Clothing

Clothing adaptation is easier in residential environments. Occupants at free will can change to lighter or heavier clothing, as it is an easy adaptive mechanism. Indraganti [4] noted clothing insulation changing with outdoor temperature and metabolic rate of the occupants in all the environments (Fig. 25.4).

In this study, clothing insulation varied from 0.18 to 0.84 clo averaging at 0.58 clo. People chose lighter clothing as the temperature moved up (Fig. 25.4), and as the metabolic activity became vigorous. A majority of the subjects were in traditional Indian attires, which allowed much better adaptation within the same ensemble [8] as is seen in Fig. 25.5. Cultural perceptions, attitudes, and fashion also could have limited the clothing adaptation.

25.6 Adaptation Through Structural Modifications

Occupants undertake many structural modifications to their tenements in order to achieve thermal comfort. In India, some of these were found to be *applying reflective paint on the roof, false ceiling, growing plants* on *window sills and balconies,* etc., as shown in Fig. 25.6. For example, u*sing false ceiling* (suspended ceiling), *extending the shades to windows*, and applying *solar control film to the windows* were found to be the most widely applied adaptations [9]. Application of suspended ceiling was found to be much higher in the RE flats.

Fig. 25.5 Clothing adaptation in women and men using traditional ensembles of *Saree* and *Lungi*. Lighter clothing was used during the midday and at high metabolic activities [4]

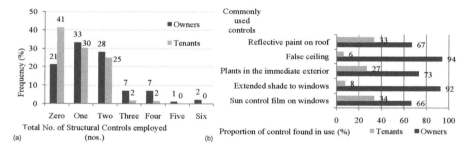

Fig. 25.6 Proportion of various controls (**a**) in use in apartments (**b**) used in tenant- and owner-occupied tenements [9]

25.6.1 Effect of Tenure on the Use of Structural Controls

Tenure was found to be impacting the decisions on the application and the use of structural controls. This is evident from Fig. 25.6b. Indraganti [9] noted a majority of the structural adaptations being done in owner-occupied apartments. In other words, owners seem to have taken additional actions especially suitable for the hot summer seasons, to make the tenement comfortable. Moreover, the apartments on rent were fitted with least number of structural controls.

The Indian code [10] does not make any of these structural environmental control provisions mandatory. This could be one of the reasons for their nonavailability in most tenant-occupied apartments. Figure 25.7 shows some of the structural interventions applied to windows, doors, and balconies as noted in a field study in apartments in India [2]. Application of these was found to be greatly enhancing the operation of the personal environmental controls. Importantly, these were mostly applied in owner-occupied tenements.

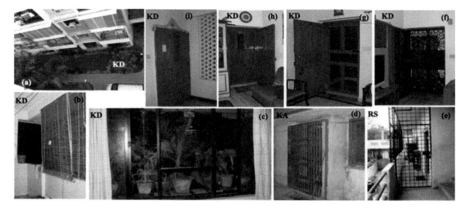

Fig. 25.7 Some interventions applied to windows, doors, and balconies to enhance adaptive operation in Indian apartments (**a, b**): bamboo blinds hung on balcony openings and (**c**) planter box extension to windows to reduce glare and sunlight and curtains to improve privacy, (**d, e, f**) metal grill gates to main doors and metal grill enclosures to balconies to improve safety, privacy, and cross ventilation, (**g, h**) additional door shutters with mosquito screens and vents for improved cross ventilation, (**i**) RCC *Jalis* (grills) applied to corridors for safety, sun control, and cross ventilation [2]

25.7 Concluding Remarks

This chapter focused on occupant adaptation in Indian apartments as observed through a field study. People in Indian apartments used traditional clothing such as saris and *chudidars* (women) and *lungi* and *salwar-kameez* (men), which offered higher adaptive opportunities within the same ensemble through change of drape. Subjects adapted to the varying thermal regime in more ways than one. These adaptations were found to be majorly through the clothing and metabolism and behavioral control actions. In addition to these, users operated windows, fans, air coolers, and ACs and achieved comfort adaptively. Many nonthermal factors were noted to be impeding occupant adaptation, some of them beyond the direct control of the inhabitant.

Moreover, some tenements were fitted with a few structural interventions to enhance occupant adaptation. Application of these in tenant occupied apartments was found to be much lower.

References

1. Heschong L (1979) Thermal delight in architecture. MIT Press, Cambridge, MA
2. Indraganti M (2010) Adaptive use of natural ventilation for thermal comfort in Indian apartments. Build Environ 45:1490–1507
3. Indraganti M (2010) Behavioural adaptation and the use of environmental controls in summer for thermal comfort in apartments in India. Energ Buildings 42(7):1019–1025

4. Indraganti M (2010) Thermal comfort in naturally ventilated apartments in summer: findings from a field study in Hyderabad, India. Appl Energy 87(3):866–883
5. Rijal HB, Honjo M, Kobayashi R, Nakaya T (2013) Investigation of comfort temperature, adaptive model and the window-opening behaviour in Japanese houses. Archit Sci Rev 56 (1):54–69
6. Nicol JF, Humphreys MA (2002) Adaptive thermal comfort and sustainable thermal standards for buildings. Energ Buildings 34:563–572
7. Brager GS, Paliaga G, de Dear RJ (2004) Operable windows, personal control. ASHRAE Trans 110(2):17–35
8. Indraganti M, Lee J, Zhang H, Arens EA (2015) Thermal adaptation and insulation opportunities provided by different drapes of Indian saris. Archit Sci Rev 58(1):87–92
9. Indraganti M, Rao KD (2010) Effect of age, gender, economic group and tenure on thermal comfort: a field study in residential buildings in hot and dry climate with seasonal variations. Energ Buildings 42(3):273–281
10. BIS (2005) National building code 2005. Bureau of Indian Standards, New Delhi

Chapter 26
Occupant Behaviour in the Various Climates of Nepal

Hom Bahadur Rijal

Abstract Surveys of the thermal environment and occupant behaviour were conducted in summer and winter for residents in the traditional houses of the Banke, Bhaktapur, Dhading, Kaski, and Solukhumbu districts in Nepal. The results show that residents adjust well to the thermal conditions of the houses. Both in summer and winter, the average clothing insulation of women is greater than that of the men. Clothing insulation is minimum in the sub-tropical climate in summer and maximum in the cool climate in winter. The clothing insulation can be predicted by the proposed regression equations.

Keywords Nepal · Traditional house · Occupant behaviour · Thermal adjustments

26.1 Introduction

In Nepal, traditional houses are adapted to the climate; similar types of house are found in locations of similar climate and culture. Houses and lifestyles vary, and various adaptive actions are used that produce comfortable thermal conditions where possible. For example, (1) there is the custom of wearing traditional clothing designed to protect both from extreme heat and extreme cold; (2) houses have cooler and warmer spaces, and residents move between them, and residents sleep in the semi-open spaces or front yards to stay cool; (3) firewood is burnt to provide heat in winter; and (4) people drink large quantities of cold water and take more cold showers to keep cool in summer, and drink large quantities of butter tea to keep warm in winter [1–3].

H. B. Rijal (✉)
Department of Restoration Ecology and Built Environment, Tokyo City University,
3-3-1 Ushikubo-nishi, Tsuzuki-ku, Yokohama, 224-8551 Japan
e-mail: rijal@tcu.ac.jp

T. Kubota et al. (eds.), *Sustainable Houses and Living in the Hot-Humid Climates of Asia*, https://doi.org/10.1007/978-981-10-8465-2_26

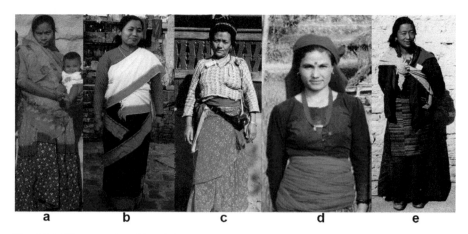

Fig. 26.1 Women and traditional clothing in winter [3]. (**a**) Banke: 1.04clo (**b**) Bhaktapur: 1.61clo (**c**) Dhading: 1.16clo (**d**) Kaski: 1.10clo (**e**) Solukhumbu: 1.79clo

Surveys of the thermal environment and occupant behaviour were conducted in summer and winter for residents in the traditional houses of five districts in Nepal. The objective of this research was to investigate the residents' methods of thermal adjustment.

26.2 Field Survey

The outdoor and indoor measurement points were at 1.5 m above ground level and 0.6 m above floor level, respectively. All data were measured at 5 min intervals and recorded using a data logger.

The factors thought to be influential for thermal adjustments, such as clothing changes, controlling apertures, burning firewood, sunbathing, taking cold showers, etc., were placed in the categories of "1, highly influential"; "2, slightly influential"; and "3, not influential". These questions were asked individually of the people who took part in the thermal comfort survey as shown in Chap. 17.

Men often wore Western-style clothing such as T-shirts, shirts, jackets, and trousers, while women often wore traditional clothing such as saris, Nepalese blouses, and shawls (Fig. 26.1). Clothing insulation values were calculated by weighing the clothes and using modern equations relating the weight of the ensemble to its thermal insulation (Hanada et al. [4, 5]). However, it is not known how well the equations estimate the thermal insulation of the traditional Nepalese clothing of these areas, and so the clo value estimates may not be precise.

Table 26.1 Outdoor and indoor air temperature [3]

Area	Climate	Daily mean outdoor air temperature T_{out} (°C)		Daily mean indoor air temperature T_{in} (°C)		
		Summer	Winter	n	Summer	Winter
1. Banke	Sub-tropical	32.4	11.3	13 (10)	32.0	13.3
2. Bhaktapur	Temperate	22.4	10.1	14 (5)	23.6	11.5
3. Dhading	Temperate	20.1	11.9	15 (8)	22.2	14.8
4. Kaski	Temperate	19.5	11.8	15 (7)	21.1	15.3
5. Solukhumbu	Cold	15.5	3.1	16 (6)	17.8	6.5

n Number of indoor measurement points (number of house)

26.3 Results

26.3.1 Thermal Environment in Each Area

The mean indoor and outdoor air temperatures are shown in Table 26.1. Summer in Banke has the highest (sub-tropical climate) mean outdoor (32.4 °C) and indoor (32.0 °C) temperatures, while in Solukhumbu (cold climate) they are lowest both in summer and winter. The indoor and outdoor temperatures have large seasonal and regional differences.

26.3.2 Methods of Thermal Adjustment by the Residents

To investigate the methods of thermal adjustments used by the residents, the mean values of each thermal adjustment are shown in Fig. 26.2. The method is similar to that used in a study in Zambia by Malama et al. [6].

To adjust the thermal environment in summer and winter, methods such as clothing, ventilation, firewood, and drink are used in each area. The most common methods of thermal adjustment in summer are "wearing thin clothing", "sleeping under a thin blanket", and "opening the opening", and in winter "staying in the sunshine", "wearing thick clothing", and "sleeping under a thick blanket". Methods unique to the sub-tropical climate in summer are "taking a nap" and "sleeping in the front yard", and to temperate and cold climates in winter, "wearing socks and shoes". To adjust to heat and cold, people move between indoor, semi-open, and front yard spaces. If they feel hot, they stay in the shade or in well-ventilated spaces, and if cold, they stay around the open hearth or seek the sun. People change their living space according to changes in the environment. In the sub-tropical climate, to get a cool atmosphere in summer, people sleep in the front yard in beds that look like a hammock made from wood and rope with mosquito nets. People also sleep in the semi-open spaces in summer in the temperate climate. However, women rarely sleep in the semi-open or front yard spaces for reasons of personal safety.

In the sub-tropical climate, people burn firewood in the semi-open and the front yard spaces for heating. Firewood combustion is one of the means to overcome the

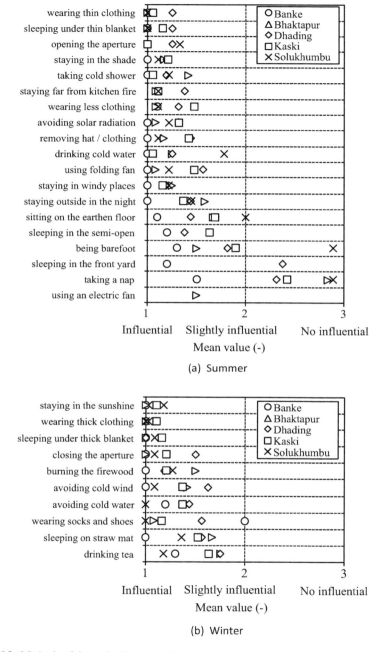

Fig. 26.2 Methods of thermal adjustment [3]

cold, and vulnerable subpopulations like children and the elderly stay near the open hearths in the morning and evening. Firewood consumption per capita per day for heating is highest in the cool climate (2.0 kg), lower in the sub-tropical climate (1.9 kg), and least in the temperate climate (0.6 kg) [7].

In the sub-tropical climate, hot summers mean that people take a shower each morning, afternoon, evening, and night, i.e. 1–9 times (mean = 4 times, n = 14) per person per day [3]. Some people do not take a shower in the afternoon because they say it makes them listless. However, most people take a shower 3–4 times in the daytime. The heat suppresses appetite; people drink large quantities of cold water which is kept naturally cold in earthenware pots.

In cool climates, to keep warm in winter, people drink large quantities of butter tea; 4–35 cups (mean = 15 cups, n = 11) per person per day [3]. They always have butter tea in the house and provide it for any guest. In summer they drink smaller quantities of butter tea, 2–23 cups (mean = 10 cups, n = 11) per person per day. To avoid using a cold outdoor toilet during the night, they drink less butter tea in the evening.

26.3.3 The Clothing Insulation

In the sub-tropical climate, people wore light clothing, and men often wore only shorts in summer. In the cool climate, people wore layers of thick clothing and, until they went to bed, shoes indoors in winter. Both in summer and winter, the average clothing insulation of women is greater than that of the men (Table 26.2). These trends are similar in Japanese houses [8]. Clothing insulation is minimum in the sub-tropical climate in summer and maximum in the cold climate in winter. These are effective to maintain the thermal comfort in each climate and season.

26.3.4 Relation Between Clothing Insulation and Temperature

Figure 26.3 shows the relation between the clothing insulation (I_{cl}) and outdoor air temperature (T_o). Each data point represents the mean value for each subject. Most of the data is from free-running mode, and some data include the heating mode (firewood burning). The following regression equations were obtained.

Indoor
$$I_{cl} = -0.046T_o + 1.7 \left(n = 85, R^2 = 0.46, \text{S.E.} = 0.005, p < 0.001 \right) \quad (26.1)$$
Semi-open
$$I_{cl} = -0.034T_o + 1.5 \left(n = 62, R^2 = 0.31, \text{S.E.} = 0.006, p < 0.001 \right) \quad (26.2)$$

Table 26.2 Clothing insulation of subjects [3]

Space	Survey area	Climate	Season	Subject Male	Subject Female	I_{cl} (clo) Male	I_{cl} (clo) Female
Indoor	1. Banke	Sub-tropical	Summer	2	4	0.30	0.64
			Winter	1	5	0.78	0.88
	2. Bhaktapur	Temperate	Summer	6	6	0.36	0.64
			Winter	8	6	0.88	1.28
	3. Dhading	Temperate	Summer	3	3	0.49	0.79
			Winter	5	3	0.67	1.15
	4. Kaski	Temperate	Summer	3	2	0.39	1.10
			Winter	2	2	0.67	1.06
	5. Solukhumbu	Cold	Summer	6	6	0.84	1.12
			Winter	6	6	1.38	2.02
	All areas		Summer	20	21	0.48	0.86
			Winter	22	22	0.88	1.28
Semi-open	1. Banke	Sub-tropical	Summer	2	6	0.33	0.40
			Winter	5	6	0.94	0.91
	2. Dhading	Temperate	Summer	5	4	0.53	0.95
			Winter	6	4	0.75	1.25
	3. Kaski	Temperate	Summer	3	5	0.56	1.15
			Winter	6	10	0.82	1.30
	All areas		Summer	10	15	0.47	0.83
			Winter	17	20	0.84	1.15

I_{cl} clothing insulation, $I_{cl\_M} = 0.000558w + 0.068$, $I_{cl\_F} = 0.00103w\text{-}0.0253$, M male, F female, w clothing weight (g)

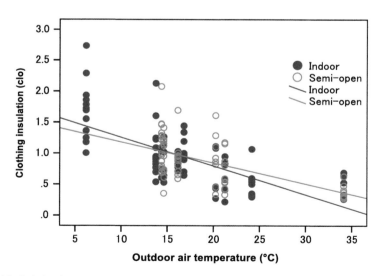

Fig. 26.3 Relation between the clothing insulation and outdoor air temperature

where n is the number of subjects; R^2 is the coefficient of determination; SE is the standard errors for regression coefficient; and p is the level of significance for the regression coefficient. The regression coefficients are negative for both equations. It shows that the clothing decreases when temperature is increased. The clothing insulation and outdoor air temperature are correlated. The clothing insulation can be predicted by the proposed regression equations.

26.4 Conclusions

Surveys of the thermal environment and occupant behaviour were conducted in summer and winter for residents in the traditional houses of the Banke, Bhaktapur, Dhading, Kaski, and Solukhumbu districts in Nepal. The results are:

1. The thermal adjustment survey shows that residents used clothing, ventilation apertures, firewood, and beverages to adjust to the thermal environment in summer and winter. The unique methods of thermal adjustment in summer are "taking a nap" and "sleeping in the front yard" in sub-tropical climate, and "wearing socks and shoes" in temperate and cool climate in winter.
2. Both in summer and winter, the average clothing insulation of women is greater than that of the men. Clothing insulation is minimum in the sub-tropical climate in summer and maximum in the cool climate in winter.
3. The clothing insulation and outdoor air temperature are correlated. The clothing insulation can be predicted by the proposed regression equations.

Acknowledgements We would like to give thanks to Prof. Harunori Yoshida, Prof. Noriko Umemiya, Prof. Michael Humphreys, and Prof. Fergus Nicol for their research guidance, to the investigated households for their cooperation, and to our families and friends for their support. This research is supported by the Japan Society for the Promotion of Science.

References

1. Rijal HB, Yoshida H, Umemiya N (2002) Investigation of the thermal comfort in Nepal. In: Proceedings of international symposium on building research and the sustainability of the built environment in the tropics, pp 243–262
2. Rijal HB, Yoshida H, Umemiya N (2003) Summer and winter thermal comfort of Nepalese in houses, J Archit Plann Environ Eng AIJ 565: 17–24 (In Japanese with English abstract)
3. Rijal HB, Yoshida H, Umemiya N (2010) Seasonal and regional differences in neutral temperatures in Nepalese traditional vernacular houses. Build Environ 45(12):2743–2753
4. Hanada K, Mihira K, Ohhata K (1981) Studies on the thermal resistance of women's underwear. J Jpn Res Assoc Text End-Users 22(10):430–437 (In Japanese with English summary)
5. Hanada K, Mihira K, Sato Y (1983) Studies on the thermal resistance of men's underwear. J Jpn Res Assoc Text End-Users 24(8):363–369 (In Japanese with English summary)

6. Malama A, Sharples S, Pitts AC, Jitkhajornwanich K (1998) An investigation of the thermal comfort adaptive model in a tropical upland climate. ASHRAE Trans 104(1):1194–1203
7. Rijal HB, Yoshida H (2002) Investigation and evaluation of firewood consumption in traditional houses in Nepal. In: Proceedings of the 9th international conference on indoor air quality and climate, vol 4, pp 1000–1005
8. Watanabe K, Rijal HB, Nakaya T (2013) Investigation of clothing insulation and thermal comfort in Japanese houses. In: PLEA2013 – 29th conference, sustainable architecture for a renewable future, Munich, Germany 10–12 September 2013

Chapter 27
Window Opening Behaviour in Japanese Dwellings

Hom Bahadur Rijal

Abstract We investigated window opening behaviour and thermal environment over a period of 4 years in the living rooms and bedrooms of dwellings in the Kanto region of Japan. We collected 36,144 data samples from 243 residents of 120 dwellings. The proportion of 'open window' in the free-running mode is significantly higher than that in the cooling and heating modes. The window opening behaviours were shown to be related to both the indoor or outdoor air temperatures. Window opening behaviour as predicted by logistic regression analysis is in agreement with the measured data. The deadband was narrower, and constraints on the window opening in the investigated dwellings were considerably smaller than had previously been found in studies of office buildings. An adaptive algorithm is developed that can be applied to predict window opening in Japanese dwellings.

Keywords Dwellings · Thermal comfort · Window opening · Comfort temperature · Deadband · Constraint

27.1 Introduction

People use various adaptive opportunities to regulate the indoor thermal environment. Window opening is one of the important behavioural adaptations for thermal comfort. Residents more freely open the window to adjust the indoor thermal environment in dwellings, as there are fewer constraints on the window opening at home than one might encounter at the office. In Japanese office buildings, the indoor thermal environment is often controlled using air conditioning. By contrast, the same is achieved in dwellings more often through window operation, as the use of air conditioning has a direct financial implication for the household. However, natural ventilation from opening windows has been decreasing in dwellings in recent years

---

H. B. Rijal (✉)
Department of Restoration Ecology and Built Environment, Tokyo City University,
3-3-1 Ushikubo-nishi, Tsuzuki-ku, Yokohama, 224-8551 Japan
e-mail: rijal@tcu.ac.jp

© Springer Nature Singapore Pte Ltd. 2018
T. Kubota et al. (eds.), *Sustainable Houses and Living in the Hot-Humid Climates of Asia*, https://doi.org/10.1007/978-981-10-8465-2_27

because of the increasing prevalence of mechanical ventilation and air conditioning. Temperature control by opening and closing windows can reduce environmental impact by minimizing the period of the year when air conditioning is needed. Therefore, it is important to introduce natural ventilation through open windows, and to minimize the use of air conditioning as much as possible in hot summers and other more moderate seasons.

People at home usually are able to control their own thermal environments, so it may be wondered what is the purpose of knowing what behaviour they choose. Models relating the occupants' behaviour to the climate are of course of scientific interest as an addition to our knowledge of human adaptive behaviour. They are useful practically too. Knowing how people open the window in winter and in summer helps towards the correct sizing of air conditioning and heating plant – oversized plant is usually less efficient. For the free-running mode of operation, the situation is different. The question is then: Can this proposed design provide the required indoor temperatures? If thermal simulation or experience suggests that it cannot, then the design can be altered, particularly with regard to window design and thermal mass, so that comfort is more likely to be obtainable. Thermal simulation packages often assume a fixed schedule of window opening, so more realistic data on window opening behaviour can be used to improve the simulations. The window opening algorithm is therefore a useful design tool.

There has been more research into window opening behaviour in offices than in dwellings [1–5]. The findings from research in offices cannot be assumed to apply to dwellings, where people's behaviour is less constrained. There is evidence that people respond differently in their own dwellings for a number of reasons, social, economic and cultural [6]. The window opening algorithm developed for work environments [7] therefore may not reflect domestic occupant behaviour adequately. Further research on residential window opening behaviour was needed to remedy this shortage.

To explore window opening behaviour and develop a window opening algorithm for Japanese dwellings, thermal measurements were made and an occupant behaviour surveys conducted for 4 years in the living rooms and bedrooms of dwellings in the Kanto region of Japan.

27.2 Field Survey

Occupant behaviour surveys and thermal measurements were conducted in 120 dwellings in Kanto region from 2010 to 2014 [8, 9]. It has hot humid summer, cold winter and mild spring and autumn. The monthly mean outdoor air temperature (relative humidity) of Tokyo is 5.7 °C (42%) in January and 28.6 °C (72%) in August.

Investigated dwellings include detected houses and apartment buildings which are typically available in Japanese urban areas. These dwellings have air conditioning which can provide heating and cooling when needed during hot or cold weather.

Indoor air temperature, globe temperature and relative humidity were measured in the living rooms and bedrooms, away from direct sunlight, at 10 min intervals using a data logger. Outdoor air temperature and relative humidity were obtained from the nearest meteorological station.

The number of subjects was 119 males and 125 females. Respondents completed the Japanese questionnaire on paper several times a day in the living rooms and twice a day in the bedrooms ('before going to bed' and 'after waking up from the bed'). Window opening behaviour was recorded by the occupant, at the time of completing the questionnaire in binary form (0, window closed; 1, window open). We collected 36,144 samples.

27.3 Results and Discussion

The collected data were divided into three groups: the FR mode (free running, where neither cooling nor heating was in use during the voting), CL mode (cooling by air conditioning) and HT mode (heating).

27.3.1 Variation of Indoor and Outdoor Temperatures During Voting

Figure 27.1 shows the monthly mean outdoor and indoor air temperature in FR mode in living room and bedroom. Because of the large number of samples, 95% confidence interval of the mean (mean $\pm$ 2 SE) is very small. The results show

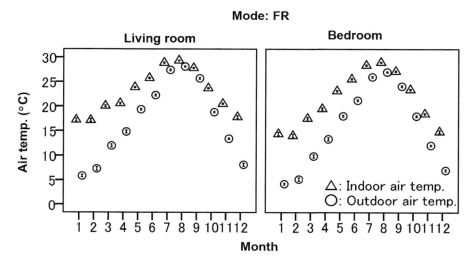

Fig. 27.1 Monthly mean outdoor and indoor air temperature with 95% confidence intervals (mean $\pm$ 2 SE)

that the seasonal difference of the indoor air temperature is quite large and that the data cover a wide range of indoor and outdoor temperatures.

27.3.2 Evaluation of Window Opening Behaviour

27.3.2.1 Status of Window Opening

To understand the window opening behaviour, the mean proportions which were calculated from the binary data (0, window closed; 1, window open) of 'window opening' are compared. The mean window opening is 0.37 for the FR mode, 0.03 for the cooling mode and 0.00 for the HT mode [9]. The result indicates that occupants did not open the window in the heating mode. The results showed that the mean windows opening is close to the Pakistan value, much lower than the UK value [1, 10]. We limit our principle analysis to the FR mode, because the windows in the other modes were so rarely open.

27.3.2.2 Season, Month and Time of the Day

Seasonal and monthly differences in proportion of windows open in FR mode are shown in Fig. 27.2. The proportion of open windows is highest in summer and lowest in winter. Window opening in autumn is significantly higher than that in spring. The mean window opening in the living room is higher than in the bedroom. Evidently, the proportion of open windows gradually increases towards the summer months (Fig. 27.2b) and gently decreases towards the winter months. The proportion of open windows gradually increases during the morning and then decreases towards the evening [8].

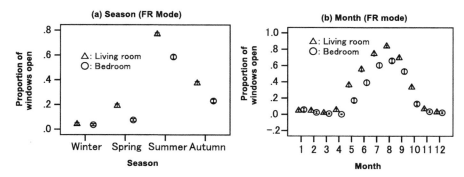

Fig. 27.2 The mean proportion of open windows with 95% confidence intervals (mean $\pm$ 2 SE) in FR mode

27.3.2.3 Relationship Between the Open Windows and Air Temperature

Figure 27.3 shows the proportion of open windows and the corresponding temperatures. The data were divided into ten groups (deciles), in an ascending order of temperature. The proportion of the window opening rises as the indoor globe or outdoor air temperature rises. The proportion of window opening in the living rooms is higher than in the bedrooms. When mean indoor air temperature is 26.7 °C, the proportion of open windows is 0.59 in living rooms and 0.47 in bedrooms (Fig. 27.3a).

However, the proportion of windows open decreases in highest outdoor temperatures. This trend is similar to the Pakistan study [10]. The reason could be that occupants close the windows to prevent the hot air entering through them. When the mean outdoor air temperature is 23.5 °C, the proportion of windows open is 0.65 in living rooms and 0.53 in bedrooms (Fig. 27.3c). These proportions are similar to the Pakistan study [10] and considerably lower than found in the UK study [1].

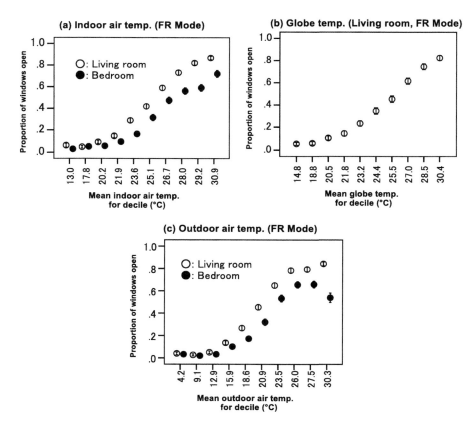

Fig. 27.3 The mean proportion of open windows with 95% confidence intervals (mean ± 2 SE) in FR mode

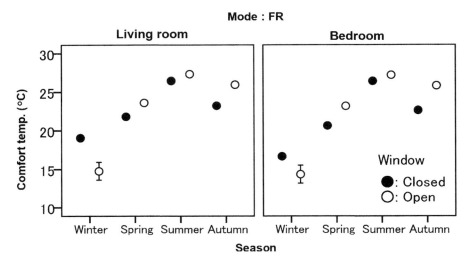

Fig. 27.4 The mean proportion of open windows with 95% confidence intervals (mean ± 2 SE) in FR mode

27.3.3 Comfort Temperature and Window Opening

Window adjustment is an important factor to enable adaptive thermal comfort to take place in buildings operating in the free-running mode. In this section, we consider how window opening is related to the manner in which the comfort temperature changes with the window opening. (In the heating and cooling modes, the comfort temperature is directly related to energy use.)

Figure 27.4 shows the seasonal variation in comfort temperature with windows open and closed. The mean comfort temperature for window open is 26.5 °C in living room which is 4.0 K higher (t-value = −69.1, $p < 0.001$) than that of the case of window closed [9]. Brager et al. [11] found in office buildings a 1.5 K higher comfort temperature for the people with an access to window operation than the group without. The temperature difference is highest in autumn. In winter the mean comfort temperature for the open window condition is significantly lower than for the window-closed condition. These trends are similar in bedroom. The results showed that window opening is effective to create the comfortable thermal environment.

27.3.4 Development of Window Opening Algorithm

27.3.4.1 Logistic Regression Curves

In the previous section, we examined the window opening behaviour based on our field data and showed some general behavioural trends. In this section we use the data to quantify and predict the general trends of the window opening behaviour in the dwellings [2]. Such predictions are of interest for the thermal simulation of dwellings.

The following regression equations using raw data were obtained for the relation between the windows open and the indoor temperature, and also for the outdoor air temperature. It is the outdoor temperature that is to be used for the prediction of the probability of the window being open in the context of thermal modelling. However, from the point of view of adaptive thermal comfort, the prediction from the indoor temperature is more fundamental, as this is the variable that, via the thermal response and consequent behaviour of the occupant, causes the window to be opened or closed. Both sets of equations are therefore presented.

Living room

$$\text{logit}(p) = 0.386T_i - 10.0 \ (n = 14{,}435, R^2 = 0.34, \text{S.E.} = 0.007, p < 0.001)$$
$$(27.1)$$
$$\text{logit}(p) = 0.362T_g - 9.4 \ (n = 10{,}979, R^2 = 0.29, \text{S.E.} = 0.008, p < 0.001)$$
$$(27.2)$$
$$\text{logit}(p) = 0.257T_o - 5.7 \ (n = 14{,}526, R^2 = 0.38, \text{S.E.} = 0.004, p < 0.001)$$
$$(27.3)$$

Bedroom

$$\text{logit}(p) = 0.302T_i - 8.4 \ (n = 10{,}633, R^2 = 0.26, \text{S.E.} = 0.007, p < 0.001)$$
$$(27.4)$$
$$\text{logit}(p) = 0.216T_o - 5.3 \ (n = 10{,}686, R^2 = 0.28, \text{S.E.} = 0.005, p < 0.001)$$
$$(27.5)$$

All data

$$\text{logit}(p) = 0.349T_i - 9.2 \ (n = 25{,}068, R^2 = 0.32, \text{S.E.} = 0.005, p < 0.001)$$
$$(27.6)$$
$$\text{logit}(p) = 0.241T_o - 5.6 \ (n = 25{,}212, R^2 = 0.35, \text{S.E.} = 0.003, p < 0.001)$$
$$(27.7)$$

T_i, indoor air temperature (°C); T_g, globe temperature (°C); T_o, outdoor air temperature (°C); n, sample size; SE, standard error of the regression coefficient; p, significance level of the regression coefficient; R^2, Cox and Snell R^2.

These logistic regression equations, based on the indoor or outdoor air temperature, are shown in Fig. 27.5. The predicted window opening is well matched with measured values. The proportion of window opening in the living rooms is higher than in the bedrooms. The relation between the proportion of windows open and indoor air temperature or globe temperature is almost the same, and so the results can be presented using the indoor air temperature alone.

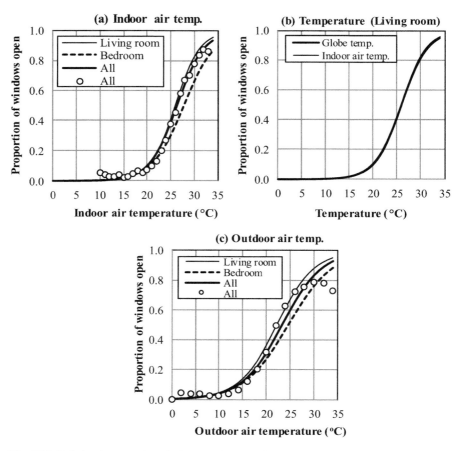

Fig. 27.5 Relation between the window opening behaviour and temperature in FR mode. Comparison of measured (open circular dots) and predicted value (curved line by equations) was shown in figure. Measured values were grouped for every 1 °C for indoor air temperature and for every 2 °C for outdoor air temperature. The grouped data for samples less than 100 are not shown

27.3.4.2 Deadband of Temperature for Window Opening Behaviour

Indoor and Outdoor Air Temperature The proportion of windows open was plotted as a scatter diagram against the indoor or outdoor air temperature as shown in Fig. 27.6. In order to obtain the sets of points for plotting, the data were sorted by dwelling and indoor or outdoor temperature and then split into groups of 25 records each. It is seen that the original regression line does not fit very well into the grouped data.

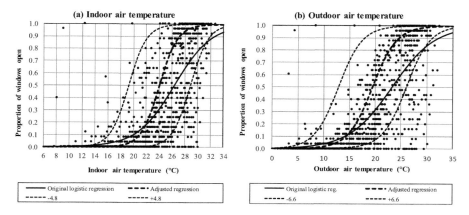

Fig. 27.6 Logistic regression curves of open window as a function of (**a**) indoor air temperature and (**b**) outdoor air temperature in all FR modes. Three lines (left, closure from the open position, centre and right, closed to open position) represent the deadband of windows open. Each data point is the mean of 25 raw data

To determine the range of temperatures over which the window opening remains unchanged, the deadband of the window opening behaviour with respect to the indoor or outdoor air temperature is established [1, 12]. This is calculated for the grouped data (mean of 25 raw data), making due allowance for the error in the logit arising from the sample size of 25. The regression equation is further adjusted using the existing procedures [2, 10, 12, 13].

As it can be seen in Fig. 27.6a, 83% of the points are within ±4.8 K range (±1.5 standard deviations) of the regression line. The deadband is 4.8 K for indoor air temperature which is the same as for Japanese dwellings in Gifu [2], bigger than that of the UK (2.1 K) and smaller than that of Pakistan (7.0 K) [10, 13]. The deadband (6.6 K) with respect to the outdoor temperature is bigger than that of Japanese dwelling of Gifu (5.4 K) [2] and the UK (5.0 K) study [1].

Temperature Departure from Comfort Temperature As we found in the previous section, the width of the deadband for windows based on indoor air temperature is wider than had been found from the UK data [1], possibly because Japan has higher seasonal differences in comfort temperatures compared with the UK. The width of this deadband thus includes a portion attributable to the seasonal shift in comfort temperature. Based on existing research, the fundamental or basic deadband can be estimated based on the temperature departure from comfort temperature ($\Delta t = T_i - T_c$) [2, 10, 14]. We have also applied the same method to establish the deadband. The following logistic regression equation using raw data was obtained for all data in between the windows open and the Δt:

$$\text{logit}(p) = 0.568\Delta t - 0.7 \ \left(n = 25,054, R^2 = 0.17, \text{S.E.} = 0.010, p < 0.001\right)$$

$$(27.8)$$

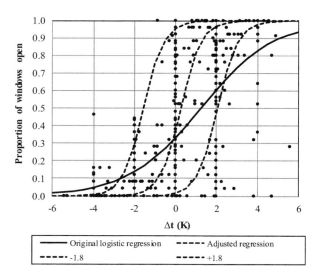

Fig. 27.7 Logistic regression curves of windows open as a function of Δt in all FR modes. Three lines (left, closure from the open position, centre and right, closed to open position) represent the deadband of windows open. Each data point is the mean of 25 raw data

Equation (27.8) is shown in Fig. 27.7 as an original logistic regression line. To obtain the 'binned' data points (mean of 25 raw data), the data were sorted by dwelling and then by the Δt and then split into groups of 25 records in order of increasing Δt. However, the original logistic regression line does not fit very well with the grouped data. The equation was calculated, and its regression gradient adjusted to make allowance for the binomial error in the predictor variable (the logits) arising from the sample size of only 25 grouped data. The regression equation is adjusted using the existing procedures [2, 10, 12].

To determine the range of Δt over which the window opening remains unchanged, the deadband of the window opening behaviour with respect to Δt is established. The deadband includes some 88% of the observations. The deadband obtained in this study is 1.8 K (± 1.5 standard deviations) which is bigger than that of Japanese dwelling of Gifu (1.4 K) and smaller than that of the UK (2.0 K), Europe (2.3 K) and Pakistan (2.8 K) studies [14]. The difference could be attributed to the greater freedom available to the residents than to office users in operating their windows.

27.3.5 Quantification of Constraints

Realistic knowledge of window opening is needed to predict the thermal comfort and energy use in dwellings. If controls such as windows are effective and easy to use, thermal discomfort can be much reduced. In practice there may be constraints that hinder the use of such controls. In any thermal simulation, these constraints need numerical values [12, 15, 16]. We explore the nature and extent of these constraints operating on the use of windows using the method already reported [2, 15]. We have reckoned all constraints from the relevant comfort temperature.

We have found that constraints on window opening in the dwellings are smaller than that of the office buildings [2, 15]. The reason might be that residents are more free to open the windows in their dwellings, and so the constraints might be smaller than those in the office buildings. However, further research is required to arrive at a firm conclusion. A constraint of 2 K is normal and unlikely to result in discomfort, constraints of around 5 K would indicate a significant problem with the use of that control, while a constraint of around 8 K would render the control redundant for controlling the thermal environment.

27.4 Conclusions

We have investigated the window opening behaviour and corresponding thermal environment for 4 years in the living rooms and bedrooms of dwellings in the Kanto region of Japan and the following results were found:

1. The proportion of the window opening in the free-running mode is significantly higher than in the cooling or heating modes.
2. The window opening is related to the indoor or outdoor air temperature in the free-running mode.
3. The window opening behaviour is predicted based on indoor and outdoor air temperature using logistic regression analysis. The predicted window opening matched well with that of the measured value.
4. The deadband of window opening in these dwellings is 1.8 K which is smaller than that for European and Pakistan office buildings.
5. The results indicate that people are more likely to open windows in their dwellings than are occupants of offices.
6. The maximum constraint of window opening in the investigated dwellings is 5.1 K which is smaller than European and Pakistan office buildings.
7. Overall window opening behaviours were shown to play a significant role in the effective achievement of comfort in Japanese dwellings, with obvious energy-saving implications from this finding.

Acknowledgements We would like to give thanks to Prof. Michael Humphreys and Prof. Fergus Nicol for their research guidance, the households who participated in the survey and the students for data entry. This research was supported by Grant-in-Aid for Scientific Research (C) 24560726.

References

1. Rijal HB, Tuohy P, Humphreys MA, Nicol JF, Samuel A, Clarke J (2007) Using results from field surveys to predict the effect of open windows on thermal comfort and energy use in buildings. Energ Buildings 39(7):823–836
2. Rijal HB, Honjo M, Kobayashi R, Nakaya T (2013) Investigation of comfort temperature, adaptive model and the window opening behaviour in Japanese houses. Archit Sci Rev 56 (1):54–69
3. Imagawa H, Rijal HB (2015) Field survey of the thermal comfort, quality of sleep and typical occupant behaviour in the bedrooms of Japanese houses during the hot and humid season. Archit Sci Rev 58(1):11–23
4. Kubota T (2007) A field survey on usage of air-conditioners and windows in apartment houses in Johor Bahru city. J Environ Eng AIJ 72(616):83–89
5. Fabi V, Andersen RV, Corgnati S, Olesen BW (2012) Occupants' window opening behaviour: a literature review of factors influencing occupant behaviour and models. Build Environ 58 (12):188–198
6. Oseland N (1995) Predicted and reported thermal sensation in climate chambers, offices and homes. Energ Buildings 23(2):105–116
7. Rijal HB, Humphreys MA, Nicol JF (2009) Understanding occupant behaviour: the use of controls in mixed-mode office buildings. Build Res Inf 37(4):381–396
8. Rijal HB, Humphreys MA, Nicol JF (2014) Study on window opening algorithm to predict occupant behaviour in Japanese houses. In: PLEA2014 – 30th conference, sustainable habitat for developing societies, Ahmedabad, India, 15–18 December, 2014
9. Rijal HB, Humphreys M, Nicol F (2015) Adaptive thermal comfort in Japanese houses during the summer season: behavioral adaptation and the effect of humidity. Buildings 5(3):1037–1054
10. Rijal HB, Tuohy P, Humphreys MA, Nicol JF, Samuel A, Raja IA, Clarke J (2008) Development of adaptive algorithms for the operation of windows, fans and doors to predict thermal comfort and energy use in Pakistani buildings. ASHRAE Trans 114(2):555–573
11. Brager GS, Paliaga G, de Dear R (2004) Operable windows, personal control, and occupant comfort. ASHRAE Trans 110(2):17–33
12. Humphreys MA, Nicol JF, Roaf S (2016) Adaptive thermal comfort: foundations and analysis. Routledge, London
13. Rijal HB, Tuohy P, Nicol F, Humphreys MA, Samuel A, Clarke J (2008) Development of an adaptive window-opening algorithm to predict the thermal comfort, energy use and overheating in buildings. J Build Perform Simul 1(1):17–30
14. Rijal HB, Tuohy P, Humphreys MA, Nicol JF, Samuel A (2011) An algorithm to represent occupant use of windows and fans including situation-specific motivations and constraints. Build Simul 4(2):117–134
15. Rijal HB, Tuohy P, Humphreys MA, Nicol JF, Samuel A (2012) Considering the impact of situation-specific motivations and constraints in the design of naturally ventilated and hybrid buildings. Archit Sci Rev 55(1):35–48
16. Nicol JF, Humphreys MA, Roaf S (2012) Adaptive thermal comfort: principles and practice. Routledge, Abingdon

Chapter 28
Occupants' Climate-Controlling Behavior in Japanese Residences

Takashi Nakaya

Abstract In this chapter, occupants' behaviors will be described from the viewpoint of reducing thermal stress, based on the field survey data in Chap. 19. Field surveys were conducted in occupied houses in Hyogo and Osaka, Japan, from August to September 2003.

In mixed mode, occupants combined air conditioning, natural ventilation, and electric fans. Especially the occupants switched between air conditioning and natural ventilation (+ fan use) controlled the indoor temperature. The use of natural ventilation and electric fan suppressed the start of air conditioning at the humidity ratio of 15 g/kg (DA) or less, at room temperature of 30 °C or less. However, when the indoor air temperature exceeds 30 °C, the start of air conditioning has no influence of natural ventilation and electric fan use, and there was no effect in suppressing air conditioning. Occupants started air conditioning at room temperature lower than wet-bulb globe temperature (WBGT) of 29 °C. Therefore, WBGT of 29 °C is considered to be hot limit of adaptation.

Keywords Field survey · Occupants' behavior · Limit of adaptation · Air conditioning · Natural ventilation · Fan

28.1 Introduction

Reducing the use of air conditioning is the essential part of reducing the energy consumption of buildings in summer. In order to reduce the air conditioning energy, it is important to reduce thermal stress by natural ventilation and electric fans and the use of air conditioners.

However, in the environment with extremely thermal stress, the risk of heat stroke increases by not using air conditioning. It is important to understand the factors that affect their behavior.

T. Nakaya (✉)
Department of Architecture, Shinshu University, Nagano, Nagano, Japan
e-mail: t-nakaya@shinshu-u.ac.jp

© Springer Nature Singapore Pte Ltd. 2018
T. Kubota et al. (eds.), *Sustainable Houses and Living in the Hot-Humid Climates of Asia*, https://doi.org/10.1007/978-981-10-8465-2_28

Moreover, the air conditioning system is mixed mode in Japanese residences. Occupants use natural ventilation, fans, and air conditioners at the discretion of occupants themselves. House occupants may stop one action and change to another or perform multiple actions simultaneously resulting in very complex behaviors. Nicol and Humphreys [1] pointed out the necessity of examining a combination of behaviors, stating that "the use of one control may change with use of another." However, there are few cases of interaction between regulating behaviors, and there are many unclear points.

First, based on field survey data [2], we will organize the occupants' behaviors. Next, the effect of natural ventilation and cooling suppression by using fan is analyzed. And to avoid heat stroke, consider the hot limit of adaptation.

28.2   Method

In this chapter, measurement data of the indoor thermal environment (air temperature, globe temperature, relative humidity) were used for discussion. Details of the measurement are described in Chap. 19. In parallel with the measurement of the thermal environment, a survey was conducted of occupants' behaviors. Focusing on the moment at which the occupants' behavior changed, we extracted data about the thermal environment in the room.

Figure 28.1 shows an example of the results. The questionnaire is formatted with time along the horizontal axis and the daily activities of the occupants in the living room along the vertical axis.

For example, Fig. 28.1 shows that the windows were opened at 5:30 pm, a fan was switched on at 6:20 pm, the windows were closed, and AC was turned on at 7:00 pm. Moreover, we asked respondents to select a reason for stopping an activity (too cold, did not get cooler, left the room, etc.). Data was obtained for a total of 1660 hours. According to this survey, it is possible to analyze the thermal environmental data at the moment at which occupant behavior changes.

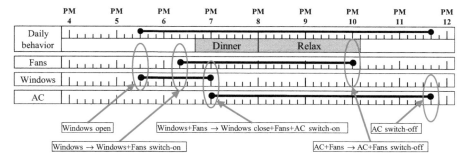

Fig. 28.1 Exemplary occupant-behavior-survey result [2]

28.3   Results

28.3.1   The Combinations of Occupant Behaviors

Table 28.1 summarizes the various combinations of occupants' behaviors. Most of the data at the moment of opening the window were [Windows open] and [AC → Windows open + AC switch-off]. Occupants often stopped air conditioning and switched to opening windows. Occupants switched between air conditioning and window opening rather than constantly running the air conditioning. Windows were closed mainly for windows close and AC start [Windows → Windows close + AC switch-on]. Since thermal stress could not be reduced by opening the window, occupants closed the window and changed to air conditioning. Most of the data at the moment of running the air conditioning belonged to the category [Windows → Windows close + AC switch-on] and [AC switch-on]. Also, there were many data points associated with a change from window opening and fan operation to air conditioning operation. Air conditioning was stopped mainly for AC stop [AC → AC off] and windows open and AC stop [AC → Windows open + AC off]. In mixed mode, occupants combined air conditioning, natural ventilation, and electric fans. Occupants who switched between air conditioning and natural ventilation (+ fan use) especially controlled the indoor temperature.

Table 28.2 shows the temperature data when the status of an AC or a window was changed. Indoor temperature and outdoor temperature at the start [AC on, window open] and stop [AC off, window close] are shown. The room temperature with the window opened (28.2 °C) is lower than that at which the window is closed (29.6 °C).

Occupants opened the window when the room temperature was low and closed the window when the room temperature was high. Also, when the window was closed or opened, the indoor temperature and outdoor temperature were at the same level. Natural ventilation cannot be expected to discharge indoor energy, as the difference between indoor and outdoor is small. It is thought that the effect of opening the window is to increase the airflow speed within the room.

The room temperature at which the AC was switched on (29.9 °C) is lower than that at which the window was closed (27.4 °C). Occupants started the air conditioning when the room temperature was high and stopped it when the room temperature was low.

28.3.2   Psychrometric Chart

28.3.2.1   Starting and Ending Occupants' Behaviors

Occupants started air conditioning when the thermal stress increased. In this chapter, the upper limit of the thermal environment that started air conditioning is defined as the hot limit of adaptation. Also, the use of natural ventilation and electric fans is

Table 28.1 Summary of combinations of occupants' behaviors [2]

Occupants' behavior		Combination of occupants' behavior	N	T_a [°C]	T_{out} [°C]
Natural ventilation	Windows open	Windows open	73	28.3	27.2
		AC → Windows open + AC off	30	27.6	28.6
		Fans on + windows open	30	28.7	28.2
		AC → Fans + windows open + AC off	5	27.6	27
		Fans → fans + windows open	4	29.5	26.5
		AC → Windows open + AC	3	28.3	30.3
	Windows close	Windows → windows close + AC on	48	30.1	29.4
		Windows → windows close	36	28.4	28.2
		Fans + windows → fans off + windows close	28	29.3	28.6
		Fans + windows → fans + windows close + AC on	14	30.8	30.1
		Fans + windows → fans off + windows close + AC on	9	31.3	30.9
		Windows → fans on + windows close + AC on	6	30.3	28.5
		Fans + windows → fans + windows close	5	29.6	29
		Windows + AC → Windows close + AC off	1	28	28
		Windows + AC → Windows close + AC	1	29	31
		Windows + AC → Fans on + windows close + AC	1	28	30
		Windows → windows close + fans on	1	29	27
Air conditioner	AC switch-on	Windows → windows close + AC on	48	30.1	29.4
		AC on	37	29.1	28.5
		Fans + windows → fans + AC on	14	30.8	30.1
		Fans + windows →AC on	9	31.3	30.9
		Fans on + AC on	6	30.7	29.2
		Windows → fans on + windows close + AC on	6	30.3	28.5
		Fans → fans + AC on	4	28.5	28.3
	AC switch-off	AC → AC off	53	27.5	27.6
		AC → Windows open + AC off	31	27.6	28.6
		AC → Fans on + AC off	12	28.1	28.3
		AC → AC(auto switch-off)	5	28	27.2
		AC → Fans on + AC off	2	26.5	28
		Fans + AC → Fans + AC off	2	26.5	28.5
		Fans + AC → Fans + AC (auto switch-off)	2	27.5	25.5
		Fans + AC → Fans + windows open+ AC off	1	28	32

T_a Air temperature at the moment of changing the behaviors
T_{out} Outdoor temperature at the moment of changing the behaviors

Table 28.2 Temperature data when the statuses of air conditioning and natural ventilation were changed [2]

Combination of occupants behavior		N	Indoor air temperature [°C]				Outdoor air temperature [°C]			
			Min.	Max.	Mean	S.D.	Min.	Max.	Mean	S.D.
Natural ventilation	Windows open	146	23.8	33.1	28.2	1.8	23.3	33.5	27.8	4.5
	Windows close	150	24.9	36.4	29.6	1.8	24.9	36.4	29.6	3.3
AC	AC switch-on	124	25.3	36.4	29.9	1.8	24.2	34.7	29.3	4.4
	AC switch-off	108	23.5	30.7	27.4	1.6	24.1	33.3	27.9	4.0

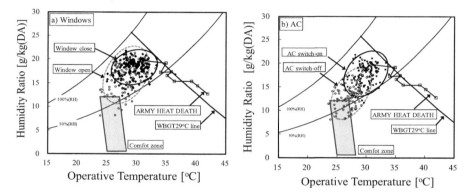

Fig. 28.2 Psychrometric chart (**a**) Windows were open/closed. (**b**) AC on/off [2]

defined as adaptive behavior. Figure 28.2 shows the data from the moments at which occupant behavior changed in a psychrometric chart. A thermal comfort zone was added to this chart. Each data point was plotted on the psychrometric chart and density ellipses ($p = 90\%$) drawn. The probability ellipse is the analysis method that expresses the distribution shape of two variables. If the data is a two-dimensional normal distribution, it includes any ratio of the area of an ellipse. The comfort zone is defined by PMV [3]. WBGT is the index used for examining the limit of the hot environment [4]. The black dots in Figure correspond to heat stroke deaths of healthy male US soldiers assigned to sedentary duties in midwestern army camp offices [5].

Figure 28.2a shows the temperature and humidity data at the moment when the status of the windows was changed. The data at the moment of opening the window showed a temperature ranging from about 24 to 32 °C and a humidity ratio ranging from 10 to 22 g/kg (DA). On the other hand, the data at the moment the window was closed showed temperatures ranging from about 27 to 34 °C and humidity ratios ranging from 15 to 22 g/kg (DA). Most changes in windows status took place when

Fig. 28.3 Pscychrometoric chart of WBGT at the moment of starting air conditioning [2]

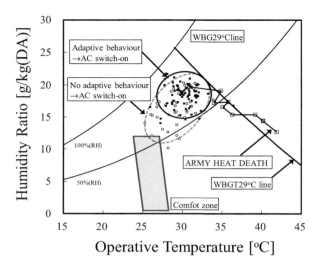

the occupant was outside of their thermal comfort zone. Window opening and closing was performed in a hotter environment than the thermal comfort zone. The upper heat boundary of the data at the moment of closing the window was WBGT 28–29 °C. The data at the moment at which air conditioning was started ranged from about 26 to 34 °C in air temperature and from 13 to 23 g/kg (DA) in humidity ratio. The data at the moment of stopping the air conditioning ranged from about 24 to 31 °C in air temperature and from about 8 to 19 g/kg (DA) in humidity ratio. The distribution of temperature and humidity data at the start and end of air conditioning was wide, and the individual differences were large. The upper heat boundary of the data at the moment of starting air conditioning was WBGT 28–29 °C, which are the same temperature and humidity levels at which the window was closed. Occupants closed the windows at lower temperature and humidity than the heatstroke-danger threshold, and started air conditioning.

28.3.2.2   The Effect of Suppressing the Start of Air Conditioning by Natural Ventilation

This study examined whether the use of air conditioning can be suppressed by natural ventilation or use of electric fans (Fig. 28.3). Figure 28.4 shows the frequency distribution of WBGT. In this chapter, the upper limit of the thermal environment that started air conditioning is defined as the hot limit of adaptation. Also, the use of natural ventilation and electric fans is defined as adaptive action. "Air-conditioning-switch-on" behavior was categorized into switching from fans or natural ventilation to air conditioning [Adaptive behavior → AC on] and simply starting air conditioning [No adaptive behavior → AC on]. The use of natural ventilation and electric fans suppressed the start of air conditioning at a humidity

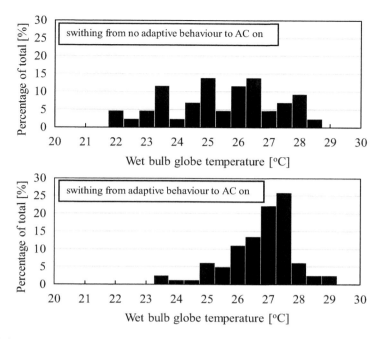

Fig. 28.4 Frequency of WBGT at the moment of starting air conditioning [2]

ratio of 15 g/kg (DA) or less, at a room temperature of 30 °C or less. However, when the indoor air temperature exceeds 30 °C, the start of air conditioning has no influence upon the natural ventilation or electric fan use, and there was no effect in suppressing air conditioning. Data on the start of air conditioning was evaluated by WBGT. Suppression of air conditioning by natural ventilation and electric fans was effective in indoor environments with WBGT 25 °C or less. Meanwhile, in the WBGT range from 28 to 29 °C, air conditioning started regardless of natural to ventilation and electric fan usage. The hot limit of adaptation was WBGT 29 °C, regardless of the adaptive behavior before the start of air conditioning. In order to increase the heat release from humans, it is effective to reduce the temperature of the indoor environment. Moreover, if the humidity of the indoor environment is low, sweat on the surface of the human body can easily evaporate and can cool the body. However, if the indoor temperature and humidity increase, the transfer potential between humans and the indoor environment becomes small. Even if wind speed and the transfer coefficient are both increased, the body does not become cool without being able to increase heat release. In other words, adaptive behaviors (windows open, fan) such as increasing the wind speed are not effective in hot environments, and the effect of reducing thermal stress is limited.

Figure 28.5 shows the conceptual diagram of occupant mixed-mode behavior during summer. The aerial diagram includes a comfort zone and the limit of adaptation. Occupants switched between natural ventilation and cooling to adjust

Fig. 28.5 Conceptual
diagram of occupant
behavior

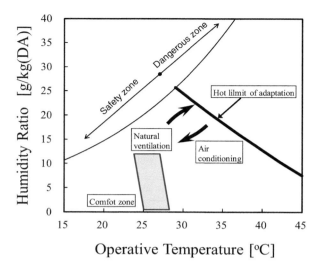

the temperature and humidity in the room. Occupants opened the windows if the indoor environment was cool and drove the air with a fan as the indoor environment became hot. Natural ventilation was done in a higher-temperature environment than the thermal comfort range. However, in order to avoid the danger of heat stroke, occupants switched from natural draft to cooling use at a lower temperature and humidity than the adaptive heat limit. In general, the thermal environment of the room is often designed to fall within the comfort zone.

In addition, prediction models of resident behavior often adopt the deviation from the neutral temperature as the predictor variable. Both of these are based on the assumption that occupants are people who require thermal comfort. However, in the example discussed in Chap. 19, the occupants lived in the heat zone outside the thermal comfort zone, but they were a high-acceptance rate. The examination of this chapter shows that there was a hot limit of adaptation. In summer, occupants in mixed mode did not necessarily give top priority to thermal comfort. The occupants had adjusted to temperature and humidity ranges that were not comfortable but acceptable and safe.

28.4 Conclusions

In this chapter, the following knowledge was obtained regarding the occupant behavior based on field-survey data taken in Japan during the summer:

- In the mixed mode, occupants combined air conditioning, natural ventilation, and electric fans. The occupants who switched between air conditioning and natural ventilation (+ fan use) especially controlled the indoor temperature. Occupants closed the windows and started air conditioning if the room temperature was high, and opened the windows and stopped air conditioning if room temperature was low.

- The use of natural ventilation and electric fans suppressed the start of air conditioning at humidity ratios of 15 g/kg (DA) or lower and at room temperatures of 30 °C or less. However, when the indoor air temperature exceeds 30 °C, the start of air conditioning has no influence of natural ventilation and electric fan use, and there was no effect in suppressing air conditioning.
- Occupants started air conditioning at room temperatures lower than WBGT 28–29 °C. In Chap. 19, occupants did not live in environments hotter than WBGT 29 °C. Also WBGT 29 °C matches the heatstroke-danger threshold of ASHRAE. Therefore, WBGT 29 °C is considered to be the hot limit of adaptation.
- In summer, occupants in mixed mode did not necessarily give top priority to thermal comfort. The occupants had adjusted to the temperature and humidity range which is not comfortable but acceptable and safe

References

1. Nicol F, Humphreys M (2004) A stochastic approach to thermal comfort, occupant behaviour and energy use in buildings. ASHRAE Trans 110(2):554–568
2. Nakaya N, Matsubara N, Kurazumi Y (2008) Use of occupant behaviour to control the indoor climate in Japanese. In: Proceedings of conference: air conditioning and the low carbon cooling challenge, Network for comfort and energy use in buildings residences
3. ASHRAE (2013) ASHRAE Standard 55 thermal environmental conditions for human occupancy. American Society of Heating, Refrigerating, and Air-conditioning Engineers
4. ISO 7243 (1989) Hot environments – Estimation of the heat stress on working man, based on the WBGT index (wet bulb globe temperature)
5. ASHRAE (2013) ASHRAE Fundamental chapter 10, Health, American Society of Heating, Refrigerating, and Air-conditioning Engineers

Box A: Occupant Adaptation in Japanese and Indian Offices

Madhavi Indraganti, Ryozo Ooka, and Hom Bahadur Rijal
Department of Architecture and Urban Planning, College of Engineering, Qatar University, Doha, Qatar
Institute of Industrial Science, The University of Tokyo, Tokyo, Japan
Department of Restoration Ecology and Built Environment, Tokyo City University, Yokohama, Japan
Email: madhavi.indragaganti@fulbrightmail.org; madhavi@qu.edu.qa; ooka@iis.u-tokyo.ac.jp; rijal@tcu.ac.jp

Abstract Designing buildings with minimal operational costs mandates the knowledge on occupant's adaptation, which is underreported now. Relying on field study reports, this box takes a closer look at various methods of user adaptation in offices. Users were found to be adaptively operating the windows, fans, and air conditioners, as the indoor conditions migrated from the comfort limits. Many barriers within and beyond the user's purview hindered adaptation. To improve occupant adaptation and thermal satisfaction, building managers may aim to mitigate the latter.

Keywords Office buildings · Thermal comfort · Personal environmental controls · Operable controls · Logistic regression · Comfort temperature

Introduction

Understanding the occupant's adaptation in buildings enables designers to expand the air movement and higro-thermal limits of indoor environments in offices. In retrospect, they impact their architecture and energy use. Research on this front is nascent in Japan and India [1–4]. Given their current energy situation [5], both Japan and India need to embrace low-energy intensive approaches to indoor environmental design and control. As a logical corollary to the discussion in Chap. 17, the following sections throw light on various adaptations that occupants undertook in offices as observed in recent field studies. These were use of fans, air conditioners (AC), windows, and behavioral adaptations. Also presented are the barriers to adaptation in offices.

T. Kubota et al. (eds.), *Sustainable Houses and Living in the Hot-Humid Climates of Asia*, https://doi.org/10.1007/978-981-10-8465-2

Fig. 1 Probability of open window, estimated through the logistic regression of proportion of open windows and indoor air temperature in India. Also shown are the actual data in 1K bins [7]

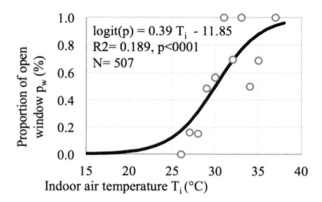

logit(p) = 0.39 T_i - 11.85
R^2= 0.189, p<0001
N= 507

Adaptive Use of Windows

Window opening is one of the immediate and economical adaptation methods of environmental control. Building occupants in Japan and India operated the windows during the temperature migrations. Window use correlated robustly with the indoor and or outdoor temperatures. It is also important to achieve higher indoor air movement in warm environments. Brager et al. [6] observed that the subjects voting uncomfortable were getting inadequate air movement from fans and windows. Interestingly, in India in naturally ventilated (NV) mode, the environments with windows open recorded higher air speeds. Mean air speed with windows open and closed was 0.2 m/s (standard deviation (S.D.): 0.3 m/s) and 0.14 m/s (S.D.: 0.25 m/s), respectively.

Window-opening behavior is analyzed through logistic regression of proportion of windows open with its stimulus, such as temperature (Fig. 1) [7]. Simulation studies of indoor environments find these equations useful. Similar relationship was noted in Japanese offices also [8]. The proportion of open windows can be estimated from these. For example in India, 80% windows would be open when indoor temperature is at 33–34 °C. The Japan study noted 60% open windows when the outdoor air temperature was 22 °C.

Adaptive Use of Fans and ACs for Enhanced Air Movement

Electrical fans are simple yet cost-effective means to enhance indoor air movement. Although they recirculate the air mass around, their usage behavior is of paramount importance to understand and provide for higher indoor air speeds at a much lower expense. Indraganti et al. [4, 9] found fan usage very high in both naturally ventilated (NV) and AC modes of building operation.

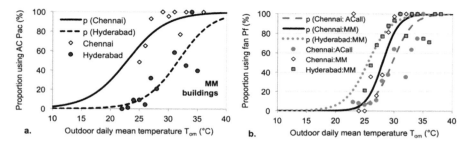

Fig. 2 Logistic regression of proportion of (**a**) AC use in MM buildings in Chennai and Hyderabad in India; (**b**) fan use in completely air-conditioned (ACall) and MM buildings in Chennai and Hyderabad. Also shown juxtaposed are the actual proportions in 1 K bins of outdoor daily mean temperature (T_{om}) [4, 9]. ($p < 0.001$ for all the relationships)

In mixed mode (MM) buildings of Japan and India, occupants continued to use fans even after switching from NV mode to AC mode during the overheated periods in summer. As a result, high indoor air speed is recorded in Japan (0.30 m/s) in MM buildings with ACs switched on. These are more than the levels mentioned in the ASHRAE standard [10]. Owing to the high operating costs of running the AC throughout the day, most offices in Japan and India were run in NV mode during the early hours adaptively switching to AC mode during the temperature migrations in summer and during the midday. This association between AC operation and temperature was found to be strong (r = -0.571 (India), r = -0.406 (Japan), $p < 0.001$) [7, 9].

Figure 2 shows the logistic regression of AC use (P_{ac}) in mixed mode buildings and fan usage in both completely air-conditioned (AC$_{all}$) and MM buildings in two cities of Chennai and Hyderabad in India. Chennai belongs to a warm-humid coastal region. Observable, it has much higher AC and fan usage even beyond the summer months possibly due to higher humidity [2]. It is important to note that the subjects desired higher indoor air movement even while voting neutral on the thermal sensation (TS) scale [6]. For example in India, 67.5% (N = 239) subjects who rated the environments *unacceptable* preferred higher air speeds, despite voting neutral on the TS scale.

Adaptation Through Clothing and Behavioral Control Actions

Occupants in offices adapt through clothing both within a day and within a long period. In a multiuser office space, clothing adaptation is well within the reach of the occupant unlike the other environmental controls like windows, fans, and ACs. Indraganti et al. [4, 9] noticed a negative correlation (at 95% confident interval) between clothing insulation and indoor air temperature. This manifests active adaptation.

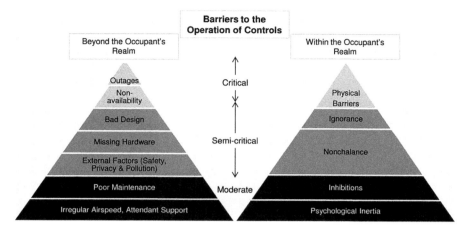

Fig. 3 Barriers in using various controls in offices categorized into three levels of interference directly under the occupant's realm and beyond [7]

Office occupants also use many behavioral control actions adaptively. *Staying in airy place* was the most voted behavioral adaptation in India throughout the year. This is possibly due to the prevalence of yearlong warm conditions. *Drinking cold/ hot beverages and changing posture* were the next most used actions.

Barriers to Occupant Adaptation

Indraganti et al. [7] noted many nonthermal factors hindering occupant adaptation. These could be categorized into two sections: those within the occupant's realm and those beyond as shown in Fig. 3 These can be categorized based on their level of interference. For example, outages and non-availability form critical barriers beyond the user's realm. On the other hand, psychological inertia creates a moderate barrier well within the user's realm.

Concluding Remarks

This box looked at various adaptation mechanisms available and in use in offices. Logistic regressions are presented to estimate the probability of fans, air conditioners, and windows in use. These would be useful in building simulations to account for the occupant behavior. Occupant adaptation and satisfaction would improve if building managers aim to remove the barriers that are beyond the user's realm.

References

[1]. Rijal HB, Humphreys MA, Nicol JF (2017) Towards an adaptive model for thermal comfort in Japanese offices. Build Res Inf https://doi.org/10.1080/09613218.2017.1288450

[2]. Indraganti M, Ooka R, Rijal HB (2015) Thermal comfort in offices in India: behavioral adaptation and the effect of age and gender. Energ Buildings 103:284–295

[3]. Indraganti M, Ooka R, Rijal HB (2014) Thermal comfort and acceptability in offices in Japan and India: a comparative analysis. In: Annual conference of the Society of Heating Air-conditioning and Sanitary Engineers of Japan (SHASE) September, Akita, Japan

[4]. Indraganti M, Ooka R, Rijal HB, Brager GS (2014) Adaptive model of thermal comfort for offices in hot and humid climates of India. Build Environ 74 (4):39–53

[5]. IEA (2015) IEA Energy Atlas. International Energy Agency

[6]. Brager GS, Paliaga G, de Dear RJ (2004) Operable Windows, Personal control. ASHRAE Trans 110(2):17–35

[7]. Indraganti M, Ooka R, Rijal HB, Brager GS (2015) Drivers and barriers to occupant adaptation in offices in India. Archit Sci Rev 58(1):77–86

[8]. Takasu M, Ooka R, Rijal HB, Indraganti M, Singh MK (2017) Study on adaptive thermal comfort in Japanese offices under various operation modes. Build Environ 118:273–288

[9]. Indraganti M, Ooka R, Rijal HB (2013) Thermal comfort in offices in summer: Findings from a field study under the 'setsuden' conditions in Tokyo, Japan. Build Environ 61:114–132

[10]. ASHRAE (2010) ANSI/ ASHRAE Standard 55-2010, Thermal environmental conditions for human occupancy. American Society of Heating, Refrigerating and Air-Conditioning Engineers, Atlanta

Part IV
Energy Consumption

Chapter 29
The Use of Energy Consumption Data

H. Takaguchi

Abstract There are two ways to build an energy consumption data. One is by downscaling the nationwide statistical data. It is often called macro data. Another way is by collecting the information of each household by questionnaire survey, and it is called micro data. Each has its own characteristics, it is important to chose the data according to the purpose.

From the viewpoint of utilization of micro energy consumption data, the opinion is divided whether actual data or design values should be collected. Both of them have advantages and disadvantages so that to collect both data carefully and order include examining its correlativity are important.

Keywords Energy statistics · Residential energy consumption

29.1 Introduction

The target of the Paris Agreement of COP21 in 2015 is to hold the increase in the global average temperature to well below 2 °C above preindustrial levels. To achieve this goal, the global society has to realize the net zero carbon society until the end of the twenty-first century. To response this situation, we have to consider to achieve zero carbon nationwide, in the city scale and buildings, and at home by understanding a real situation of carbon emission, that is, energy consumption and implementation of the functioning measures.

For example, the energy consumption of the building sector in Japan has occupied 30% of national energy consumption, and a half of it is at home. To reduce the energy consumption at home, researchers all over the world have been measuring the energy consumption and tried to understand the structure of it. Now these data and research have been used to make the building standard for insulation and energy benchmark.

H. Takaguchi (✉)
Department of Architecture, Waseda University, Tokyo, Japan
e-mail: takaguchi@waseda.jp

On the other hand, collecting an energy consumption data becomes easier and easier because of the spread of a smart meter which has occurred in all over the world simultaneously. The building engineer will be required to develop the plan of the building and city development based on the evidence and to propose the idea to balance reducing energy consumption and improvement of the quality of life.

29.2 Macro Data and Micro Data

There are two ways to build an energy consumption data. One is by downscaling the nationwide statistical data. For example, the energy consumption per household can be calculated from the volume of sales of an energy company and the number of households. It is often called macro data. Another way is by collecting the information of each household by questionnaire survey, and it is called micro data. Both of them have advantages and disadvantages and are used properly according to the purpose.

Usually, macro data is developed from industry statistics, trade statistics, construction statistics, population census, and so on. In each country macro data are organized by the United Nations or IEA (International Energy Agency), so that they are comparatively easy to obtain. But macro data cannot respond to the requirement to do the detailed cross analysis.

On the other hand, micro data is developed on each individual data. Therefore it is possible to make a group of particular characteristic buildings and obtain more detailed information. But because of this reason, it is necessary to consider carefully its statistical reliability and representativeness.

29.3 Examples in Major Countries

The US Department of Energy has started the Commercial Buildings Energy Consumption Survey (CBECS) because of the oil shock in 1973 and 1979. The number of samples is about 5000 buildings chosen according to the condition of the real building stock in the USA. Ultimately, those data is published as an extended statistics as whole building stock. In addition to this survey, some states have surveyed their own survey. For example, the state of California surveyed the "California Commercial Building End-Use Survey" in 2002.

European countries had not carried out a survey as the European Union for a long time. Only a few countries such as Denmark and Norway had their own survey and database. But in 2003, EPBD (Energy Performance of Buildings Directive) was enforced by the European Union, and the movement to create a mass database has been activated. EPBD is a scheme to open the energy performance data of building when the building construction is completed and its transaction happens. EU required each member of countries to develop a calculation method of energy performance, benchmark, certification system, and human resource development. Each country could design their own code and scheme according to their circumstance. Because of this EPBD, various projects in cooperation with neighboring countries were implemented. IMPACT (improving energy performance assessment and certification schemes by tests) and EL-TERTIARY for the commercial building were the major projects.

29.4 Examples of the Use of Energy Data

The most popular way of use of energy consumption data of building would be an application as a benchmark. The labeling sysytem of Energy Star in the US certificates TOP 25% buildings as a high-performance building according to the actual energy consumption data. CASBEE in Japan has a similar scheme in its evaluation system. Now many national governments and local governments have tried to develop the standard or code for buildings and used these data as a foundation. From the building owner's point of view, they can compare the energy consumption with other buildings, analyze the consumption trends in their buildings and conduct energy diagnosis, identify facilities to be repaired, and plan for maintenance. Air conditioner manufacturers use the detailed energy consumption data for understanding an actual way of their product use. They analyze an hourly or minutely data for understanding the actual performance, lifetime of the product, and so on.

From the viewpoint of utilization of energy consumption data, the opinion is divided whether actual data or design values should be collected. The actual data is affected by the business content of the tenant, the vacant room, and the grade of common space. Meanwhile, the design values seem fair as a comparison under the certain circumstance. But the design values always have a concern whether a building surely is built the same as the design value and managed properly. It can be said that the misfit risk between the design value and actual value. And as a practical matter, it is difficult to recalculate the design value of the existing building which occupies the majority of buildings so that the only actual data can be collected. Therefore the real estate investment company usually collects the actual data to sort out the investment.

- The actual data and the design value of the building energy consumption are the two wheels to consider the building performance. Both of them have advantages and disadvantages so that to collect both data carefully and order include examining its correlativity are important.

Chapter 30
Overview of Energy Consumption in Hot-Humid Climates of Asia

Kazuhiro Fukuyo

Abstract Energy demand in the ASEAN member countries has increased by 2.6 times between 1990 and 2014, while the world's energy demand has increased by 56% in the same period. However, the energy consumption profiles in the ASEAN countries are not uniform. In Brunei and Singapore, electricity is a main residential energy resource, while in other countries, biofuels are important energy resources. Residential electricity consumption increases steadily with the economic development in the ASEAN member countries. Generally, a 1000-dollar increase in GDP per capita results in a 30~40 kWh increase. In the least developed countries, such as Cambodia, Lao PDR, and Myanmar, a severe inequality in electricity accessibility exists.

Keywords Energy statistics · Residential energy consumption

30.1 Introduction

Energy demand in the ASEAN member countries, i.e., countries with hot-humid climates, has rapidly increased in the past decades. Figure 30.1 shows the total primary energy supply (TPES) in the world in 1990 and 2014. It shows remarkable increase in energy demand in China, India, and the ASEAN. The ASEAN's energy demand has increased by 2.6 times between 1990 and 2014, while the world's energy demand has increased by 56% in the same period.

This rapid increase in the ASEAN's energy demand is mainly driven by the socioeconomic development in the ASEAN member countries. As the ASEAN emerges as a hub of the global economy, the region consumes energy more. The international energy agency (IEA) predicted that the ASEAN's energy demand will grow by 80% from today to about 46 EJ (1100 Mtoe) in 2040, accompanying with

K. Fukuyo (✉)
Graduate School of Innovation and Technology Management, Yamaguchi University, Yamaguchi, Japan
e-mail: fukuyo@yamaguchi-u.ac.jp

© Springer Nature Singapore Pte Ltd. 2018
T. Kubota et al. (eds.), *Sustainable Houses and Living in the Hot-Humid Climates of Asia*, https://doi.org/10.1007/978-981-10-8465-2_30

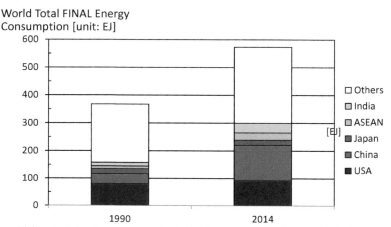

All data excluding Laos are provided by IEA. Laos energy data is provided by UNSD.

Fig. 30.1 Total primary energy supply (TPES) in the world in 1990 and 2014. All data excluding Laos are provided by IEA. Laos energy data is provided by UNSD

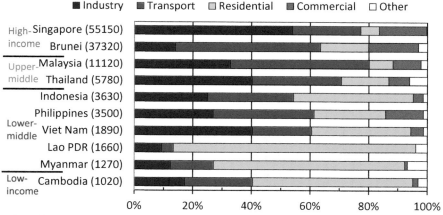

The values given in parentheses indicate the Gross National Income (GNI) per capita in 2014 in current US dollars (World Bank)

Fig. 30.2 Composition of final energy consumption for the ASEAN countries. The values given in parentheses indicate the Gross National Income (GNI) per capita in 2014 in current US dollars (World Bank)

the regional economic development [1]. Energy policies in the ASEAN countries are of growing importance in global energy security and prevention of global warming.

The energy consumption profiles in the ASEAN countries are not uniform. Patterns of their energy use vary depending on their socioeconomic conditions, climates, etc. Figure 30.2 shows the sectoral composition of energy consumption in the ASEAN member countries in 2014 (only the data of Lao PDR is provided by United Nations Statistics Division [2], while others are provided by IEA) [3]. The numbers given in parentheses indicate the Gross National Income (GNI) per capita in 2014 in current US dollars [4]). In the high-income (Singapore and Brunei Darussalam) and upper-middle-income countries (Malaysia and Thailand), the share of the residential sector is less than 16%; in the lower-middle-income (Indonesia, Philippines, Vietnam, Laos, and Myanmar) and low-income countries (Cambodia), the share of the residential sector is more than 24%. This figure indicates that energy policies for the residential sector are of higher importance in the countries with lower GNI per capita. Energy policies in the ASEAN countries should be established and implemented in accordance with the actual conditions in each country.

30.2   Residential Energy Consumption

Figure 30.3 shows the nationwide residential energy consumption per household in 2014 (The same as in Fig. 30.2, the data of Lao PDR is provided by United Nations Statistics Division [2], and the others are provided by IEA [3]). These data represent the nationwide average values which cover urban and rural areas in each country. The numbers given in parentheses indicate the average size of household in each country. The energy consumption patterns vary a lot from country to country. In Brunei and Singapore, electricity is a main residential energy resource. In other countries, biofuels, i.e., firewood and charcoal, are important energy resources, mainly for cooking as used to be. Biofuels have an environmental impact but will continue to be important energy resources for many years to come. It is necessary to improve the energy efficiency of biofuels to reduce energy consumption and protect forest resources.

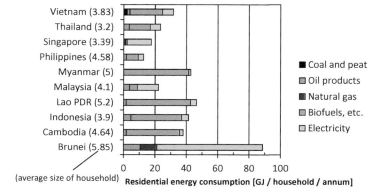

Fig. 30.3 Residential energy consumption per household, 2014

30.3   Stead Increase in Electricity Consumption

Electricity consumption increases steadily with the economic development in the ASEAN member countries. Figure 30.4 shows the gross domestic product (GDP) based on purchasing power parity (PPP) per capita in current international dollars based on the 2011 ICP round [4] and yearly residential electricity consumption per capita [2, 3] in each country from 1990 to 2014. There is a significant correlation between the economic development and electricity consumption. Although the causality between them should be carefully examined, it is believed that the economic development leads first to rise in the living standard, second to widespread use of electric appliances, and then to increase in energy consumption. In general, the

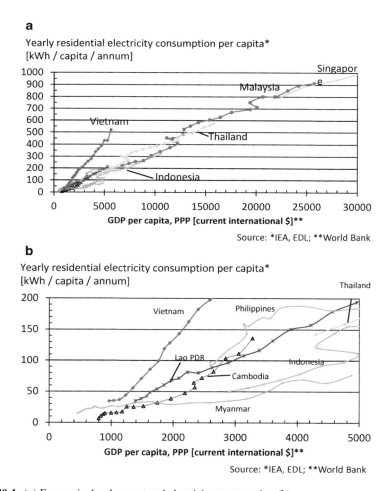

Fig. 30.4 (a) Economic development and electricity consumption (b)

figure shows that a 1000-dollar increase in GDP per capita results in a 30~40 kWh increase in electricity consumption per capita per annum. Exceptionally, Vietnam shows rapid increase in residential electricity consumption at an above-average pace. This is thought to be causally related to a high accessibility to electricity in Vietnam. According to the World Development Indicators [4], the access to electricity in Vietnam is 99% of its population in 2012, while the access to electricity is significantly lower in the other lower-middle-income and low-income countries.

It is important issue not only how to generate and distribute electricity to meet the rise of the living standard but also how to conserve the electricity for prevention of global warming.

30.4   Intra-Country Inequality of the Access to Electricity

In the least developed countries of the ASEAN members, a severe inequality in electricity accessibility exists. Figure 30.5 shows the access to electricity in rural, urban, and whole area of each country [4]. There is a big gap of accessibility between rural and urban area in Cambodia, Lao PDR, and Myanmar. Figure 30.6 shows the electricity consumption by province in Cambodia in 2012 [5]. It presents the electricity consumption per capita per annum in each province. This figure exhibits electricity inequality within the country. The people in Phnom Penh city (the capital city) and Kandal Province (the neighboring province to the capital) enjoy the benefits of electricity, while the people in the rural provinces such as Preah Vihear are hard to access the electricity. Concentration of the electricity grid in the urban area leads to the widening of the intra-country inequality in electricity accessibility and living standards.

There is a possibility that the inequality in electricity consumption will be gradually resolved in the populated provinces in Cambodia; a higher concentration of people encourages economic activity and investment to infrastructures including

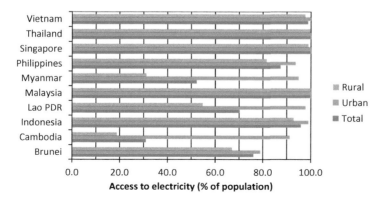

Fig. 30.5 Access to electricity: Intra-country inequality

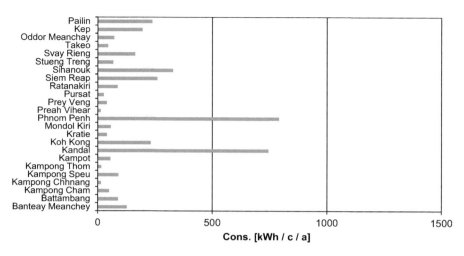

Fig. 30.6 Electricity consumption by Province in Cambodia (2012)

electricity grids. On the contrary, the inequality in electricity consumption may not be resolved in sparsely populated areas, unless some measures are taken.

In the sparsely populated areas, the construction and connection of the national electricity grids may not be economical. Establishment of cross-border electricity grids, construction of local micro-grids, and introduction of renewable energy technologies will be effective measures. Electrifications by biomass [6] and solar power [7] are higher potential measures in Cambodia. The rural electrification promotion project by introducing PV cells in Laos, which was supported by the Japan International Cooperation Agency (JICA), offers a useful example in solving the inequality in electricity use in Cambodia.

30.5   Conclusion

Energy demand in the ASEAN countries has rapidly increased in the past decades. Especially residential electricity consumption increases steadily with the economic development in these countries. It is an important issue, not only how to generate and distribute electricity to meet the demand but also how to conserve the electricity for preventing global warming. With the rise of the electricity consumption, the intra-country inequality in electricity accessibility has emerged in the least developed ASEAN countries, i.e., Cambodia, Laos, and Myanmar. Some measures for solving the inequality between rural and urban areas, such as establishment of the cross-border electricity grid, construction of local micro-grids, and introduction of renewable energy, may be required.

Note on the Macr-Energy Data

We need comprehensive macro-energy data that are superior in terms of reliability, consistency, and usability. Only the limited numbers of the organizations, such as IEA and UNSD, provide them. However, they are not perfect. For example, the IEA doesn't cover the energy statistics of Lao PDR. The UNSD provides the energy statistics in short periods. The author thus used these data in combination with the other statistics provided by each country or energy supplier.

References

1. International Energy Agency (2015) Energy demand prospects. In: Southeast Asia energy outlook 2015. International Energy Agency, Paris, p 30
2. United Nations Statistics Division (2015) 2014 Energy balances. United Nations, New York
3. Statistics (2017) International Energy Agency, Paris. http://www.iea.org/statistics/. Accessed 15 Feb 2017
4. World Development Indicators (2017) World Bank, Washington, DC. http://data.worldbank.org/. Accessed 25 June 2017
5. Fukuyo K (2015) Regional differences in electricity accessibility among the Asian least developed countries. Paper presented at the International joint-conference, SENVAR-iNTA-AVAN 2015, Universiti Teknologi Malaysia, Johor Bahru, 24–26 Nov 2015
6. Abe H et al (2007) Potential for rural electrification based on biomass gasification in Cambodia. Biomass Bioenergy 31:656–664. https://doi.org/10.1016/j.biombioe.2007.06.023
7. Janjai S et al (2011) Estimation of solar radiation over Cambodia from long-term satellite data. Renew Energy 36:1214–1220. https://doi.org/10.1016/j.renene.2010.09.023

Chapter 31
Energy Consumption and Indoor Temperature in Cambodian Houses

Hiroto Takaguchi

Abstract Cambodia is located almost at the center of the Mekong Delta. The largest lake in Southeast Asia, named the Tonle Sap, is located at the center of Cambodia. The Khmer dynasty had flourished in Cambodia, and economic growth has continued following the end of the civil war. There are four main types of housing in Cambodia. They include "town houses" (which are often called "shop houses"), "detached houses," "traditional houses," and "condominiums." In this chapter, the energy consumption of these four types of houses is reported by use and by household income. The results are from several interview surveys that were conducted in Phnom Penh, Siem Reap, Kandal, and Takeo.

The energy consumption of Cambodian houses shows a strong correlation with household income, location, and the style of the house. The average annual energy consumption in urban areas was 13.3 GJ per household. In rural areas, the average annual energy consumption was 22.3 GJ per household. Low-income families depended highly on biomass energy and used more than high-income households because the combustion efficiency of biomass is low. As a result, the energy consumption in rural areas was larger than that in urban areas. When the income of household exceeded approximately $300 per year, the shift from using biomass energy to LPG and electricity progressed rapidly. Corresponding to this shift in energy source, energy efficiency improved, and energy consumption decreased. Ultimately, as household income increased even more, energy consumption also increased.

Keywords Cambodia · House · Energy consumption · Indoor temperature

H. Takaguchi (✉)
Department of Architecture, Waseda University, Tokyo, Japan
e-mail: takaguchi@waseda.jp

© Springer Nature Singapore Pte Ltd. 2018
T. Kubota et al. (eds.), *Sustainable Houses and Living in the Hot-Humid Climates of Asia*, https://doi.org/10.1007/978-981-10-8465-2_31

Fig. 31.1 Town house

31.1 Introduction

Cambodia is located near the center of the Mekong Delta. The largest lake in Southeast Asia, named the Tonle Sap, is located in the center of Cambodia. Cambodia was the protectorate of France from 1867 to 1953 and was administered as a part of the colony of French Indochina. Therefore, there is an apparent influence of French style in many buildings.

There are four main types of housing in Cambodia. A "town house" shares walls with other units, has a shop space on the ground floor, and has living spaces on the second floor and above (Fig. 31.1). A town house is made of a bent structure, and the walls are filled with locally made bricks. A "detached house" and "traditional house" are both independent, single-unit houses (Fig. 31.2 and 31.3). Traditional houses are built on stilts and are made with local materials such as wood, palm tree leaves, and local tile or zinc for the roof. The fourth style is an apartment house that is called a "condominium" (modern apartment houses) (Fig. 31.4).

The abundance of each of these types of houses depends on the character of each region in Cambodia. Many areas are rapidly changing, especially in Phnom Penh and surrounding suburbs. According to trends in economic growth, the style of the houses has been changing rapidly. For example, as river management systems have been improved, the underfloor space of the stilt homes has been increasingly renovated into usable indoor space.

Generally, the lifetimes of buildings in countries that do not experience significant natural disasters, such as earthquakes, are assumed to be relatively long.

Fig. 31.2 Detached house

Fig. 31.3 Traditional house

Because of this, the performance of the building at the time of construction and ease of renovation are very important for future energy efficiency. Currently, the usage of air conditioners in Cambodia is low. Recent rates were 15% in Phnom Penh and 11% in Kandal. According to projections of economic growth and improvement to the quality of life, the use of air conditioners and other electric equipment is predicted to

Fig. 31.4 Condominium (under construction)

increase rapidly. For this reason, it is very important to understand building performance, energy consumption in the building, and occupant behavior.

31.2 Energy Consumption of Houses in Urban Areas

In 2015, we conducted a field survey to investigate the energy consumption of town houses and detached houses in Phnom Penh and the suburban area of Kandal. We did not survey the energy consumption in condominiums because at the time, condominiums were not common living spaces for the people of Cambodia. The surveyor visited the targeted houses and interviewed inhabitants about their family structure, characteristics of the building, and energy consumption. A total of 300 households were surveyed, 200 of which were in Phnom Penh and 100 in Kandal.

The average of annual energy consumption of houses in urban areas was 13.3 GJ/ year. Figure 31.5 shows the annual energy consumption divided by resources and use. Every surveyed household was connected to the regional electric grid, and many houses used LPG for cooking. However, many houses in suburban areas used firewood for cooking. Electricity and LPG account for 84% of the total energy consumption, whereas biomass fuels account for 16%. At the time of the survey, the households located in the center of the urban areas consumed more electricity than those in suburban areas. Figure 31.6 shows the typical interior of the ground floor and kitchen in a town house.

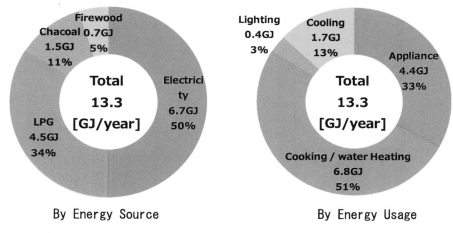

By Energy Source   By Energy Usage

Fig. 31.5 Energy consumption in urban houses in 2015

Ground Floor   Kitchen

Fig. 31.6 Interior space of a typical town house

31.3   Energy Consumption in Rural Houses (Traditional Houses)

In 2014, we conducted a field survey in the rural area of Takeo. The purpose of the survey was to determine the energy consumption in a traditional house. We conducted interviews at 23 households. The average annual energy consumption per household was 22.3 GJ/year. Figure 31.7 shows the annual energy consumption divided by energy sources and use. Charcoal and firewood account for approximately 83% of total energy sources, and LPG and electricity account for less than 17% of the total. Some households generated electricity by producing a biogas that comes from fermented manure. In Takeo, such biomass power generation systems are supported by international support organizations such as JICA. Considering energy usage, cooking accounts for approximately 95% of the total. None of the

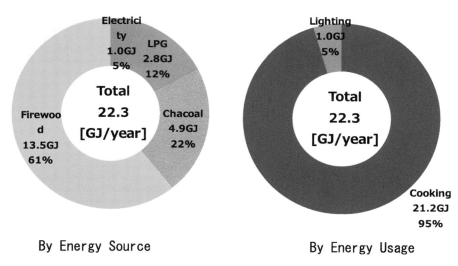

By Energy Source   By Energy Usage

Fig. 31.7 Energy consumption in rural houses in 2014

investigated houses used air conditioners, so their electricity consumption was limited to lighting and electric appliances. As such, energy consumption for air conditioning was smaller than in urban areas.

The energy consumption in rural areas was 68% larger than in an urban area. This is because of the low efficiency of energy output by firewood and charcoal. The actual use of fuel energy and electricity were much smaller. Considering CO_2 emissions, households in rural areas are much larger than those in urban areas, but if biomass energy could be developed as a renewable or carbon-neutral energy source, emissions in rural areas would also be much smaller.

Figure 31.8 shows an underfloor space of a stilt house. The space is shaded from direct sunlight, so it is comparatively more comfortable than spaces inside the house. Figure 31.9 shows a fireplace in the kitchen. To prevent the heat from the fireplace entering the house, the fireplace is often located outside the house. Figure 31.10 shows a cooking stove that uses biogas.

31.4   Energy Sources for Cooking

The energy sources for cooking in Cambodia vary. In urban areas, LPG cylinders (Fig. 31.11) are very popular. The cylinders are sold in energy shops or at gasoline stands in towns and are also deliverable. Portable gas stoves (Fig. 31.12) are also used in urban and rural areas. Cartridges for the gas stoves are can be refilled repeatedly.

Fig. 31.8   Under-floor space

Fig. 31.9   Kitchen using firewood

In rural areas, LPG is usually used on rainy days or when cooking needs to be done quickly. Biomass fuels, such as firewood (Fig. 31.13) or charcoal, are most commonly used for cooking. According to the Cambodia Socio-Economic Survey,

Fig. 31.10 Kitchen using
biomass gas

Fig. 31.11 LPG cylinder

Fig. 31.12 Portable gas stove and refillable cartridges

Fig. 31.13 Firewood

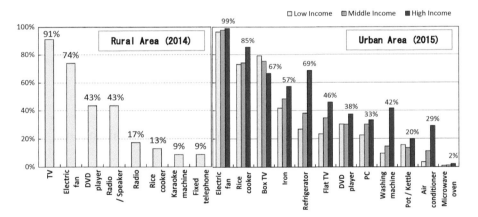

Fig. 31.14 Penetration rate in rural (n = 23) and urban areas (n = 297)

the share of LPG in households has increased from 0.7% to 3.3% in the past 10 years. The shift from using biomass to fuel for energy is projected to continue.

31.5 Penetration of Home Appliances

Figure 31.14 shows the penetration rate of home appliances in rural and urban areas. Even in rural areas, the ownership of TVs has already spread 91%. No rural households that were surveyed owned an air conditioner, but 74% did own an

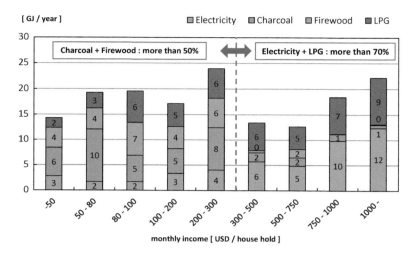

Fig. 31.15 Energy consumption based on family income in 2010

electric fan. The ownership of electric fans has spread almost 100% in urban areas. In high-income households, the penetration rate of rice cookers was 85%, refrigerators was 69%, and washing machines was 42%. The penetration rate of air conditioners has reached almost 30% in urban areas. In contrast, the penetration rate of washing machines and air conditioners was very low for low-income households.

31.6   Energy Consumption of Households Based on Income

Household energy consumption has a strong correlation with family income. In addition, energy sources vary according to family income. Figure 31.15 shows the relationship between family income and energy consumption. These results are based on a questionnaire survey that was done in Phnom Penh and Siem Reap in 2010. Of the surveys completed, 151 samples are from Phnom Penh, and 56 are from Siem Reap [1].

In households with less than $300 monthly income, firewood and charcoal accounted for over 50% of total energy consumption. In households with more than $300 monthly income, electricity and LPG accounted for over 70% of total energy consumption. A monthly income of $300 was a threshold at which the preferred energy source changed greatly. The change follows a conversion in cooking energy source from biomass to other sources, such as LPG.

Because the thermal efficiency of LPG and electricity is higher than that of biomass energy, energy consumption in the lowest-income family was higher than in households with an income ranging from $300 to $750. Considering CO_2 emissions, however, emissions by low-income families were much lower than high-income families.

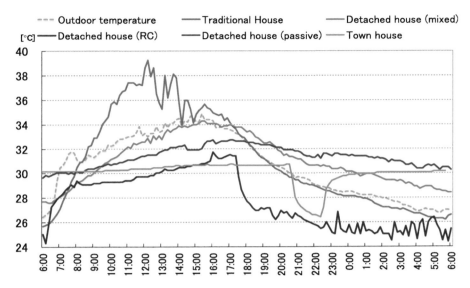

Fig. 31.16 Comparison of the Indoor thermal environments of five housing types

31.7   Indoor Air Temperature

At the end of this chapter, I discuss the thermal environment of each type of house. Figure 31.16 shows the hourly change of temperature in a representative living room in each style of house. On the day of data collection, the ambient outdoor temperature was 26 °C in the morning and 35 °C in the afternoon. Surprisingly, the temperature inside the traditional house surpassed the outdoor temperature and reached 39 °C at noon. The traditional house has recently upgraded the roof material from palm leaves to tin, and as such, the temperature of the living room increased because of radiation from the tin roof which became very hot in the direct sunlight. In the RC house that had an air conditioner, the effects of the cooling was observed. The indoor temperature remained relatively stable and was lower than the detached house which had an RC ground floor and wooden second floor.

31.8   Conclusions

The energy consumption of Cambodian houses is strongly correlated with household income, location, and the style of the house. Low-income families depend highly on biomass energy and use more than high-income households because the combustion efficiency of biomass is low. Considering CO_2 emissions, however, if biomass energies are considered to be carbon neutral, low-income households have already achieved a low carbon footprint. When the income of household exceeds approximately \$300/year, an energy shift from biomass to LPG and electricity is abrupt. Corresponding the change, energy efficiency is higher, and energy consumption decreases, but CO_2 emissions increase. Finally, when household income increases even more, both energy consumption and CO_2 emissions continue to increase. We must find ways to combine the economic growth, energy consumption, and CO_2 reduction to improve quality of life.

In addition, house style and methods of renovation are very important for energy efficiency. In traditional houses, for example, indoor air temperatures were hotter than other house styles because of the chosen roof material. In the chosen traditional house, the roof material had been changed from palm leaves to tin in order to diminish workloads associated with regularly replacing palm leaves. As a result, radiation from the tin roof generates a poor indoor environment.

Reference

1. Miyazaki K, Yamamoto Y, Washiya S, Takaguchi H (2012) Research on mitigation methods of energy consumption increase of Cambodian houses – proposal of energy conservation model houses. J Environ Eng 77(673):193–202

Chapter 32
Household Energy Consumption and CO_2 Emissions for Residential Buildings in Jakarta and Bandung of Indonesia

Usep Surahman and Tetsu Kubota

Abstract Indonesia has been experiencing population and high economic growth in line with rapid urbanization. As a consequence, the need for living areas increased, and an enormous number of residential buildings have been developed in major cities, such as Jakarta and Bandung. The household sector contributes to the nationwide final energy consumption by 29% in 2011, and the household energy consumption is expected to increase dramatically in the near future. The objective of this study is to analyze household energy consumption and CO_2 emission profiles for urban houses in major cities of Indonesia. Three surveys investigating household energy consumption within individual houses and apartments were conducted in Bandung (2011 and 2014) and Jakarta (2012). The results show that, overall, the average annual energy consumption of all samples in Jakarta was approximately 20.6 GJ, which is 5.0 and 7.1 GJ larger than individual houses and apartments in Bandung, respectively. The difference was primarily attributed to a difference in the ownership and use of air-conditioning between the two cities due to different altitudes. The profiles of CO_2 emissions are similar to those of energy. It is important to reduce the use of air-conditioning for operational energy for urban houses in the future.

Keywords Household energy consumption · CO_2 emissions · Individual houses · Apartments · Indonesia

U. Surahman (✉)
Universitas Pendidikan Indonesia (UPI), Bandung, Indonesia
e-mail: usep@upi.edu

T. Kubota
Graduate School for International Development and Cooperation (IDEC), Hiroshima University, Hiroshima, Japan

© Springer Nature Singapore Pte Ltd. 2018
T. Kubota et al. (eds.), *Sustainable Houses and Living in the Hot-Humid Climates of Asia*, https://doi.org/10.1007/978-981-10-8465-2_32

325

32.1   Introduction

The ultimate purpose of this study is to propose low-energy and low-carbon residential buildings in major cities of Indonesia. Over the last few decades, Indonesia has been experiencing high economic growth in line with rapid urbanization and population growth. The percentage of people living in urban areas reached approximately 53% in 2014 [1]. Consequently, the need for living spaces increased rapidly, and an enormous number of residential buildings have been developed, especially in major cities. This tremendous urbanization found in the major cities sees a large increase in urban energy consumption.

In Indonesia, the household sector contributed 33.2% of the nationwide final energy consumption during the period of 2000–2013 [2]. The household energy consumption is expected to increase dramatically as the middle class in urban areas rises in the near future [3]. Energy-saving strategies are, therefore, essential to be introduced further to make the cities more sustainable.

A few studies of household energy consumption were conducted in Indonesia. For instance, Utama and Ghewala [4] evaluated on the life cycle energy consumption of high-rise apartments in Jakarta. Furthermore, Utama and Ghewala [5] investigated the life cycle energy of single landed houses with different materials of walls in Semarang. Kurdi [6] estimated the life cycle energy and CO_2 emissions of planned houses in seven large cities in Indonesia. The above studies provide rare and useful results of energy consumption in residential buildings of Indonesia. However, these studies only focused on mass/planned houses. In Indonesia, individual/unplanned houses are typical of residences in major cities rather than the said mass houses, as discussed in Sect. 32.2.

This study, which focuses on individual houses in major cities of Indonesia, aims to assess the household energy consumption and CO_2 emissions of urban residential buildings comprising three house categories (simple, medium, and luxurious houses). Three surveys were conducted in the cities of Bandung ($n = 247$ for individual houses and $n = 300$ for apartment units) in 2011 and 2014 and Jakarta ($n = 297$) in 2012 to obtain household energy consumption profiles of these buildings. The results of this analysis will provide useful insights for policy making in energy policy for achieving low-energy and low-carbon societies in Indonesia.

32.2   Methodology

32.2.1   Case Study Houses

Jakarta and Bandung were selected as the case study cities, which represent rapidly developing cities. Both cities are located in the same region of Java in Indonesia. Jakarta, the capital city of Indonesia, had a population of 9.99 million [7], whereas Bandung, the capital of West Java Province, had a population of 2.45 million in

2012 [8]. Both experience hot and humid tropical climates. However, the monthly average temperature in Bandung (22.9–23.9 °C) is not as high as that in Jakarta (27.1–28.9 °C) because of the former's relatively high altitude. On average, Bandung and Jakarta are situated 700–800 and 5–10 m above the sea level, respectively.

In most major cities in Indonesia, individual houses, called *Kampungs*, account for the largest proportion of the existing housing stocks. These dwellings are settled in unplanned and overcrowded urban villages without being provided with proper, basic urban infrastructure and services [9]. These individual houses accounted for approximately 74% of total housing stocks in Jakarta in 2012 [7] and approximately 89% of those in Bandung [8]. In contrast, mass houses are defined as houses constructed in a proper modern urban planning. The recent mass developments comprising terraced houses are included in this type. This accounted for another 26% in Jakarta and 11% in Bandung. Moreover, these houses can be further classified into three house categories based on their construction cost and lot size, namely, simple, medium, and luxurious houses (Fig. 32.1) [10]. These houses have average technical life spans of approximately 20, 35, and 50 years, respectively [11].

We assumed that subsets of the population for respective house categories (i.e., simple, medium, and luxurious houses) were homogeneous in Indonesia. Therefore, the disproportional stratified sampling was applied; thus, a large sample size was not necessary to represent the entire population (2.2 million and 510 thousand houses in Jakarta and Bandung, respectively). Several typical residential neighborhoods were selected from the cities of Jakarta (14 areas) and Bandung (6 areas), respectively. A total of 297 and 247 residential buildings were then chosen randomly in the selected neighborhoods in the two cities by considering the distance from city center and their establishment years, respectively (see Table 32.1). These samples were selected by considering the abovementioned existing ratios of unplanned and planned houses in Jakarta. The whole sample consisted of mass houses of 23% and individual houses

Fig. 32.1 Views of sample residential buildings. (**a**) Simple house; (**b**) medium house; (**c**) luxurious house; (**d**) simple apartment; (**e**) medium apartment; (**f**) luxurious apartment

Table 32.1 Brief profiles of sample houses in Jakarta and Bandung

	Jakarta (individual houses)				Bandung (individual houses)				Bandung (apartment)			
	S	M	L	W	S	M	L	W	S	M	L	W
Sample size (individual/mass)	125 (125/0)	115 (75/40)	57 (29/28)	297 (229/68)	120 (120/0)	99 (99/0)	28 (28/0)	247 (247/0)	30 (0/30)	230 (0/230)	40 (0/40)	300 (0/300)
Household size (persons)	4.3	4.5	5.3	4.5	4.7	4.7	5.6	4.8	3.6	1.7	1.6	1.9
Monthly household income (%)												
<100 (USD)	4.8	1.7	1.8	3.0	10.0	0.0	0.0	4.5	24.1	0.9	0.0	3.2
100–499	76.8	59.1	19.2	58.9	75.8	58.6	7.1	61.5	69.0	36.1	41.0	40.2
500–1000	16.8	31.3	38.6	26.6	14.2	38.4	57.2	28.7	6.9	34.3	56.4	34.5
>1000	1.6	7.9	40.4	11.5	0.0	3.0	35.7	5.3	0	28.7	2.6	22.1
Total	100	100	100	100	100	100	100	100	100	100	100	100
Average (USD)	353.9	553.1	1047.9	572.3	330.5	550.8	1105.2	518.1	272.4	596.7	492.3	548.7
Total floor area (%)												
<50 (m²)	71.2	9.6	0.0	33.7	50.8	6.1	0.0	25.9	100.0	99.1	95.0	98.7
50–99	20.0	51.3	0.0	28.3	39.2	34.3	3.6	32.4	0.0	0.9	5.0	1.3
100–300	8.8	36.5	84.2	34.0	10.0	58.6	64.3	37.7	0.0	0.0	0.0	0.0
>300	0.0	2.6	15.8	4.0	0.0	1.0	32.1	4.0	0.0	0.0	0.0	0.0
Total	100	100	100	100	100	100	100	100	100	100	100	100
Average (m²)	44.2	107.2	213.6	101.1	59.6	124.3	283.1	110.9	23.0	31.8	38.4	31.8
Housing age (%)												
<10 (year)	22.4	25.2	17.5	22.6	13.3	26.3	57.1	23.5	0.0	0.0	0.0	0.0
10–20	40.0	17.4	21.0	27.6	14.2	28.3	28.6	21.5	0.0	65.7	5.0	49.0
21–30	17.6	29.6	31.6	24.9	21.7	15.1	14.3	18.2	100.0	34.3	95.0	51.0
31–40	12.8	15.6	21.1	15.5	8.3	9.1	0	7.7	0.0	0.0	0.0	0.0
>40	7.2	12.2	8.8	9.4	42.5	21.2	0	29.1	0.0	0.0	0.0	0.0
Total	100	100	100	100	100	100	100	100	100	100	100	100
Average (year(s))	19.6	24.1	24.6	22.0	34.5	23.8	10	27.5	24.0	17.4	23.5	18.9

S, Simple houses, *M* Medium houses, *L* Luxurious houses, *W* Whole samples

of 77%. Since the proportion of individual houses is very large, the survey focused only on these houses in Bandung.

As shown in Table 32.1, the average household size was about 4–5 persons for individual houses and 2–4 persons for apartments with a small variation between the three categories. The monthly average household income was also investigated by a multiple-choice question. As expected, the average income increases with house category from simple to luxurious houses. As shown, the total floor area also increases with house category. The largest percentage of total floor area was less than 50 m^2 (71% in Jakarta and 51% in Bandung) for simple houses and almost 100% for apartment units, 50–99 m^2 (51%) in Jakarta and 100–300 m^2 (58%) in Bandung for medium houses, and 100–300 m^2 (84% in Jakarta and 64% in Bandung) for luxurious houses. These differences are due to the difference of land cost where that of Jakarta is more expensive than that of Bandung. The age distribution of the surveyed houses (actual ages) varies slightly between the two cities. The average age of luxurious houses in Bandung is exceptionally low.

Meanwhile, the detailed household energy consumption data are necessary for the analysis of operational energy. However, there are few previous relevant investigations in Indonesia [4–6]. Since the energy consumption data were not available in both cities, the detailed interviews and measurement of appliance capacity by using watt checkers (MWC01, OSAKI) were conducted in order to obtain the data.

32.2.2   Operational Energy

Energy consumption for respective household appliances was estimated through multiplying the number of appliances by their usage time and electric capacity, which were acquired through the interviews and measurements. The annual average household energy consumption was then calculated by combining consumption for all the appliances. As described earlier, the seasonal variation in climatic conditions is not large in Jakarta and Bandung. Therefore, the usage time of appliances was assumed to be constant throughout the year. Nevertheless, the small seasonal changes of air temperature and humidity were considered in the estimation of energy consumption of air-conditioners and refrigerators, although the resultant changes were found to be negligible.

32.3   Results and Discussion

32.3.1   Operational Energy

Figure 32.2 presents the ownership levels of major household appliances in respective case studies. As shown, light bulbs, televisions, and refrigerators recorded high ownership levels similarly in the two cities among three house categories. In the case

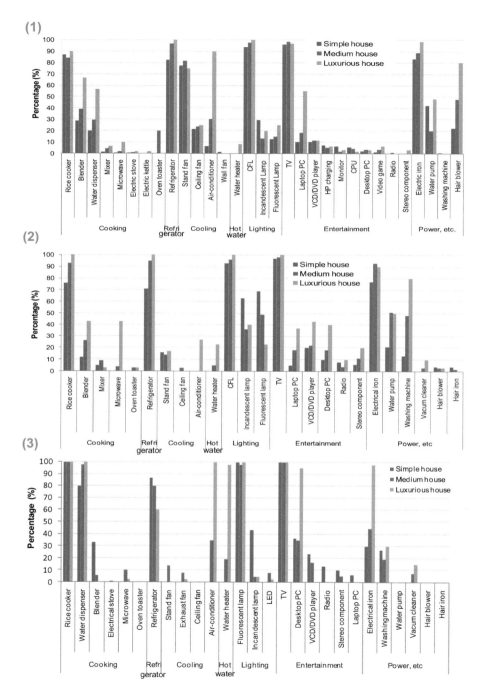

Fig. 32.2 Ownership level of appliances in (**1**) individual houses in Jakarta, (**2**) individual houses Bandung and (**3**) apartments in Bandung

of Jakarta (Fig. 32.2(1)), standing fans also recorded high ownership levels of 75–83% reflecting the severe hot climatic condition of the city. The ownership level of air-conditioners significantly differs between the two cities: it is 6–89% in Jakarta and 0–29% for individual houses as well as 0–100% for apartments in Bandung depending on house types. In general, the ownership levels of other appliances increase from simple to luxurious houses, respectively.

Figure 32.3 shows the annual household energy consumption averaged in respective house categories. Figure 32.3a indicates the energy consumption by different energy sources, and Fig. 32.3b shows those by different end-use categories. Overall, the average annual energy consumption of all samples in Jakarta is approximately 20.6 GJ, which is 5.0 G and 7.1 GJ larger than individual houses and apartments in

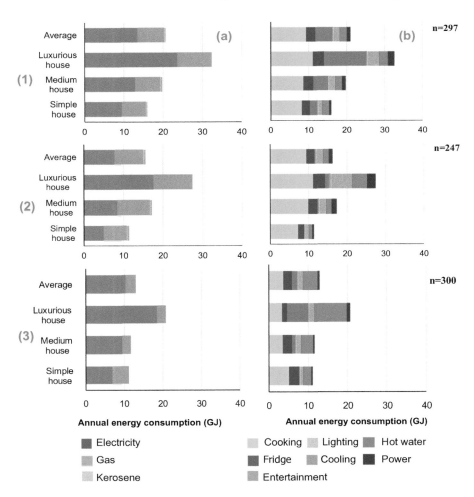

Fig. 32.3 Annual household energy consumption by house category in (**1**) Jakarta, (**2**) individual houses in Bandung and (**3**) apartments in Bandung, (**a**) by energy sources and (**b**) by end uses.

Bandung (15.6 GJ and 12.9 GJ), respectively. The difference is mainly attributed to the use of air-conditioning between the two cities. As shown, the energy consumption for cooling accounts for 27.8% in Jakarta on average (Fig. 32.3(1b)), whereas the corresponding percentage is only 1.8% and 10.5% in Bandung (Figs. 32.3(2b) and (3b)). Hence, in the case of Jakarta, basically, the average household energy consumption of house categories increases with the increase in ownership and use of air-conditioning, and the entertainment largely influences the increase in the overall energy consumption (Fig. 32.3(1b)). Since the average household size did not vary largely among the three house categories, the above difference of ownership and usage levels of cooling appliances in Jakarta, especially air-conditioner, and those of cooking and lighting in Bandung is directly reflected in the large difference of annual energy consumption among three house categories in both cities. Energy consumption caused by electricity use is larger than by LPG: 61–73% in Jakarta (Fig. 32.3(1a)) as well as 47–65% for individual houses and 63–90% for apartments in Bandung (Figs. 32.3(2a) and (3a)).

32.3.2 Operational CO_2 Emissions

The annual household CO_2 emissions were estimated through multiplying the energy consumption for each fuel type by its corresponding CO_2 emission factor [12]. The profiles of CO_2 emissions are similar to those of energy. The average annual CO_2 emission in Jakarta is estimated at 7.8 ton CO_2-equivalent, while that of Bandung is 4.8 ton CO_2-equivalent for individual houses and 5.9 ton CO_2-equivalent for apartments. The major contributors in Jakarta are cooling (2.4 ton (31%)), cooking (1.6 ton (20%)), and refrigerator (1.3 ton (17%)), while those in Bandung are cooking (1.2 ton (26%)), refrigerator (1.1 ton (23%)), and lighting (1.0 ton (21%)). If the amount of CO_2 emissions caused by cooling is excluded, then the difference of total CO_2 emissions between the two cities would be insignificant (5.4 ton in Jakarta, 4.7 ton (individual houses) and 5.7 ton (apartments) in Bandung). This clearly indicates that the increase in the use of air-conditioning in the future would dramatically increase the household energy consumption and therefore their CO_2 emissions.

32.4 Conclusions

Three case studies, which investigated household energy consumption profiles, in Bandung and Jakarta, were analyzed in order to identify the profiles of household energy consumption and CO_2 emissions in major cities of Indonesia.

The average annual energy consumption of all samples in Jakarta was approximately 20.6 GJ, which was 5.0 GJ and 7.1 GJ larger than individual houses and apartments in Bandung, respectively. The difference was mainly attributed to the use of air-conditioning between the two cities. Hence, in the case of Jakarta, basically, the average household energy consumption increased with the increase in ownership and use of air-conditioning. Meanwhile, in the case of Bandung, the energy consumption for cooking, lighting, and entertainment largely influenced the increase in the overall energy consumption. Accordingly, the average annual CO_2 emission in Jakarta was estimated at 7.8 ton CO_2-equivalent, while that of Bandung was 4.8 ton CO_2-equivalent for individual houses and 5.9 ton CO_2-equivalent for apartments.

In conclusion, it is important to reduce energy consumption/CO_2 emissions caused by air-conditioner through potential energy-saving strategies such as adopting passive cooling techniques, encouraging efficiency improvement of air-conditioner usage such as using better insulation, changing setting point temperature of air-conditioner, etc.

Acknowledgments   This research was supported by a JSPS KAKENHI (Grant No. JP 15KK0210). We would also like to thank Mr. Yohei Ito and Dr. Ari Wijaya of Universitas Persada Indonesia, Dr. Hanson E. Kusuma of Institut Teknologi Bandung, and the students who kindly administered our survey.

References

1. United Nation (UN) (2014) World urbanization prospect. UN, New York
2. Indonesia (2014) Handbook of energy and economic statistics of Indonesia. Ministry of Energy and Mineral Resources of Indonesia, Jakarta
3. JETRO (2011) Market investment attraction—Indonesia (in Japanese). Japan External Trade Organization, Tokyo
4. Utama NA et al (2009) Indonesian high rise buildings: a life cycle energy assessment, energy and. Building 41:1263–1268
5. Utama NA et al (2008) Life cycle energy of single landed houses in Indonesia. Energy Build 40:1911–1916
6. Kurdi Z (2006) Determining factors of CO_2 emission in housing and settlement of urban area in Indonesia. Ministry of Public Work of Indonesia and National Institute for land, Infrastructure and Management of Japan, Bandung
7. Jakarta (2013) Jakarta in figures. Statistical Centre of Jakarta, Jakarta
8. Bandung (2013) Bandung in figures. Agency for the Centre of Statistic of West Java, Bandung
9. World Bank (1995) Indonesia impact evaluation report. Enhancing the quality of life in Urban Indonesia: The legacy of Kampung improvement Program. Report No. 14747-IND. Operation Evaluation Department, the World Bank, Washington, DC
10. Indonesia (1995) Housing categories based on lot area and construction cost; The ministerial decree of public work no. 45. Jakarta (In Bahasa, Indonesia)
11. Indonesia (2002) Planning procedures for developing earthquake resistant on buildings; Indonesian Nasional Standard (SNI) 03–1726. Ministry of Housing, Jakarta (In Bahasa, Indonesia)
12. Nansai K et al (2002)Embodied energy and emission intensity data for Japan using input–output table (3EID) -Inventory data for LCA-. National Institute for Environmental Studies, Japan

Chapter 33
Firewood Consumption in Nepal

Hom Bahadur Rijal

Abstract Firewood consumption and air temperature were investigated in winter and summer in traditional houses in the Banke, Bhaktapur, Dhading, Kaski, and Solukhumbu districts of Nepal. The firewood consumption rate was 235–1130 kg/capita/year. The results showed that the households in the temperate climate used less firewood than those in the subtropical climate. The indoor and outdoor temperature difference (7.8 °C), the vertical temperature difference (7.1 °C), and the maximum indoor air temperature (42 °C) were most significant in the kitchen. The results demonstrated a waste of energy in winter and an uncomfortable thermal environment in summer. If thermal storage on the wall were introduced in winter as well as airtight openings and improvements in fireplaces, we could reduce the usage of firewood, and the thermal and air environment would be improved.

Keywords Firewood · Energy · Thermal environment · Regional difference · Nepal

33.1 Introduction

Firewood is the most primitive natural energy resource for human beings. Possibly, people started to burn firewood for cooking and heating after discovering how to create fire. In most developing countries, firewood is used as the main energy source. Especially in the rural areas, people are fully dependent on firewood without being able to access any artificial energy. Firewood could become one of the sustainable energy sources which is available locally, if managed properly. However, a shortage of firewood is increasingly reported in different parts of the world due to deforestation and population growth. It is also anticipated that firewood consumption will increase after oil supplies expire. Thus, it is essential to know the firewood consumption in the different regions, climates, and seasons for sustainable development

H. B. Rijal (✉)
Department of Restoration Ecology and Built Environment, Tokyo City University
3-3-1 Ushikubo-nishi, Tsuzuki-ku, Yokohama, 224-8551 Japan
e-mail: rijal@tcu.ac.jp

© Springer Nature Singapore Pte Ltd. 2018
T. Kubota et al. (eds.), *Sustainable Houses and Living in the Hot-Humid Climates of Asia*, https://doi.org/10.1007/978-981-10-8465-2_33

of firewood by better forest management. If we establish firewood combustion systems in buildings, we can use waste wood including old building materials. In this chapter, we will explain about the firewood consumption by relating this to the thermal environment in traditional houses in Nepal.

In rural areas of Nepal, firewood is the main source of energy for cooking, heating, and lighting. In the past, firewood was abundant and it could be easily collected in the vicinity of the villages. However, collecting firewood is becoming more difficult due to overuse and the shrinking of forest areas. Therefore, the reduction of firewood consumption is presently one of the most important issues in Nepal, with restrictions imposed on quantities of wood gathered. Another important issue is the improvement of the thermal environment and of air quality. In Nepal, firewood is burned in an open fireplace, is very energy inefficient, and creates high indoor air temperatures and low indoor air quality (Fig. 33.1). In Nepal, there have been a few studies conducted on firewood consumption and indoor air quality [1–4], but no such research has been found on the thermal environment in relation to firewood consumption. In this research, we focus on energy consumption and the thermal environment in terms of firewood usage. We try to quantify these elements by considering the regional and seasonal differences of Nepal.

In order to evaluate and improve the energy consumption and thermal environment in the traditional houses of five districts of Nepal, the firewood consumption rates and air temperatures were measured and investigated in winter and summer. The objectives of this research are (1) to estimate the energy consumption, such as firewood consumption for cooking and heating, and (2) to measure the relationship between firewood consumption and the thermal environment, such as indoor and outdoor temperature differences, and the vertical temperature differences in the kitchen [5].

Fig. 33.1 Indoor air pollution in Dhading district

33.2   Method

33.2.1   The Study Areas

The research is conducted in five districts of Nepal (see Chap. 6). The climatic zone of Banke is subtropical; Bhaktapur, Dhading, and Kaski are temperate; and Solukhumbu has a cool climate [6]. They were chosen according to the altitude, climate, topography, ethnicity, energy, and housing conditions in Nepal. Bhaktapur is an urban area and the other districts are rural. In the rural areas, agriculture and livestock are the main occupations and firewood is used for cooking and heating. In Bhaktapur, straw, twigs, and rice husks are also used as cooking fuel. Firewood was burned in an open-hearth stove and a semi-open stove. In Banke, people also burned firewood in living rooms and semipublic spaces for heating. The kitchen room was used for both dining and living. In Dhading and Kaski, the kitchen was also used for sleeping. In Solukhumbu, because of the cold climate, people spent most of their time in the kitchen.

33.2.2   Measurements

Investigations were conducted in winter (December 2000 to January 2001) and summer (April to May 2000). In the 5 districts, the daily firewood consumption rate was measured in 133 households for 34 days (in Dhading 3 households for 486 days from 18 August 1999 to 27 November 2000) (Fig. 33.2). The short

Fig. 33.2 Measurement of firewood [5]

measurement period was 3–7 days, and the long measurement period was 468 days. The amount of firewood consumed in a day was determined by first allocating each family a specific amount of firewood each day, which was weighed using a scale. In the morning of the next day, the amount of firewood leftover was weighed and recorded, and the families were then given more firewood. This process was repeated for 468 consecutive days. Family size was divided into 3 types: (a) large (more than 9 persons), (b) medium (5–8 persons), and (c) small (1–4 persons). The number of family members and guests was recorded. Children below 5 years of age were counted as 0.5 person, because of their lower amount of food consumption. The weight of firewood included twigs, crop residues, bamboo, straw, and rice husks. Only well-dried firewood was measured. Outdoor (above 1.5 m from ground level) and kitchen air temperatures (above 0.6 m from floor level) were measured in 5 min intervals and recorded by a data logger. Measured data were calibrated.

33.3 Results

Average firewood consumption is shown in Table 33.1. Firewood consumption per capita per day is calculated by dividing the total firewood consumption by the total number of consumers. Firewood consumption is classified as cooking and heating. Firewood for cooking per year was estimated by taking the firewood consumption per capita per day in the summer and multiplying it by 365 (days). For heating, the amount was determined by subtracting the amount of firewood used per capita per day in the winter from the amount of firewood per capita per day in the summer and multiplying it by 90 (heating days), the length of the heating season being similar in all regions. In Dhading, summer and winter firewood consumption was 1.4 kg/capita/day; therefore, we could not estimate the firewood for heating. Total firewood consumption is the sum of cooking and heating firewood.

33.3.1 The Regional and Seasonal Difference of Firewood Consumption

Figures 33.3 and 33.4 show the daily firewood consumption in short and long measurement periods. The average firewood consumption rates were 0.6–4.6 and 0.8–2.6 kg/capita/day in winter and summer (Table 33.1). Firewood consumption was higher in winter because people use extra firewood for heating. Firewood consumption was highly correlated to the outdoor air temperature during the cooking and heating period (r = 0.87, Fig. 33.5). Because the cool climate (Solukhumbu) had the lowest outdoor air temperature, the rate of firewood consumption in that area was the highest in both winter and summer. In winter, the outdoor air temperatures are similar in the subtropical (Banke) and temperate climates (Dhading, Kaski);

Table 33.1 Average firewood consumption in each district (winter/summer) [5]

Items (winter/summer)	Ba	Bh++	Dh	Dh[a]	Ka	So	Total
No. of households in survey	41/2	4	39/3	3	28/2	21/1	133
No. of measurement days	7/4	5	7/4	468	7/4	6/3	511
Total no. of consumers	1890/ 100	143	1521/ 93	9240	1228/ 67	636/ 32	14,857
Family size	6.6/ 12.5	8.4	5.7/ 7.8	6.6	6.4/ 8.4	5.0/ 10.5	6.5
Total firewood consumption [kg]	5464/ 100	92	2139/ 132	13,482	1744/ 51	2933/ 83	25,854
Firewood consumption [kg/household/day]	19.2/ 12.3	5.4	8.0/ 11.0	9.6	9.1/ 6.4	23.3/ 27.7	12.4
Firewood consumption [kg/capita/day]	2.9/ 1.0	0.6	1.4/ 1.4	1.5	1.4/ 0.8	4.6/ 2.6	2.1
Firewood for cooking+ [kg/capita/year]	365	235	511	–	292	949	470
Firewood for heating+ [kg/capita/year]	170	–	–	–	56	181	68
Total firewood consumption+ [kg/capita/year]	535	235	511	561[b]	348	1130	553

Ba Banke, *Bh* Bhaktapur, *Dh* Dhading, *Ka* Kaski, *So* Solukhumbu, +: estimated value, ++: winter data only (mainly straw burn)
[a]Long measurement period
[b]Accumulated value in a year

Fig. 33.3 Daily firewood consumption per capita per day in winter [5]

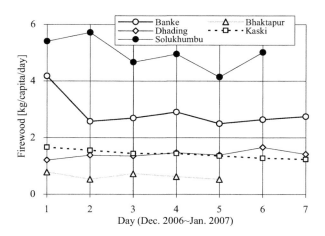

however, the firewood consumption rate was 1.5 kg/capita/day higher in the subtropical climate. The reason for this is that people burn firewood not only in the kitchen but also in the living room and the front yard for winter heating, while in summer, firewood consumption in the subtropical climate was 0.4 kg less than that of the temperate climate (Dhading). This is because the maximum outdoor air

Fig. 33.4 Monthly mean firewood consumption per person per day in Dhading [5]

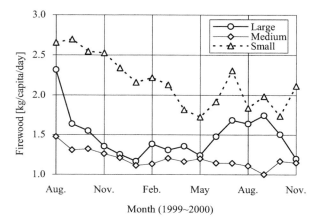

Month (1999~2000)

Fig. 33.5 Relation between firewood consumption and outdoor air temperature during the cooking and heating period [5]

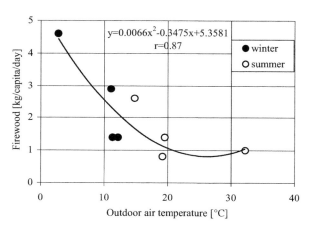

temperature in subtropical climate reaches 42 °C in the summer; therefore, people put out their fires soon after cooking. The climates, topographies, and ethnic groups of Dhading and Kaski are similar; the firewood consumption in these areas was 1.4 kg/capita/day in winter. But in summer firewood consumption of Kaski was 0.6 kg/capita/day lower than Dhading. This could be due to the fact that Kaski has insufficient firewood compared to Dhading and its people might have used their firewood more sparingly. The firewood consumption of Bhaktapur was lowest because people often used other forms of fuel for cooking only, including straw, rice husks, twigs, and devices such as electric heaters and kerosene stoves. Firewood was used 2.1, 5.2, and 5.2 times more for cooking than for heating in the Banke, Kaski, and Solukhumbu areas, respectively. Total firewood consumption rates were 235–1130 kg/capita/year (Table 33.1). In Dhading, the difference between measured and estimated firewood consumption was 50 kg/capita/year. It can be said that measured and estimated data are well matched.

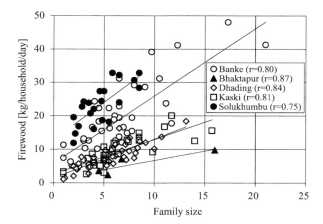

Fig. 33.6 Relation between family size and daily mean firewood consumption in winter [5]

The results showed that firewood consumption has regional and seasonal differences. The regional difference was 4.0 kg/capita/day in winter and 1.8 kg/capita/day in summer. The maximum seasonal difference was 2.0 kg in the cool climate (Solukhumbu). The temperate climate used the least amount of firewood followed by the subtropical climate using more and the cold climate using the most. It was unexpected to find that the temperate climate used less firewood than the subtropical climate. This could be due to the fact the subtropical climate has a greater annual temperature range. Over the winter, the people in the subtropical climate, who are less resilient to cooler temperature, unnecessarily use more firewood than the people in the temperate climate. Paradoxically, they also use more firewood in the summer in order to build fires for cooking more rapidly. After quickly cooking their food, they would extinguish the fires, wasting some of the firewood in the process.

33.3.2 Relation Between Firewood Combustion and Thermal Environment

Figures 33.6 and 33.7 show the relation between family size and firewood consumption in the short and long measurement period. Firewood consumption was highly correlated to the family size (r = 0.75–0.87). Because of the great amount of food cooked in large families, firewood consumption per day was higher than that of small- or medium-sized families. However, the volume and openings of the kitchens in each climatic zone were similar, irrespective of family size. This meant that the large family kitchen created higher indoor air temperatures than did small- or medium-sized family kitchens and caused greater discomfort.

Because of lesser firewood consumption, the indoor and outdoor temperature differences were smaller in summer than in winter (Table 33.2). In winter, because of lower firewood consumption, the indoor and outdoor temperature differences and the

Fig. 33.7 Relation between family size and daily firewood consumption in Dhading. Each point represents the monthly mean value [5]

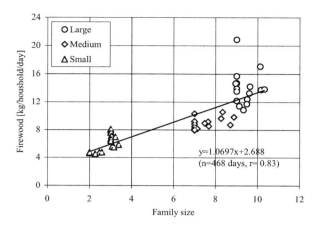

Table 33.2 Thermal environment in the kitchen during cooking and heating [5]

Items	Season	Banke	Bhaktapur	Dhading	Kaski	Solukhumbu
T_{out} [°C]	Winter	11.1	9.2	11.3	12.2	2.8
	Summer	32.2	21.0	19.5	19.2	14.8
T_{in} [°C]	Sample days[a]	8/6	5/5	6/4	7/7	5/5
	Winter	14.9	11.1	19.1	16.3	6.7
	Summer	32.2	23.6	25.6	22.2	17.8
T_{in}-T_{out} [K]	Winter	3.8	1.9	7.8	4.2	4.0
	Summer	0.0	2.6	6.0	3.0	3.0
T_c-T_f [K]	Sample days	5	2	3	4	2
	Winter	4.3	2.9	7.1	5.8	5.0

T_{out} and T_{in} Outdoor and indoor air temp., T_c and T_f Air temp. around ceiling and floor level
[a]Winter/summer; winter, 7:00–10:00 and 17:00–20:00; summer, 6:00–9:00 and 18:00–21:00

vertical temperature differences of Bhaktapur were the smallest. However, Dhading consumed less firewood than Banke and Solukhumbu; the temperature differences were highest in winter and summer. This might be due to the volume of the kitchen. Dhading was lesser in volume compared to Banke and Solukhumbu. In Dhading, the vertical temperature difference was 6.5 °C in winter, which is similar to present research [7]. The vertical temperature difference was greater than the ASHRAE ST-55 thermal comfort standard (3 °C).

33.3.3 Discussion

In the present research, the firewood consumption rates were 0.6–4.6 kg/capita/day. In Bhogteni, the firewood consumption rate was 0.96–1.75 kg/capita/day [1]. Bhogteni is similar to Dhading and Kaski in climate, topography, and ethnic group, and as a result it is close to 1.4 kg/capita/day. In Bhogteni the firewood

Fig. 33.8 Proposed and
installed brick-iron stove [7]

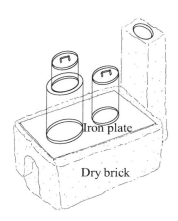

Iron plate

Dry brick

consumption was 0.95 m³/capita/year (570 kg), which is similar to Dhading (561 kg). In high elevations (2700–3900 m) and low elevations (1300 m), firewood consumption was 8.2 and 2.8 kg/capita/day [2]. Compared to high elevations, Solukhumbu had 3.6 kg/capita/day lower firewood consumption. The reason for this could be that it is located at 100–1300 m lower than the areas of high elevation studied by Davidson et al. [1], and consequently outdoor air temperature is 10 °C higher. As a result Solukhumbu needed less firewood than places at higher elevations. Furthermore, compared to lower elevations, Dhading and Kaski had a 1.4 kg lower firewood consumption. Compared with existing research, the firewood consumption rate was found to be similar in areas with similar climates. It can be assumed that the condition of the indoor thermal environment in Bhogteni and lower elevations would be similar to Dhading and Kaski, while the temperature distribution in higher elevations would be higher than Solukhumbu.

Firewood was burned in open fireplaces in excessively ventilated rooms, which accelerated the firewood combustion speed. As a result, more firewood for cooking and heating was required. Because of inadequate firewood usage, low thermal efficiency and high temperature distribution were observed. In winter, the heat loss was greater because cold air entered inside and hot air went outside from the opening of the kitchen. In summer, there was no system for expelling the hot indoor air, and as a result, the temperature rose substantially, which created an uncomfortable indoor environment. Therefore, a waste of energy in winter and an uncomfortable thermal environment in summer were observed. If use were made of thermal storage in the walls as well as airtight openings and improvements in fireplaces in winter, we could reduce the usage of firewood, and the thermal and air environment would be improved.

For the purposes of reducing cost and firewood consumption, strengthening the pot hole, and improving the indoor air quality, a brick-iron stove was proposed and installed in the 44 houses in the Salle village, Dhading District (Fig. 33.8, $30 US, January 2003) [7].

33.4 Conclusions

Firewood consumption and air temperature were investigated in winter and summer in traditional houses in the Banke, Bhaktapur, Dhading, Kaski, and Solukhumbu districts of Nepal. The results are as follows:

1. The firewood consumption rate was 235–1130 kg/capita/year. The results showed that the temperate climate used less firewood than the subtropical climate. Regional differences depended on the method of firewood usage. If better firewood usage methods were introduced, we could reduce the regional differences of firewood consumption.
2. The indoor and outdoor temperature difference (7.8 °C), the vertical temperature difference (7.1 °C), and maximum indoor air temperature (42 °C) were most significant in the kitchen. The results demonstrated a waste of energy in winter and an uncomfortable thermal environment in summer. If use were made of thermal storage in the walls as well as airtight openings and improvements in fireplaces in winter, we could reduce the usage of firewood, and the thermal and air environment would be improved.

Acknowledgments   We would like to give thanks to Prof. Harunori Yoshida and Prof. Noriko Umemiya for their research guidance and valuable advice, to the investigated households for their cooperation, and to our families and friends for their support. This research is supported by the Kurata Memorial Hitachi Science and Technology Foundation.

References

1. Davidson CI, Lin SF, Osborn JF (1986) Indoor and outdoor air pollution in the Himalayas. Environ Sci Technol 20(6):561–567
2. Fox J (1984) Firewood consumption in a Nepali village. Environ Manag 8(3):243–250
3. Panday RK (1995) Development disorders in the Himalayan heights challenges and strategies for environment and development altitude geography. Ratna Pustak Bhandar, Baghbazar, Kathmandu Nepal, pp 53, 290
4. Pandey MR, Neupane RP, Gautam A, Shrestha IB (1990) The effectiveness of smokeless stoves in reducing indoor air pollution in a rural hill region of Nepal. Mt Res Dev 10(4):313–320
5. Rijal HB, Yoshida H (2002) Investigation and evaluation of firewood consumption in traditional houses in Nepal. Proceedings of the 9th International conference on indoor air quality and climate (Monterey), vol. 4, pp 1000–1005
6. Rijal HB, Yoshida H, Umemiya N (2001) Investigation of winter thermal environment in traditional vernacular houses in a mountain area of Nepal. J Archit Plann Environ Eng AIJ 546:37–44 (in Japanese with English abstract)
7. Rijal HB, Yoshida H, Miyazaki T, Uchiyama I (2005), Indoor air pollution from firewood combustion in traditional houses of Nepal. Proceedings of the 10th International Conference on Indoor Air Quality and Climate (Beijing), 4–9 September, pp 3625–3629

Chapter 34
Thermal Environment and Energy Use of Houses in Bangkok, Thailand

Weijun Gao, Suapphong Kritsanawonghong, Pawinee Iamtrakul, and Chanachok Pratchayawutthirat

Abstract The Bangkok Metropolitan Region is the largest city of Thailand. The residential sector electricity consumption has risen steadily for the past year. This study has a questionnaire collected in Bangkok to investigate household electricity consumption behavior and field measurement on the residential indoor and outdoor thermal environment related to electricity consumption behavior.

The results of analysis indicated that the influence of climatic factor has resulted in higher temperature than thermal comfort standard. The findings also revealed that Thai people have to adapt their own lifestyle to make good use of fan and air-conditioner.

Keywords Thailand · Bangkok · Residential building · Energy consumption · Questionnaire survey

34.1 Introduction

The growth of electricity demand in Thailand is strongly influenced by the rapid increasing population and improving living standard (as measured by per capita gross domestic product, GDP). Recent data indicates that electricity consumption in Thailand is rising at around 4.5% per year, and this rate of forecast continues to the medium term with annual increases of 4–7.5% up to 2016 [1]. It is anticipated that

W. Gao (✉)
Faculty of Environmental Engineering, Department of Architecture, The University of Kitakyushu, Fukuoka, Japan
e-mail: weijun@kitakyu-u.ac.jp

S. Kritsanawonghong
Business Development, Renewable Energy Division, Prime Road Group, Bangkok, Thailand

P. Iamtrakul
Faculty of Agriculture and Planning, Thammasat University, Bangkok, Thailand

C. Pratchayawutthirat
The Eco Knowledge Program, Univentures Public Company Limited, Bangkok, Thailand

© Springer Nature Singapore Pte Ltd. 2018
T. Kubota et al. (eds.), *Sustainable Houses and Living in the Hot-Humid Climates of Asia*, https://doi.org/10.1007/978-981-10-8465-2_34

345

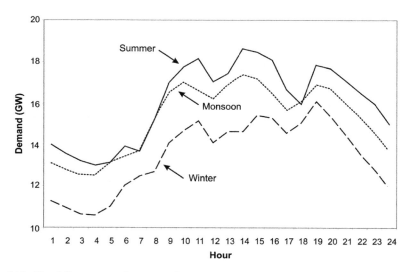

Fig. 34.1 The daily consumption pattern in the three seasons (Source: [2])

this and longer-term growth will be affected by the changes in weather patterns brought by climate change.

Thailand features a hot, humid climate with the average temperature being 31 °C within an annual range of 22–39 °C. The seasonality can be seen clearly in Fig. 34.1 which shows the daily consumption pattern in the three seasons: winter, summer, and monsoon. The load pattern is closely related to the daily temperature profile, power demand starting to increase around 8 am up to the peak around 2 pm before falling back and then picking up again in the evening. With a hot, humid climate, these differences reflect the hotter summer temperatures that lead people to spend more time indoors increasing in-house demand for air-conditioning and refrigerator.

34.2 Questionnaire Survey on Indoor Environment and Energy Consumption Behavior in Bangkok

34.2.1 Outline of the Survey

The survey was conducted during November 2011 among 100 selected households. The type of building involved in the survey ranges from low-rise to high-rise housing.

Table 34.1 shows the detail of the questionnaire which consists of five parts: family characteristic, building characteristic, housing equipment, lifestyles related to pattern of electricity consumption, and effect for energy saving and environment.

Table 34.1 Content of the questionnaire survey

Item	Content
Family characteristic	Number of residents, income, electricity consumption information, water consumption information
Building characteristic	Living house total area, construction year, structure, architecture area, type of building, renovation
Housing equipment	Air-condition control and ventilation
Lifestyle related to electricity	Time to open window, time to turn on air-condition, bath and cooking activity
Effect for energy saving and environment	Concerned about environmental program, management of waste and garbage, using the advance technology, and how to reduce the energy

Fig. 34.2 Construction year of the building

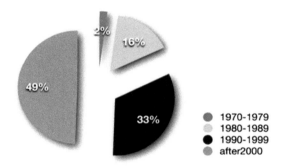

34.2.2   Results of the Survey

34.2.2.1   Building Characteristic

Construction Year and Structure
2% of surveyed buildings were constructed during 1970–1979, 16% during 1980–1989, 33% during 1990–1999, and 49% after 2000, which is illustrated in Fig. 34.2. It can be seen that most of the residential building was constructed after the year of 2000.

10% of surveyed buildings were built of wood, brick, and concrete and 90% of buildings were built of reinforced concrete.

Family Characteristic
The number of family member will influence the space and energy utilization. Figure 34.3 shows the trend of the family characteristics from the survey. There are 1% of families with 6 members, 6% with 5 members, 22% with 4 member, 35%

Fig. 34.3 Number of residents

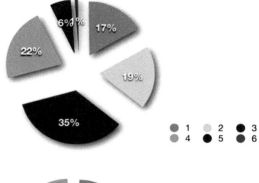

Fig. 34.4 Average income (Baht/month)

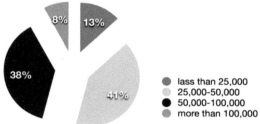

with 3 members, 19% with 2 members, and 17% with 1 member. The average number of people per family is 2.84persons/household.

Average Income (Baht/Month)

Figure 34.4 shows the average income of the people in Bangkok, which ranges between 25,000 and 75,000 baht. There are 13% of families with less than 25,000 baht, 41% between 25,000 and 50,000, 38% between 50,000 and 100,000, and 8% with more than 100,000.

34.2.2.2 Housing Equipment

Cooling Equipment

In the survey, most of the households have air-conditioners. 8% of the total households only used electric fan. Also from the survey, 92% of families used electric fan on daytime and air-conditioner on nighttime. And about 11% of the total households have used air-conditioner all day. Most of households in condominium use air-conditioner all day.

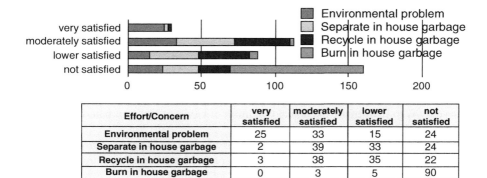

Effort/Concern	very satisfied	moderately satisfied	lower satisfied	not satisfied
Environmental problem	25	33	15	24
Separate in house garbage	2	39	33	24
Recycle in house garbage	3	38	35	22
Burn in house garbage	0	3	5	90

Fig. 34.5   Concerns about environmental problem issue

Equipment of Hot Water Supply
About 92% of the households in the survey use hot water for shower and 100% of hot water was produced by electricity. The average duration time spent in shower room in the survey is around 18.5 min.

34.2.2.3   Effort for Energy Saving and Environment

Behavior and Attitudes
In the survey, most of Thai families are only moderately concerned about environmental problem issue. Most Thai families are not interested in the garbage separation, which can be shown in the survey result of moderately satisfied, lowly satisfied, and not satisfied. This portion can be taken into account for 39%, 33%, and 24%, respectively. Also recycling in households is not much of a concern in which only 3% represent those who usually separate the garbage, 73% those who do so in a few times, and 22% those who never mind to recycle. And a few Thai families (about 5%) burn the house garbage (Fig. 34.5).

Advance Technology
In this questionnaire survey, 96% of Bangkok residents does not pay consideration on using advance technology. Only 4% use the technology to save energy, e.g., LED lamp.

How to Reduce the Energy
Only 9% of the population in Bangkok do not care about energy saving. Most people attempt to turn off the lights every time after use and also try to use less hot water. Some of Thai people stop using air-conditioner and decrease bath time.

34.3   Field Measurement on Indoor and Outdoor Thermal Environment in Bangkok

In order to estimate the future trend of residential energy consumption and indoor environment requirement in Thailand, it is necessary to know the situation of the usage of household equipment and the indoor thermal conditions. The purpose of this study is to investigate the actual condition of environment in urban dwelling in Bangkok, Thailand.

In this survey we investigated houses in the Bangkok Metropolitan Region by short-term investigation during November 2011 (winter) and long-term investigation during November 2011 to April 2012 (winter and summer).

The houses under investigation were located in the urban areas of Bangkok metropolis.

34.3.1   Short-Term Investigation in October to November 2011

Table 34.2 shows the number of measured houses, cooling system, and outdoor condition during investigation (Fig. 34.6). The questionnaire survey together with the measurements of indoor temperature and humidity was done in November 2011 for Bangkok metropolis.

As for construction year of residence, two buildings were built before the year 2000 which are House No. 2 and House No. 5; the other three buildings were built after the year 2000. The structure of the building was built of reinforced concrete (Fig. 34.7).

Table 34.2 The number of measured houses

Location of housing	Type of housing	Survey period	Cooling system	Average outdoor conditions during investigation
1	Duplex detached	2011/10	Individual space cooling	30.5 °C, 65.9%RH
2	Side detached	2011/11	Individual space cooling	31.2 °C, 52.1%RH
3	Mid-rise apartment	2011/11	Individual space cooling	29.3 °C, 56.4%RH
4	Duplex detached	2011/11	Individual space cooling	30.7 °C, 52.3%RH
5	Side detached	2011/11	Individual space cooling	29.3 °C, 46.1%RH

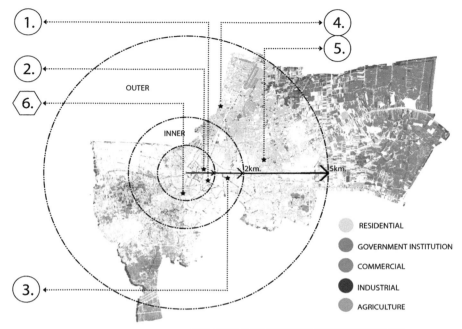

RISK AREAS IN BANGKOK TO USE HIGH ELECTRICITY

Fig. 34.6 Number of residents

34.3.2 The Time Variation of Temperature and Humidity in the Measured Houses

The measurement period is from 16 October 2011 to 26 November 2011. Due to the lack of meters, each house will be measured for 10 days.

In the field studies, occupants use individual space cooling system to low down their house's temperature in Bangkok climate.

It is found that the indoor thermal condition was relatively uncomfortable. The relative humidity and temperature were high inside of houses. From five houses investigated, two houses have no air-conditioner in the living room, and indoor thermal environments are not comfortable. On the other hand, in houses with air-conditioner in the living room, the residents turn on the air-conditioner when they do some activities indoor. (Indoor thermal conditions are relatively stable and comfort). From five houses, the bedroom are equipped with air-conditioner. Generally, the residents operate air-conditioner on sleeping time. The indoor thermal condition during cold and hot seasons is affected by the outdoor condition. The indoor temperature during cold season remains at relatively stable level; therefore the residents use only fan for the whole day. During hot season, the indoor temperature and humidity were higher than the comfort zone. In that case, occupants

HOUSE NO.1   HOUSE NO.2   HOUSE NO.3

HOUSE NO.4   HOUSE NO.5

Fig. 34.7 Investigated houses

use cooling system in their bedroom on evening time instead of staying in the living room, where there is no air-conditioner in some houses.

In House No. 1 (Fig. 34.8), air-conditioning was not used during the measurement period for the living room and bedroom (only last 3 days, in bedroom used air-conditioner). The temperature of the living room and the bedroom is remarkably stable at around 28 °C during the measurement period. The relative humidity is very high, ranging from 50 to 84%RH in living room, and occupants would prefer opening their windows instead of air-conditioner use because the indoor temperature and humidity are almost equal to the outdoor conditions.

The other houses measured from November have shown the same tendency for the variation of temperature. As an example, the time variation of temperature in House No. 4 is shown in Fig. 34.9.

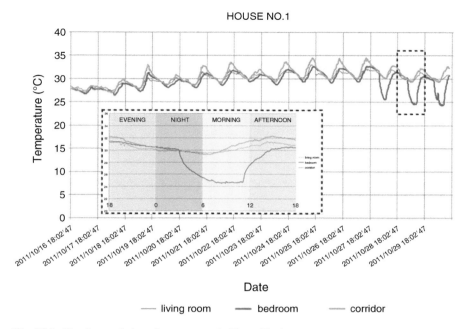

Fig. 34.8 The time variation of temperature in House No.1

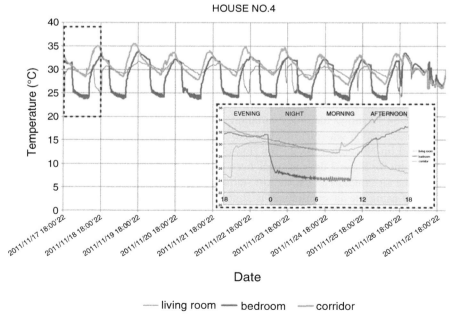

Fig. 34.9 The time variation of temperature in House No.4

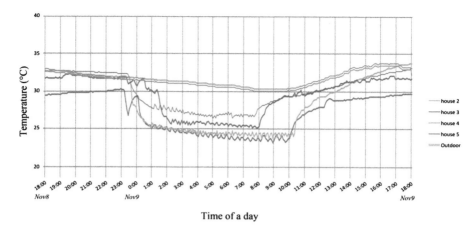

Fig. 34.10 The temperature variation on November 8–9(House 2,3) and on November 19–20 (House 4,5)

Table 34.3 The number of subject house

Location of housing	Type of housing	Survey period	Cooling system	Average outdoor conditions during investigation
6	Side detached	2011/ 1–2012/06	Individual space cooling	Dec 2011–Feb 2012 = 28.6 °C, 65.7%RH
				Mar 2012–Jun 2012 = 30.9 °C, 65.6%RH

Air-conditioning was used in the living room between 15:00 and 18:00 when occupants may have activities such as eating, watching television, and entertaining guests. From 23:00, air-conditioning would be used for the whole night. The indoor temperature would be distributed in the ranges of 24–34 °C.

Figure 34.10 shows the 1-day variation of indoor temperature in the bedroom for four houses, which was measured on November 8–9 for House No. 2 and House No. 3 and on November 19–29 for House No. 4 and No. 5. The average outdoor temperature for those 2 days also has been shown in the figure. From the figure, we can see the air-conditioner use almost starts from sleeping time in the bedroom. The pattern of using the air-conditioner is almost the same for the four houses.

34.3.3 Long-Term Investigation on Winter and Summer

Table 34.3 shows the house (Fig. 34.11) for long-term investigation. The questionnaire survey together with the measurements of indoor and outdoor temperature and humidity was done from December 2011 to June 2012 for Bangkok metropolis.

Fig. 34.11 Measured houses (House No.6)

As for construction year of residence, this building was built in 2007. The structure of the building was built of reinforced concrete.

From long-term investigation, it can be divided into three seasons, which are winter (first of December 2011 to the end of February 2012) (in fact, it is not real winter as most people usually think), summer (first of March 2012 to the end of June 2012), and monsoon (raining season).

House No. 6 only used electric fan in winter season. In the summer season, House No. 6 uses electric fan on daytime and air-conditioner on nighttime, which start from 11 pm to morning. Household in the survey used hot-water for shower activity and the average of duration time in bathroom is about 20 min.

34.3.4 Temperature Variation of Cooling System Use in Season

Figure 34.12 shows the average, maximum and minimum value of the week temperatures from December 2011 to June 2012, which living room temperature were tilted up follow the outdoor conditions. The bedroom temperature was slightly constant during the measurement period because occupants used the methods such as opening some windows for ventilation in winter time or turning on an air-conditioning in summer time.

In this study, occupants use individual space cooling system to low down their house's temperature in Bangkok metropolis. It is found that the indoor thermal condition was relatively uncomfortable. The indoor thermal condition during cold (winter) and hot (summer) seasons is affected by the outdoor condition.

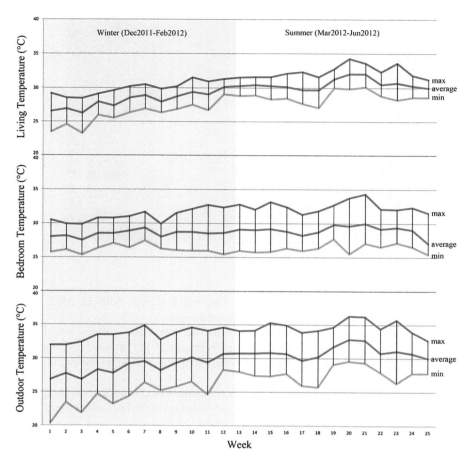

Fig. 34.12 The average temperature per week during December 2011 till June 2012

From Fig. 34.13, the indoor temperature during cold season remains at relatively stable level. Therefore, the residents only use fan or open windows for the whole day to adjust the room temperature.

During the hot season, the indoor temperature and humidity were higher than the comfort zone. In that case, occupants use cooling system in their bedroom in the evening time.

On the top of Fig. 34.13, the general lifestyle observed in Thai houses has also been shown. In the hot season, the residents have a tendency to turn on the air-conditioner before sleep and turn it off when they go to work/school. In some other season, the Thailand people have a custom to make good use of fan to get a better environment.

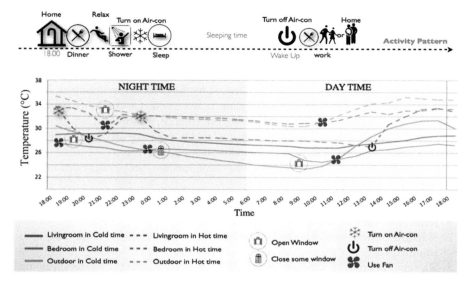

Fig. 34.13 Summary of seasonal temperature variations related with residential behavior and use of cooling system during November 2011–December 2012

34.4 Conclusion

Energy consumption, indoor thermal environment, and lifestyle are very important data required for basic understanding and to draw future consideration about energy saving policy in the tropical climate country.

The conclusion for questionnaire survey on residential indoor environment and electricity consumption in Bangkok can be made as follows:

1. About half of the buildings that have been investigated were constructed after 2000. 10% of surveyed buildings were built of wood, brick, and concrete and 90% of buildings were built of reinforced concrete.
2. 57% of households have three to four family members, which consist of parents and children. The average number of people per family is 2.84 persons/household.
3. Most of households in the survey have at least one air-conditioner. About 92% households use electric fan on daytime and air-conditioner on nighttime. About 11% of the total households used air-conditioner all day.
4. Most of the surveyed families care about environmental problem at some extent. But a few households in the questionnaire use the advance energy-saving technology.

In the field studies, occupants use individual space cooling system to low down their house's temperature in hot season. It is found that the indoor thermal condition was relatively uncomfortable. The humidity and temperature were high inside of houses. From the measurement of five houses, we found the bedroom in many houses was equipped with an air-conditioner which was turn on in the sleeping time generally.

The indoor temperature during winter season remains at relatively stable level, and the residents use only fan for the whole day.

During hot season, the indoor temperature and humidity were higher than the comfort zone, and occupants use cooling system mainly in the bedroom during sleeping time.

The indoor environment is greatly influenced by the outdoor environment. It may mean the insulation property of the building is not good. On the other hand, the shielding of solar radiation also shows low performance. However it is possible to improve the indoor thermal environment quality by reducing the amount of solar radiation through improvement of the window component.

Specifically, the following energy conservation method can be suggested in Thai housing with hot climate.

1. In this climate air-conditioning will always be required but can be greatly reduced if building design minimizes overheating.
2. Window overhangs (designed for this latitude) or operable sunshades (extend in summer, retract in winter) can reduce or eliminate air-conditioning.
3. Raising the indoor comfort temperature limit will reduce air-conditioning energy consumption (raise thermostat cooling set point).
4. Minimize or eliminate west-facing glazing to reduce summer and fall afternoon heat gain.
5. Orient most of the glass to the north, shaded by vertical fins, in very hot climates, if there are essentially no passive solar needs.
6. Traditional homes in hot humid climates used lightweight construction with operable walls and shaded outdoor porches, raised above ground.
7. High-efficiency air-conditioner (at least Energy Star) should prove cost-effective.
8. Use plant materials (ivy, bushes, trees) especially on the west to shade the structure (if summer rains support native plant growth).
9. Keep the building small (right-sized) because excessive floor area wastes heating and cooling energy.
10. A whole-house fan or natural ventilation can store nighttime "coolth" in high mass interior surfaces, thus reducing or eliminating air-conditioning.
11. Use light-colored building materials and cool roofs (with high emissivity) to minimize conducted heat gain.
12. On hot days ceiling fans or indoor air motion can make it seem cooler by at least 5 degrees F (2.8C); thus less air-conditioning is needed.

References

1. Energy Policy and Planning Office (2012) Summary of Thailand power development plan 2012–2030. Ministry of Energy
2. Parkpoom GHS (2008) Using weather sensitivity to forecast Thailand's electricity demand. Int Energy J 9:237–242

Chapter 35
Energy Consumption of Residential Buildings in China

Hiroshi Yoshino and Qingyuan Zhang

Abstract Based on Chinese statistics, unit energy consumption for 277 Chinese cities was clarified over the period of 2002–2012. The meter reading of energy consumption showed that annual energy consumption was recorded from 10 to 30 GJ, excluding the energy consumption by the district heating. Roughly one third of the energy consumption was used for cooking, and another one third was used for space heating, cooling, and hot water supply, and the final one third was used for the other purposes. The results of the detailed measurement indicated that in the eight apartments without district heating, the annual energy consumption varied from 8 to 32 GJ. Also, it was observed that there are large variations in the structure of end usage. That implies that there is room for further improvement in saving energy.

Keywords China · Energy consumption · Residential building · Statistical analysis · Measurement

35.1 Introduction

Along with the rapid economic development in recent years in China, energy consumption has also increased significantly. Over the last 12 years, from 2000 to 2012, the GDP has tripled and the energy consumption is doubled. Residential energy consumption accounts for 11% to the total in 2010, and this proportion is expected to further increase because it accounts for 30–40% in other developed countries.

China is divided into five regions in accordance with the climatic conditions, namely, severely cold region, cold region, hot-summer-cold-winter region,

H. Yoshino (✉)
Institute of Liberal Arts and Science, Tohoku University, Sendai, Japan
e-mail: yoshino@sabine.pln.archi.tohoku.ac.jp

Q. Zhang
Institute of Urban Innovation, Yokohama National University, Yokohama, Japan

© Springer Nature Singapore Pte Ltd. 2018
T. Kubota et al. (eds.), *Sustainable Houses and Living in the Hot-Humid Climates of Asia*, https://doi.org/10.1007/978-981-10-8465-2_35

hot-summer-warm-winter region, and mild region. In the cities with cold climate conditions, district heating system is adopted so that the room temperature remains at a comfortable level during the winter. On the other hand, in the southern area, either individual heating equipment is used or no heating equipment is utilized. Thus, a very large difference in terms of energy consumption for space heating between the two regions is found. It does not make sense to discuss the residential energy consumption in China using one lump.

In addition, energy consumption between rural and urban houses is significantly different. Biomass is normally used as a fuel in rural areas. Because of extremely poor efficiency of biomass energy, energy consumption in rural area is about three times as much as that of urban areas.

In this chapter, a brief review of China's residential energy consumption is firstly given based on the analysis from statistical data published in China. Then the results of investigation on energy consumption by meter reading and questionnaire as well as indoor temperature measurement by liquid crystal thermometers are reported. Finally, detailed measurement results of energy consumption using power meter and other instruments are described.

When talking about energy consumption, it may be either based on primary energy or secondary energy. In this chapter, the secondary energy is used.

35.2 Analysis of Residential Energy Consumption in Urban Area on the Basis of Statistical Data

In this section, based on Chinese statistics [1–3], the annual energy consumption of an apartment in 277 cities of China is clarified during the period of 2002–2012.

35.2.1 Energy Consumption Per Household by Energy Source

The energy consumption per household is calculated by multiplying the consumption of each type of energy source per capita and the number of family members. The consumption of electricity, coal gas and natural gas, and LPG per person of China Urban Statistical Yearbook [1] is used here. Data for the family members of each household can be found in the China Statistical Yearbook [2], in which the average of family members in China has decreased from 3.43 to 3.14 in the period of 2002–2012.

The total secondary energy consumed yearly by a household is called the unit energy consumption (UEC) in this chapter. The UEC is calculated by summing up all of the energies mentioned above.

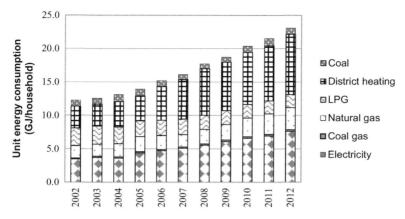

Fig. 35.1 Change of annual energy consumption per household

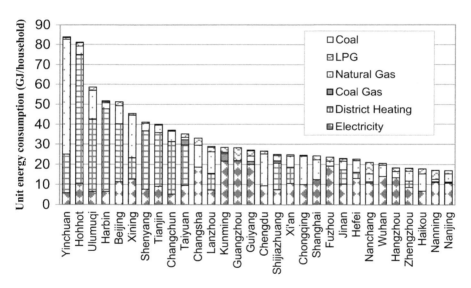

Fig. 35.2 Annual energy consumption per household in capital cities of each province in 2012

Figure 35.1 shows the total annual energy consumption per household by summing up all kinds of the energy. Energy consumption for space heating supplied by district heating system is included. The method of calculation is described in the next section. The annual energy consumption increased from 12.3 GJ/household to 23.1 GJ/household in the period of 2002–2012, with the improvement of living standard in urban areas in China. Energy for district heating and electricity accounts for large proportion among various kinds of energy.

The annual energy consumption per household in the capital cities of each province in China is shown in Fig. 35.2. The maximum annual energy consumption,

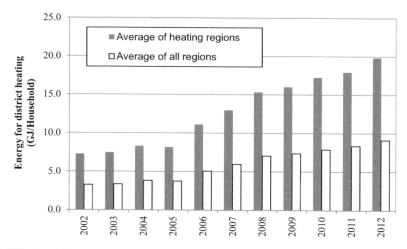

Fig. 35.3 Annual energy consumption for district heating per household

81 GJ/household, among all the capital cities is found in Yinchuan, whereas the minimum, 17 GJ/household, is in Nanjing. The proportion of natural gas for Yinchuan, Xining, Chongqing, etc. is large because these locations are close to natural gas fields. The average of these 30 cities is 33.2 GJ/household, which is 1.44 times as much as the average value of the 277 cities. This implies that more energy is consumed in capital cities than noncapital cities.

35.2.2 Energy Consumption for District Heating

According to the thermal design code for civil buildings, China is classified into five regions as described already. District heating systems are only equipped in the severely cold region and cold region; therefore, these regions are also called heating regions.

Energy consumption for district heating has been increasing significantly in recent years. As shown in Fig. 35.3, the energy consumption per household averaged over all of the 277 cities increases from 3.3 to 9.1 GJ, while the average of the cities in the heating regions alone increases from 7.0 to 20.0 GJ in the period of 2002–2012. In the cities other than the heating regions, energy consumed for space heating in winter includes electricity, gas, coal, etc., without an energy consumption for district heating.

35.3 Survey of Residential Energy Consumption in the Main Cities [4]

35.3.1 Energy Consumption Using Questionnaire and Meter Reading

35.3.1.1 Survey Overview

Research target is seven Chinese cities located in different climate conditions. Severely cold region (Harbin) and cold region (Beijing) are referred to heating region, and these areas account for about 70% of land area of the whole country. In hot-summer-cold-winter region (Nanjing, Shanghai, Chongqing, Changsha), population is concentrated and economic development is significant. The hot-summer-warm-winter region (Guangzhou) is characterized by high temperature and high humidity. Questionnaire survey was conducted in the winter (March) and the summer (September) of 2006. The overall response rate was 75.4% and 69.1% in winter and summer, respectively. The total number of questionnaire collected was 500. These houses belong to the middle class in terms of income and the number of family members is mainly three to four. The meter readings of electricity and gas were conducted once a month in each house, and the energy consumption per month was calculated. In Harbin and Beijing where the district heating system is used, the energy consumption for them is not included in this survey because no information is available. Out of 500, the number of houses which obtained the data of the annual energy consumption is 218. The investigation period was 1 year from January 2006 to January 2007 (in the case of Shanghai, it was from April 2006 to January 2007). Collaborators in China contributed for distribution and collection of the questionnaire and meter reading.

35.3.1.2 Ways of Usage of Space Heating and Cooling

1. Space Heating

In Harbin and Beijing which belong to the region with cold climate, district heating is utilized. However, electrical individual space heater was used in 18% and 5% of the houses in Harbin and Beijing, respectively. On the other hand, in the hot-summer-cold-winter region covering Nanjing, Shanghai, Changsha, and Chongqing, a local space heater is used. In Guangzhou, which falls into the hot-summer-warm-winter climate area, there is generally no heating equipped in most of the houses. Daily changes of heating usage rate are almost similar to that of Fig. 42.5 of Chap. 42. In Harbin and Beijing, all of the houses provided by district heating system were heated for 24 hours, whereas most of the houses in Nanjing, Shanghai, Changsha, and Chongqing were operated only during evening time (18:00 to 23:00).

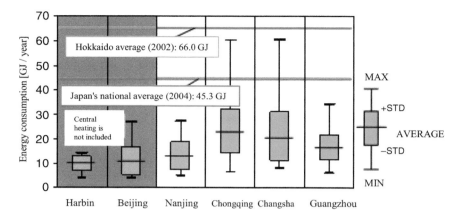

Fig. 35.4 Annual energy consumption per apartment in six Chinese cities and comparison with Japan

2. Air Conditioning

The prevalence of the air conditioner is more than 80% with the exception of the Harbin (24%). Daily changes of air conditioning usage rate are almost similar to that of Fig. 42.13 of Chap. 42. Two times of peak can be seen in the cities with the exception of Guangzhou and Harbin. The major peak appeared in the evening time (20:00–21:00), and cooling was used near dawn in Shanghai and Chongqing. In addition, the peak can be seen between lunch time and afternoon break (around 12:00 to 15:00). In Guangzhou, unlike other cities, there is no peak during the day, and cooling was used from the time for bed until dawn in many houses.

3. Annual Energy Consumption

Figure 35.4 shows the annual energy consumption per household in different cities including the values of mean, minimum and maximum, and ±standard deviation. The mean values of annual energy consumption in Chongqing and Changsha are 23.3 GJ and 21.2 GJ, respectively. The mean values are higher, and the standard deviations are larger than those of the other cities. The consumptions in Harbin and Beijing, i.e., cold region, are 10 GJ and 11 GJ, respectively. These values do not include the district heating energy consumption. Also, in Harbin, comparing with the other cities, the standard deviation is small. In comparison with Japan, energy consumption in Chongqing, which is the highest, is 51% of Japan's mean value in 2004 [5] and is 35% of the Hokkaido's mean value in 2002 [6].

4. Monthly Energy Consumption

Monthly electricity and gas consumptions and the total in each city are shown in Fig. 35.5, with the outdoor air temperature of each month

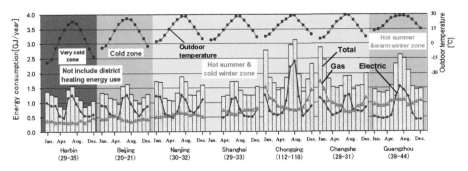

Fig. 35.5 Monthly energy consumption for a year of seven cities

1) Electricity Consumption

In every city, under strong influence of the outside air temperature change, energy consumption has varied greatly depending on the season. The average electricity consumption in winter is slightly more than the interim period due to heating equipment used.

In Chongqing located in the basin, electricity consumption in the summer is higher than the other cities because space cooling is used very often. In Guangzhou with warm winter, the consumption during the winter is almost the same as the interim period because of no use of space heating.

2) Gas Consumption

In southern region such as Chongqing, Changsha, and Guangzhou, gas consumption is higher than the northern region because most of the people prefer stir-fry cooking and cook dishes, and also they use gas for long-time cooking. In Chongqing and Changsha, in order to use hot water and cooking in the winter, gas consumption is slightly more than the interim period.

5. Analysis of Energy Consumption for Various Kinds of End Use in Nanjing, Chongqing

Based on the energy consumption obtained of each month, the energy consumed for various kinds of end use such as heating, cooling, hot water, cooking, and lighting/others was analyzed. There are two groups of houses in terms of usage of gas. In one group (Type 1, 113 houses), the gas is used for both cooking and hot water supply. In the other group (Type 2, 66 houses), the gas is only used for cooking. They use electricity to other types of applications. Then it is assumed that the average value of the energy consumption for hot water supply of Type 1 was the same as Type 2 housing. Annual energy consumptions for heating and cooling were estimated as follows. Firstly, electricity consumption in the interim period, used for purposes other than heating and cooling, was assumed to be constant throughout the

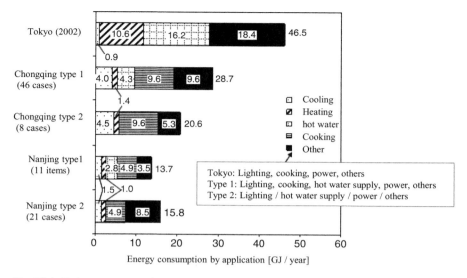

Fig. 35.6 End-use structure of annual energy consumption in two Chinese cities and Tokyo

year. Heating energy consumption of the winter was obtained by subtracting the consumption of non-heating from the total electricity consumption in each month throughout the winter. Cooling energy consumption was also determined in the same manner.

Figure 35.6 shows the end use of energy consumption of the two types of housing in Nanjing and Chongqing. The annual energy consumption in Chongqing is more than that of Nanjing. In Chongqing, the consumption of Type 1 is 28 GJ/year, which is 1.3 times more than that of Type 2. In Nanjing, Type 1 is 13.7 GJ/year, which is a little less than 90% of Type 2. Compared to Japan, the consumption of Chongqing Type 1 is 60% of Tokyo. Since the residential average floor area is 88m^2 and 64m^2 in Chongqing and Tokyo, respectively, the difference between two cities is greater when comparing with consumption per floor area. Energy consumption for cooling in Chongqing is three times more than that in Nanjing. Nanjing and Tokyo have no great difference in the cooling energy consumption. For energy consumption for space heating, the value is small, and there is no great difference between Chongqing and Nanjing. That is nearly 10% of Tokyo.

The energy consumption for cooking is large in the both cities. The ratio to the total in Nanjing and Chongqing is 30.6% and 46.4%, respectively. The absolute value of Chongqing is twice that of Nanjing.

The energy consumption for hot water could be estimated in Type 1. In Nanjing, it is 60% of Chongqing and approximately 20% of Tokyo.

35.4   Long-Term Detailed Measurement of Energy Consumption and Indoor Thermal Environment [7]

Eighteen apartments in nine cities in China, i.e., Harbin, Shenyang, Dalian, Beijing, Changsha, Maanshan, Shanghai, Guangzhou and Kunming were selected for the measurement of energy consumption for different kinds of end uses and indoor and outdoor environment during 1 year. These houses were introduced by research collaborators.

35.4.1   Methods of Measuring the Energy Consumption and the Indoor Environment

1. Power Consumption
   Power consumption was measured at the circuit breaker panel for major end use and at each outlet of the room for different end uses. The system was composed of the measuring instrument, a transmitter and a data logger with the receiver for accumulating the power consumption. The data was stored at 1-minute intervals.
2. Gas Consumption
   Gas consumption was recorded at 5-minute interval by a camera installed in the front of the meter display window.
3. Temperature and Humidity
   Temperature and humidity were recorded at every 30 min intervals by a small data logger with the sensors (measurement accuracy of temperature: ± 0.3 °C, relative humidity: $\pm 5\%$). The data logger was set in the living room, the master bedroom, and the outside of rooms.

35.4.2   Survey Results

1. Annual Energy Consumption

Figure 35.7 shows the annual energy consumption with the breakdown of the 18 apartments. The first nine apartments in the upper part of Fig. 35.7 with no district heating are arranged from the larger of the energy consumption. The average value of China annual energy consumption was approximately 20 GJ. The apartments of Beijing 07 and Shanghai 01 consumed 50% more of the energy than the average. The former was provided with a central heating system using the gas. The energy consumption for space heating was about 18 GJ and accounted for more than half of the total. Since there is a lack of information of measured energy consumption of a house with the district heating, this data are valuable because all rooms are heated. In addition, the space heating energy consumption in 20 apartments with floor heating provided by the district heating system was measured in Shenyang. In

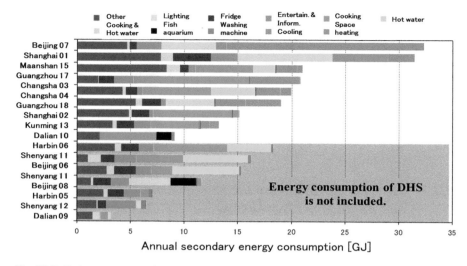

Fig. 35.7 End-use structure of annual energy consumption of 18 apartments in different cities

this case, for obtaining the energy consumption, the surface temperature of both the supply and the return pipes and the flow rate of hot water header for floor heating were measured. The average value in 20 apartments was 32.2 GJ.

The house of Shanghai 01 with five family members consumed a lot of energy for hot water, cooking, and air conditioning due to the high living standard. The houses of Maanshan 15, Guangzhou 17, and Changsha 03 and 04 consumed about 20 GJ. But the structure of energy consumption for end use was quite different. For example, it was found that the energy consumption for cooking in Guangzhou 17, hot water supply in Changsha 03, and space heating in Changsha 04 was relatively large. In the houses of Guangzhou 18, Shanghai 02, and Kunming 13 in which the total energy consumption was less than 15 GJ, a large proportion of energy for cooking was found. In the house of Kunming, space heating and cooling were not in use, and electrical energy consumed for water tank with tropical fish accounted for about 20%.

Seven houses of the lower part of the figure do not include energy consumption for the district heating. In Dalian 10, Harbin 06, and Shenyang 11, which consumed around 15 GJ, the proportion of hot water supply and cooking was noticeable. In Beijing 06, energy consumption for a water tank with tropical fish accounted for 30%. In Harbin 05 and Shenyang 12 with around 7 GJ, energy used for audio equipment was noticeable. Dalian 09 consumed extremely small amount as 3 GJ, implying that the people were living frugally.

It is seen that the energy consumption of houses in different cities is largely different from each other and depends upon the lifestyle of each house.

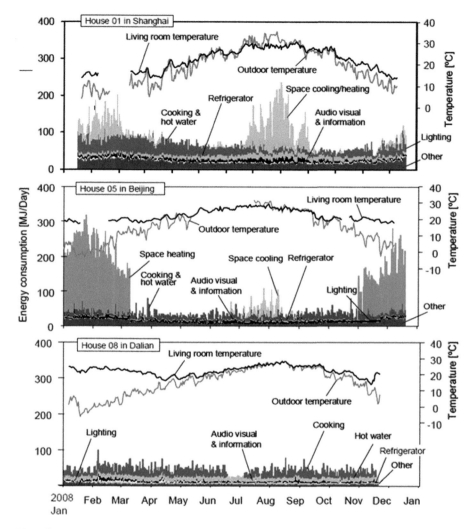

Fig. 35.8 Energy consumption change for different usage during a year in three apartments in Shanghai, Beijing, and Dalian

2. Long-Term Change of Energy Consumption

Figure 35.8 shows the change in energy consumption and temperature of 1 year in Shanghai 01, Beijing 07, and Dalian 08. In Shanghai 01, clear peaks for winter heating and summer cooling were dominant. Energy consumption for cooking and hot water supply had a large proportion and in winter was clearly greater than in summer. In addition, energy for audio instrument, refrigerator, and lighting has been used over the years. The amount for the refrigerator used in the summer was relatively large. Maximum value was found in summer with space cooling of the value about 250 MJ. Although the room temperature was a daily mean value, it ranged from 13 to 18 °C during the winter and from 28 to 30 °C during the summer.

In Beijing 07, it can be seen that the energy consumption for winter heating accounts for large proportion. The energy consumed for summer cooling was small. In other usages, energy consumed for cooking and hot water supply and audio instrument was noticeable, but the total amount was small in the winter when compared to space heating. The maximum value, which was found during the winter, was more than 300 MJ. Comparing to Shanghai 01, the space heating energy consumption was very large.

In Dalian 08, which did not include the energy consumption by district heating, there was no major change throughout the year. The consumption for cooking was noticeable.

35.5 Summary

1. Analysis by Statistical Material
   Based on Chinese statistics, unit energy consumption for 277 Chinese cities was clarified over the period of 2002–2012. Energy consumption per urban household increased from 12.3 to 23.1 GJ. Energy for district heating and electricity accounted for large proportion among all kinds of energies.

2. Energy Consumption by the Meter Reading
   Excluding the energy consumption by the district heating, annual energy consumption was recorded from 10 to 30 GJ. Roughly one third of the energy consumption was used for cooking, and another one third was used for space heating, cooling, and hot water supply, and the final one third was used for the other purposes. In the southern region, the electricity consumption during the summer increased due to operation of air conditioner. The ratio of annual cooling energy consumption is not so much at the present. But it is expected that the cooling energy consumption will increase rapidly because of the need for indoor thermal comfort in the near future.

3. The Results of the Detailed Measurement
   In the eight apartments without district heating, the annual energy consumption varied from 8 to 32 GJ. Also, it was observed that there are large variations in the structure of end usage. However, it is commonly deemed that the percentage of energy consumption for cooking is large. In an apartment installed with the central heating in Beijing, energy consumption for space heating was 18 GJ and shared more than half of the total. According to another survey in Shenyang, the average value of energy consumption for space heating in apartments with space heating was 32.2 GJ. Energy consumption excluding for space heating in apartments with district heating was a large difference from 3 to 18 GJ. In conclusion, there exists a wide variation in the energy consumption, which implies that there is room for further improvement in saving energy.

References

1. National Bureau of Statistics of China: China urban statistical yearbook, 2002–2013, China Statistics Press
2. National Bureau of Statistics of China: China Statistical Yearbook, 2002–2013, China Statistics Press
3. National Bureau of Statistics of China: Energy statistics yearbook, 2002–2013, China Statistics Press
4. Jian Z, Yoshino H et al (2007) The survey of the use of cooling & heating systems and energy consumption of urban apartments in China. Technical papers of annual meeting, The society of heating, air-conditioning and sanitary engineers of Japan, pp 1451–1454 (In Japanese)
5. The Energy Data and Modeling Center (2006) Report of Energy and economic statistics. The Institute of Energy Economics, Japan (In Japanese)
6. Jyukankyo Research Institute Inc.(2002) Annual report of household energy consumption (In Japanese)
7. Hu T, Yoshino H, Zhou J (2012) Field measurements of residential energy consumption and indoor thermal environment in six Chinese cities. Energies 5(6):1927–1942

Chapter 36
Household Energy Consumption in Slum Areas: A Case Study of Tacloban City, Philippines

Eric Casimero Oliva and Tetsu Kubota

Abstract This chapter presents the results of a survey on household energy consumption in existing slum settlements and the resettlement site in the city of Tacloban, Philippines. The results showed that the household income level in the resettlement site was surprisingly not improved compared to that of the existing slums. It was found that both groups shared almost similar household energy consumption pattern and their annual energy consumption was approximately 19–20 GJ including out-of-home energy consumption. Among all the uses, cooking consumed the highest energy, corresponding to 71% in the existing slums and 74% in the resettlement site. Further results revealed that the households in the resettlement site became more of biomass users (wood fuel) as compared to those in the existing slums.

Keywords Household energy consumption · Urban slum · Developing countries · Philippines

36.1 Introduction

Although the proportion of the urban population residing in slums today (2014) is lower than it was some two decades ago, the absolute number of slum dwellers continues to increase [1]. Currently, one in eight people across the world lives in slums. In 2000, 39% of the urban population in developing countries resided in slums; this declined to 30% in 2014 [1]. Though household energy consumption in slums is considered small at the moment, the energy consumption of this large population is projected to be increased dramatically with higher living standards

E. C. Oliva
National Housing Authority, Regional Office No. 8, Tacloban, Philippines

T. Kubota (✉)
Graduate School for International Development and Cooperation (IDEC), Hiroshima University, Hiroshima, Japan
e-mail: tetsu@hiroshima-u.ac.jp

© Springer Nature Singapore Pte Ltd. 2018
T. Kubota et al. (eds.), *Sustainable Houses and Living in the Hot-Humid Climates of Asia*, https://doi.org/10.1007/978-981-10-8465-2_36

and disposal income as twin drivers in the near future towards becoming middle class. Hence, it is important to investigate the energy consumption patterns of existing slums in order to project the future energy demand of the growing middle class. This chapter presents the results of a survey on household energy consumption in the existing slum settlements and the resettlement site in the city of Tacloban, Philippines.

36.2 Case Study Area

In the Philippines where slum prevalence was at 38.3% as of 2014 [1], the government implements the resettlement program as a strategy in averting poverty and homelessness [2]. As provided under the Philippine law on housing, relocation or resettlement for slum or squatter settlements entails the provision of decent yet affordable house and lot (or lot only) together with basic services and employment opportunities to significantly uplift the conditions of the underprivileged beneficiaries. Such development could make not only improvement in family income but likewise changes in the behaviour and lifestyle of the beneficiaries [3].

Past and present studies on slums and related services such as housing mainly focused on affordability, financial and structural policies which are basically intended to come up with solutions how to reach out more beneficiaries. Studies assessing the impact of present housing programs on energy consumption, however, are scarce if not none. It is therefore the intent of this chapter to shed light on the impact of the resettlement program on the energy consumption of its beneficiaries considering the lifestyle change brought about by distant relocation.

A survey was conducted in the city of Tacloban, Philippines, from August to October 2010 (Fig. 36.1). The city is located on the northeast part of Leyte Island, situated 580 km southwest of Manila. Tacloban City has a total population of 221.2 thousands as of 2010 census having an annual growth rate of 2.16% [4]. Urbanization is at 87% indicating that majority of its inhabitants live in the city urban districts [5]. Slum dwellers consisted around 10,000 households or about one-third of the urban inhabitants based on 2010 estimate [6] (Figs. 36.1 and 36.2a). In 2003, a resettlement housing project named "Tacloban Resettlement Project" was established some 5 km northwest of the city to serve as a relocation site for the evicted slum dwellers occupying the city centre (Figs. 36.1 and 36.2b). Tacloban City was selected as the subject area considering the following reasons:

- Being the regional capital of Eastern Visayas region, it can represent most of the cities in terms of population and urban development, especially those outside of the mega Manila. Resettlement program in the country is implemented almost similarly throughout the nation.
- Location and condition of resettlement project of Tacloban City follow the overall trend of resettlement development in the country, i.e. moving farther from the original location of the beneficiaries or the city centre where they come from.

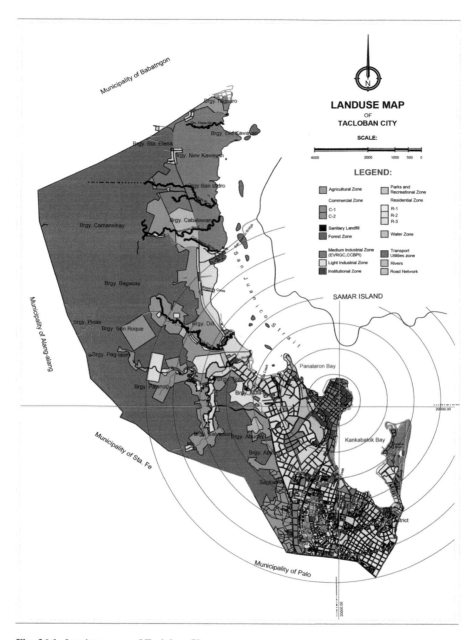

Fig. 36.1 Land use map of Tacloban City

Fig. 36.2 Houses in (**a**) slum settlement and (**b**) resettlement site

36.3 Methodology

To ensure comparability, two distinct groups of dwellers were interviewed using prepared survey questionnaires designed to capture information concerning the following key categories: (1) demography and livelihood, (2) housing condition and (3) energy consumption. The first group consisted of sample households from the existing slums (Group 1), and the second group consisted of former slum dwellers now relocated at the Tacloban Resettlement Project (Group 2). A total sample size of 371 households was determined, and actual samples were identified on site systematically at every 12-house interval. For Group 2, on the other hand, considering its smaller population size which is just 287 households, a standard sample size of 100 households was secured. At the end of the survey, actual samples collected were 285 for Group 1 and 107 for Group 2 with a response rate of 67%. Energy consumption was investigated in detail through on-site measurements and interviews. The detailed methodology is described in the previous paper [7].

36.4 Results and Discussion

36.4.1 Changes in Living Conditions

The characteristics of sample households in both groups can be briefly described in Table 36.1. Households in Groups 1 and 2 have on average 5.3 members in the family. Though samples examined in this chapter are unrelated due to the fact that previous circumstances of Group 2 cannot be readily ascertained, it is assumed that living conditions of Group 2 before they were relocated are similar with that of Group 1, the existing slum dwellers. This is mainly because sample households in Group 1 come from various slum colonies of the city including the remaining slums adjacent to the previous location of the households of Group 2. Meanwhile, durations of stay in the present sites are significantly different between the two groups. The duration among households in Group 1 varies widely from a few months to more than 40 years with an average of 17.9 years, while almost all the households of Group 2 relocated to the site after its establishment, i.e. 2003, and thus the duration ranges up to 7–8 years with an average of 5.6 years.

Relatively, more household members in Group 1 (1.7) compared to Group 2 (1.4) are able to find work or any other source of income, which is suggestive, though not conclusive, of a slightly higher household income of the former vis-à-vis the latter as indicated in Table 36.1. This result contradicts the initial assumption that household income in Group 2 must be higher than in Group 1 considering them to be beneficiaries of the government program on housing and related services. One possible reason for this, as per interviews conducted with the representatives of the proponent, is that no appropriate official livelihood intervention was ever implemented for the resettlement's beneficiaries since the establishment of the project. This could mean that households in Group 2 were left on their own right after relocation and some may still be dependent until this point in time from their previous livelihoods in the city centre where they came from. Another possible reason, which was confirmed by some residents who expressed their personal comments, is the shifting from old livelihoods by some relocatees into merely subsistence farming. Thus, more families in Group 2 fall below the poverty threshold (47.5%).

Table 36.1 Brief profile of sampled households

	Group 1 (Existing slums)	Group 2 (Resettlement)	Whole
Sample size (response rate)	385 (69%)	107 (60%)	492 (67%)
Household size (persons)	5.3	5.3	5.3
Age distribution (%)			
Children	38.1	44.6	39.5
Working age	59.0	52.8	57.7
Elderly	2.9	2.6	2.9
No. of working members (persons)	1.7	1.4	1.6
Monthly household income (%)			
< 110 USD	34.8	40.4	36.0
110–220	36.5	40.4	37.3
220–330	14.5	6.1	12.7
>330	14.2	13.2	14.0
Average (USD)	194.8	172.2	189.9
Poverty ratio (%)	41.5	47.5	42.8
Types of livelihoods (%)			
Business	8.2	2.8	7.1
Employed	41.2	51.7	43.3
Self-employed	47.5	43.5	46.7
Others	3.1	2.1	2.9
Housing materials used (%)			
Light, temporary	38.2	29.0	36.2
Mixed, but predominantly light	35.8	31.8	34.9
Strong, permanent	8.1	14.0	9.4
Mixed, but predominantly strong	17.9	25.2	19.5
Floor area (%)			
< 25 m^2	52.5	34.6	48.6
25–45	25.5	37.4	28.0
45–65	11.7	15.0	12.4
>65	10.4	13.1	11.0
Average (m^2)	36.0	38.6	36.5
Sanitation facilities (%)			
Toilet with septic tank	59.3	87.0	65.3
Shared toilet with septic tank	7.8	10.2	8.3
Public toilet with septic tank	1.6	0.9	1.4
Toilet without septic tank	12.2	0.9	9.7
None	19.2	0.9	15.2

In Group 1, majority of the working members are self-employed (48%), followed by employed (41%), either regular or temporary in private and public establishments, and the rest are either having their own business or other forms of livelihood (Table 36.1). Contrary to Group 1, the majority of workers in Group 2 are employed either temporary or permanent in private and public establishments (52%), followed by self-employed (43%) and the others.

In terms of type of housing, households belonging to Group 2 improved their houses using permanent and stronger materials (39%) as compared to households in Group 1 (26%) (Table 36.1). Though there is still considerable proportion of households using light, temporary and salvaged materials of about 61%, it is actually still lower than the proportion in Group 1, having about 74%. Such kind of condition, however, doesn't last so long and is expected to drastically improve in several years. It is possible since the resettlement program is in fact intended to provide beneficiaries with security of tenure. This feeling of security among the beneficiaries motivates households to invest on the construction of their houses despite limitations in financial resources. As shown, sanitation has greatly improved for households in Group 2. Other improvements brought about by the relocation in the living environment found in this survey include total floor area, drainage system, better road network within the settlement area and better air quality.

36.4.2   Energy Consumption

The annual household energy consumption profiles of the two groups are summarized in Fig. 36.3. Both in-home and out-of-home direct energy consumption were investigated in this study. In-home energy consumption consisted of cooking energy and electricity consumptions, while the out-of-home energy consumption consisted of the proportionate petrol or diesel consumption for transportation.

In terms of proportion, Groups 1 and 2 share almost similar consumption pattern. The least share is on electricity comprising only 12.1% in Group 1 and much lower share of 8.8% in Group 2. Electricity consumption remains very minimal due to the fact that appliance ownership is very low. Figure 36.4 indicates that most of the appliances owned by the households in both groups are just the basic and necessary ones. The share of consumption for transportation over the total is similar at 17% (3.4 GJ) in both groups, although travelling households in Group 2 takes more than three times the distance being travelled by the households in Group 1 (Table 36.2). This is mainly due to the mode of public transportation used. From Table 36.2, it shows that approximately 32% of the travellers in Group 1 use motorized tricycles having carrying capacity of up to four passengers only and multicabs with carrying capacity of up to 12 passengers only. This means that the effective consumption per km per capita for these modes is much higher than that for jeepneys which are able to accommodate 18 or more passengers, so that even if the average travel distance of households in Group 1 is much lesser than that of Group 2, the energy consumed is almost the same.

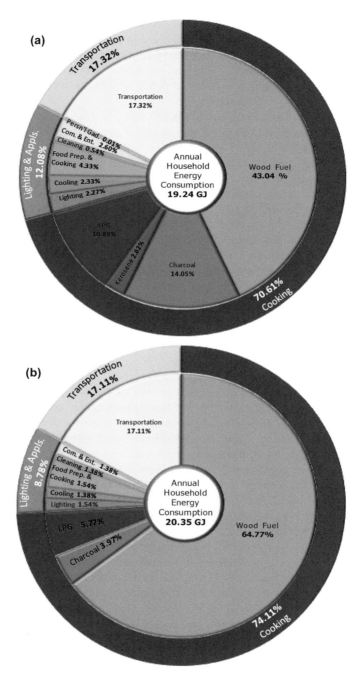

Fig. 36.3 Annual average household energy consumption profile (Secondary energy) (**a**) Group 1 (existing slums); (**b**) Group 2 (resettlement site)

Fig. 36.4 Ownership level of household appliances

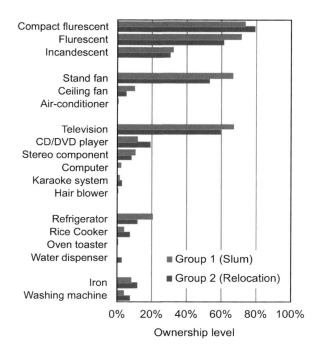

Ownership level

Table 36.2 Details of household energy consumption profile

	Group 1 (Existing slums)	Group 2 (Resettlement)	Whole
Daily average travel distance (km)	13.7	43.7	20.4
Transportation modes (%)			
Jeepney	53.1	98.1	62.5
Motorized tricycle	31.9	0.6	25.4
Multicab	13.3	0.6	10.7
Own vehicle	0.8	0.6	0.8
Passenger bus	0.8	0	0.7
Energy expense to total household income (average) (%)	24.9	26.1	25.1
Cooking	10.3	8.8	10.0
Lighting and appliance (electricity)	4.5	3.4	4.3
Transportation	10.0	13.9	10.9

Among all the uses, cooking consumes the highest energy of all corresponding to 71% (13.7 GJ) in Group 1 and 74% (15.1 GJ) in Group 2 (Fig. 36.3). Though there is barely any difference in the frequency of cooking between households in the two groups, it can easily be conceived that the total energy consumption in both groups is primarily determined by the cooking means (Fig. 36.5). The further results showed that the relocated households became more of biomass users as compared to the

Fig. 36.5 Typical cooking
means

existing slum dwellers. As shown in Fig. 36.3, the energy consumption using wood
fuel accounts for 65% in Group 2, followed by LPG (5.8%) and charcoal (4.0%),
while the wood fuel in Group 1 accounts only for 43%, followed by charcoal (14%),
LPG (11%) and kerosene (2.6%). Several reasons can be attributable to this shift.
One is that the resettlement site is surrounded by thick vegetation and that wood fuel
is freely accessible. Second, the unrelenting increases in petroleum product prices
discourage households from using petroleum-based cooking fuel products such as
LPG and kerosene and therefore prefer the use of cheaper wood fuel. Third is the
personal choice of other households to use wood for cooking especially in cooking
rice. Some respondents said that rice tastes better when cooked using wood.

Given the fact that the resettled households become more wood fuel users, several
problems may eventually arise on top of the apparent increase in their absolute
energy consumption. Household and community safety may be the utmost concern.
Considering that the typical method of cooking using wood is in open air, the danger
of breaking up fire disaster is likely. Continued exposure to smoke could likewise
pose health problems to the household members. Such practice when sustained by
more people may eventually cause deforestation. A simple but possible remedy by
promoting the use of improved cook stoves might prove useful during the transition
stage of the resettled households.

In terms of energy expense, Table 36.2 shows that transportation energy has the
highest share of expense in Group 2 (more than half of overall energy budget). In

Group 1, on the other hand, cooking and transportation share almost the same proportion, with electricity as the lowest in both groups. When compared to the average income of households in both Groups 1 and 2, result showed that about 25% of household income is spent for energy in Group 1, while a slightly higher proportion of 26% of their household income is being spent by households in Group 2, that is, considering that roughly 45% of firewood used by wood fuel users are collected for free. Therefore, reduction of inefficient fuel use would be the key to reduce the household expenses for energy and therefore would mean more savings of their income while achieving the energy saving objectives in the resettlement sites as well as in the existing slum areas.

36.5   Conclusions

The survey was conducted in the city of Tacloban in 2010 in order to compare the household energy consumption patterns between the existing slum settlements (Group 1) and resettlement site (Group 2). The main findings are as follows:

- The household income level was expected to be higher in Group 2 than in Group 1, but the result was otherwise. This was partly because the number of workers in household was larger in Group 1 than in Group 2. It was also reported that no appropriate official livelihood intervention was ever implemented by the local authority since the establishment of the relocation project.
- However, some improvements were seen in Group 2 especially in terms of the living conditions, such as building construction, floor area, sanitation facilities, road networks, etc., as compared to Group 1.
- Groups 1 and 2 shared almost similar household energy consumption pattern. Among all the uses, cooking consumed the highest energy of all, corresponding to 71% in Group 1 and 74% in Group 2. Further results revealed that the households in Group 2 became more of biomass users (wood fuel) as compared to those in Group 1. Since the fuel efficiency of wood fuel is lower than the others, the above difference of fuel source resulted in the larger portion and therefore larger energy consumption for cooking in Group 2. One possible reason was that the relocation site in Tacloban City is surrounded by thick vegetation and that wood fuel is freely accessible.

Acknowledgements We would like to extend our special thanks to Ms. Doris Secreto of the National Housing Authority (NHA) in Tacloban City for her unsurpassed assistance in doing the legwork at the site while the authors were still in Japan and also to Ms. Rochie of NHA-Main Office for her assistance in gathering data and related information concerning the resettlement projects in the Philippines. After this survey was conducted, Tacloban City was critically devastated by the typhoon, Yolanda, in 2013, causing the deaths of more than six thousands. We sincerely extend our heartfelt condolence for the loss of people in the city.

References

1. UN-Habitat, Urbanization and Development: Emerging Futures, World Cities Report 2016, United Nations Human Settlements Programme, 2016
2. Joseph EF (2003) An evaluation of the resettlement program in the Philippines: the case of Towerville resettlement project. A country Report on the New Town Development and Urban Renewal, National Housing Authority
3. Munarriz MT (1986) Traditional vs modern behavior in a housing development project: the Tondo foreshore community. Habitat Int 10(3):395–402
4. PSA (2015) Philippine statistical yearbook, Philippine statistics authority (PSA), Republic of the Philippines, 2015
5. City Planning & Development Office, Tacloban City, Tacloban City Ecological Profile (2009) City Planning & Development Office, Tacloban City, Philippines, 2009
6. Unpublished Report, Presidential Commission for the Urban Poor, Tacloban City
7. Oliva EC, Kubota T (2011) Analysis of lifestyle changes in the relocated slum dwellers of Tacloban city, Philippines specially focusing on the energy consumption pattern. In: Proceedings of 11th international congress of Asian planning schools association (APSA 2011), September, Tokyo, 19–21

Part V
Indoor Thermal Environment

Chapter 37
Thermal Function of Internal Courtyards in Traditional Chinese Shophouses in Malaysia

Mohd Azuan Zakaria, Tetsu Kubota, and Doris Hooi Chyee Toe

Abstract This study aims to identify the thermal functions of internal courtyards in traditional Chinese shophouses (CSHs) located in the hot-humid climate of Malaysia with the aim of providing useful passive cooling strategies for modern urban houses. This chapter investigates the detailed thermal environments of the selected two traditional CSHs with different courtyard types to discuss the thermal function of the courtyards. As a result, it was suggested that closed, cross-ventilated courtyards be embedded to achieve indoor thermal comfort and avoid excessive humidity in hot-humid climates. Meanwhile, it was also recommended that a staggered form of courtyard with V-shaped roofs should be designed as a nocturnal cooling source.

Keywords Thermal comfort · Courtyard · Vernacular architecture · Passive cooling · Natural ventilation · Hot-humid climate

37.1 Introduction

Unlike detached houses, an elongated row house such as the CSHs typically has only a few openings on the external walls, thus making cross-ventilation often unsatisfactory. As most of the studies on the environmental effects of courtyards only analysed the resulting ventilation improvements, there are relatively few studies that examined the effects of courtyards on ventilation performance and the resulting indoor thermal comfort. In a hot-humid region, such as in Malaysia, indoor air

M. A. Zakaria
Faculty of Civil and Environmental Engineering, Universiti Tun Hussein Onn Malaysia, Batu Pahat, Johor, Malaysia

T. Kubota (✉)
Graduate School for International Development and Cooperation (IDEC), Hiroshima University, Hiroshima, Japan
e-mail: tetsu@hiroshima-u.ac.jp

D. H. C. Toe
Faculty of Built Environment, Universiti Teknologi Malaysia, Skudai, Johor, Malaysia

© Springer Nature Singapore Pte Ltd. 2018
T. Kubota et al. (eds.), *Sustainable Houses and Living in the Hot-Humid Climates of Asia*, https://doi.org/10.1007/978-981-10-8465-2_37

387

temperature can be several degrees lower than the corresponding outdoor temperature during the day in a high thermal mass building [1]. In these circumstances, there is a major trade-off in achieving indoor thermal comfort by maintaining lower indoor air temperatures through reducing air changes with the outdoors, i.e. night ventilation, and increasing indoor wind speeds for sweat evaporation by improving natural ventilation while allowing an increase in temperature during daytime.

This chapter aims to discuss the thermal effects of internal courtyards in Malaysian CSHs and focuses especially on the above-mentioned trade-off in hot-humid climates [2]. The detailed thermal environments of the selected two traditional CSHs with different courtyard types are investigated to analyse the thermal functions of courtyards as well as the resulting thermal comfort conditions. The design recommendations for internal courtyards are then made based on the results of the analyses.

37.2 The Case Study Chinese Shophouses

The detailed field measurements were conducted in the two selected CSHs that were analysed in [2], namely, CSH 1 and CSH 2. The measurements were conducted in October 2011 (CSH 1) and in September 2014 (CSH 2) (Figs. 37.1 and 37.2, respectively).

The CSH 1 was originally constructed around the time of the Dutch colonial era in the nineteenth century, and it was restored in 2004 by the National University of Singapore (NUS) to function as an academic centre. In contrast, the CSH 2 was originally built between the seventeenth and nineteenth centuries and was strongly influenced by Dutch architecture. It was recently restored by the Heritage of Malaysia Trust and currently used as an exhibition facility for tourists. During the period of operation, the external windows were open only on the first floor in CSH 1 (10:00 to 17:00), while in CSH 2, the external doors and windows were open on both floor (11:00 to 16:00). Ceiling fans in both CSHs were not used during the measurement period. There was no air conditioner installed in either of the CSHs.

The detailed descriptions of CSHs 1 and 2 are provided in [2]. The building structures of both CSHs were of load-bearing lime-plastered brick walls with thicknesses of 250–350 mm (CSH 1) and 150–300 mm (CSH 2), concrete and timber frame. The U-values of walls in CSH 1 range from 1.6 to 1.9 W/(m^{2}K), while those in CSH 2 vary from 1.7 to 2.5 W/(m^{2}K). The thermal masses of their living halls are high at about 1270–1300 kg/m^2 in CSH 1 and about 470–490 kg/m^2 in CSH 2.

As presented in Figs. 37.1 and 37.2, each of the CSHs has two internal courtyards. The front courtyards of CSH 1 (hereafter, CY 1-1), CY 1-2 and CY 2-1 of CSH 2 were classified as "deep and closed" courtyards, while CY 2-2 as a "small and staggered" courtyard. An important difference between CSHs 1 and 2 is that the two courtyards in CSH 1 were separated by a partition, though there were permanent ventilation openings in the partition, whereas the courtyards in CSH 2 did not have any partitions between them. Therefore, in CSH 2, sufficient cross-ventilation was allowed between the two courtyards.

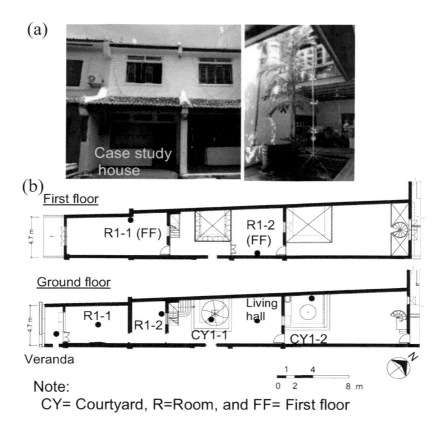

Fig. 37.1   CSH 1. (**a**) Front exterior view and courtyard; (**b**) floor plans [2]

37.3   Methods

The detailed methods of measurement in CSH 1 are presented in [3]. In the recent measurements (CSH 2), the air temperature and relative humidity at 1.5 m above the floor level were measured in the centre of each space (T&D TR-72) (Fig. 37.2b). In the living hall, however, the other thermal parameters including globe temperature and wind speed were also measured for thermal comfort evaluation. Meanwhile, outdoor weather conditions were recorded from veranda (T&D TR73U) as well as at the weather station, which was located in a small open space approximately 500 m from the measurement site.

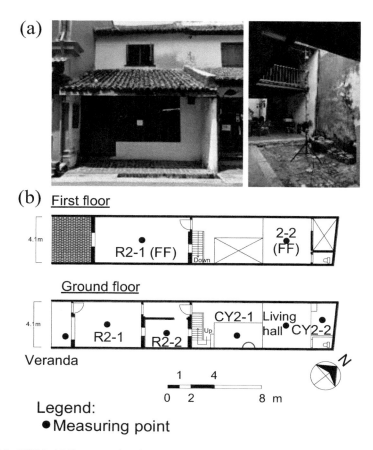

Fig. 37.2 CSH 2. (**a**) Front exterior view and courtyard; (**b**) floor plans [2]

37.4 Variations in Thermal Environments of Whole House

Figure 37.3a illustrates the average air temperatures in indoor spaces in both CSHs during the day (15:00) and the night (0:00). Meanwhile, Fig. 37.3b presents the statistical summary of air temperatures and degrees of humidity in these spaces. As indicated in Fig. 37.3b, outdoor air temperatures ranged from 26.5 to 35.0 °C with an average of 30.0 °C in CSH 1 during the fair-weather days, while those of CSH 2 ranged from 25.5 to 33.5 °C with an average of 29.5 °C. Meanwhile, the outdoor relative humidity during the same days ranged from 54 to 85% in CSH 1, whereas that in CSH 2 ranged from 53 to 89%. In both cases, the prevailing wind directions were from SSW to SW during the daytime (winds blew towards the front façade at (almost) a right angle) and from NE to ENE during the night with average wind speeds of 0.7–1.0 m/s.

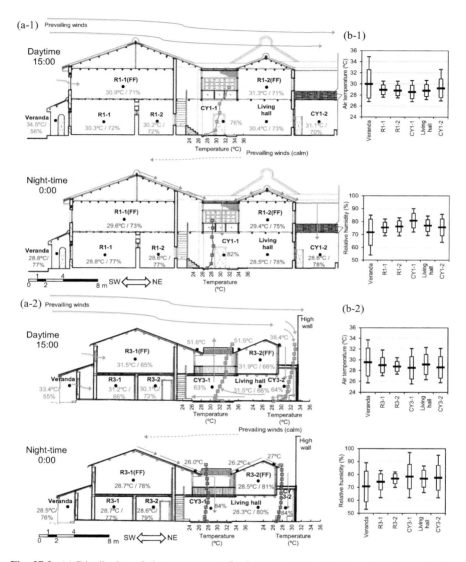

Fig. 37.3 (**a**) Distribution of air temperatures. (**b**) Statistical summary (5th and 95th percentiles, mean and ± one S.D) of measurement at 1.5m above floor in CSH 1 and CSH 2 (Note: Sectional drawings with permission of TTCLC, National University of Singapore and the Heritage of Malaysia Trust [2])

In CSH 1, CY 1-1 recorded a steep temperature gradient (thermal stratification) during the day, particularly from the ceiling level of ground floor and above (Fig. 37.3a-1). At this time, the average daily maximum air temperature in CY 1-1 and the surrounding indoor spaces were approximately 4.0 °C lower than the corresponding outdoor temperature (Fig. 37.3b-1). In contrast, the average daily maximum temperature in CY 1-2 was approximately 2.0 °C higher than that in the living hall. As the living hall was situated next to the CY 1-1 with no partition, its air temperature was nearly the same as that of the courtyard. In contrast, the daily minimum air temperatures were lowest in these two courtyards and were almost the same as the corresponding outdoor temperature (Figs. 37.3a-1 and b-1). Moreover, the temperature gradient was almost absent in the courtyard at night, which clearly indicates that the cool outdoor air sufficiently entered the building through the courtyard as all external windows and doors were closed. Thus, it is concluded that the courtyards serve as cooling sources to the surrounding spaces during the night.

In CSH 2, the lowest daytime air temperatures were observed not in the courtyards but rather in the room located on the ground floor (R2-2) (Figs. 37.3a-2 and b-2). This is mainly due to the structural cooling effect of the high thermal mass building. Meanwhile, the daily maximum air temperatures in the three spaces, i.e. CY 2-1, living hall and CY 2-2, recorded almost the same values, which were only 1.5 °C lower than the corresponding outdoor temperature. As presented in Fig. 37.3a-2, a steep temperature gradient is evident, especially in the CY 2-1 during daytime hours, but the clear gradient is found only at the lower levels, i.e. below the ground floor ceiling level. This indicates that the stratum of relatively cooled air in the CY 2-1 is thinner than that of the CY 1-1. This is primarily because the two courtyards (CY 2-1 and CY 2-2) were connected without any partition, and thus cross-ventilation (i.e. inflow of outdoor air) was improved by the two courtyards in CSH 2. CY 2-2 was situated at the end of the building, and the height of the adjacent building was 3.4 m higher than the original CSH, i.e. staggered form (see Fig. 37.3a-2). Furthermore, it was observed that the prevailing winds hit the high wall of the adjacent building and created downward winds that then flowed into CY 2-2 as illustrated in Fig. 37.3a-2. As the airflow went downward, it travelled through the living hall towards CY 2-1, thus generating upward winds in the courtyard. The inflow of outdoor warm air increased the indoor air temperature and equalized air temperatures between the two courtyards as well as the air temperatures in the living hall. The nocturnal air temperature profiles for CSH 2 were similar to those for CSH 1.

37.5 Thermal Comfort in Living Halls

Figure 37.4 presents the temporal variations of measured major thermal parameters in the living halls of CSHs 1 and 2. As a result of the difference in the locations of the living halls, the daytime air temperatures in the living hall of CSH 2 were approximately 1.5 °C higher than those of CSH 1, although the outdoor air temperatures of CSH 1 were approximately 1 °C higher than those of CSH 2. Moreover, outdoor wind speeds were almost the same during the two measurements, ranging from 0.1 to 3.0 m/s. However, indoor wind speeds in the living halls varied, particularly during the day with the indoor wind speed in CSH 2 being nearly twice as high as that in

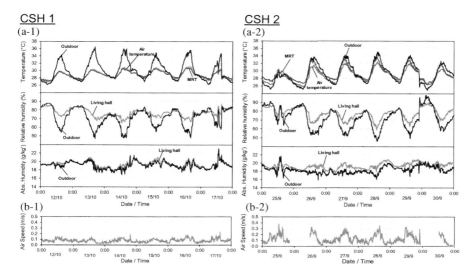

Fig. 37.4 (**a**) Air temperature, MRT, relative and absolute humidity in the living halls of case study CSHs with the corresponding outdoor conditions. (**b**) Wind speed in the living halls [2]

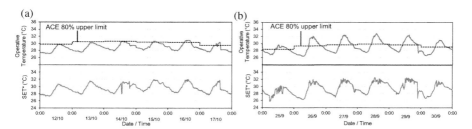

Fig. 37.5 Operative temperature (OT) with corresponding ACE 80% upper limits and SET* in the living halls. (**a**) CSH 1; (**b**) CSH 2 [2]

CSH 1 (Fig. 37.4b). This indicates that the courtyards of the two CSHs performed different thermal functions. In CSH 1, the narrow, deep internal courtyard (CY 1-1) created "cool but still" indoor thermal conditions, whereas in CSH 2, the connected two courtyards (CY 2-1 and CY 2-2) led to "warm but breezy" conditions.

In this study, operative temperature (OT) and SET* were used as indices to evaluate thermal comfort in the living halls (Fig. 37.5). In the upper part of the figures, the thermal comfort was evaluated on the basis of an adaptive comfort equation (ACE) developed by the authors for naturally ventilated buildings in hot-humid climates [4]. The 80% adaptive comfort upper limit of the OT was used for the evaluation. Meanwhile, a metabolic rate of 1.0 met and a clothing value of 0.4 clo were applied in the calculation of SET*. As presented in Fig. 37.5, with respect to the OT, the living hall of CSH 1 recorded lower values (27.0–31.0 °C) than the living hall of CSH 2 (26.5–32.5 °C) during daytime hours. The calculated OT of

CSH 1 generally fell within the limits; however, the OT of CSH 2 exceeded the limit over 40% of the time during the measurement period. When SET* was used for the evaluation, however, both of the living halls recorded similar maximum SET* values during the day (approximately 32.0 °C SET*). This clearly indicates that the indoor thermal comfort in the living hall of CSH 2 was improved by the increased wind speed caused by the connected two courtyards.

Conversely, the RH exceeded 70% throughout most of the measurement period in the living hall of CSH 1, whereas it dropped to as low as 60% during the daytime in CSH 2 (Figs. 37.4a-1 and a-2). While the effects of RH on all aspects of human comfort have yet to be established [5, 6], it is generally recognized that extreme levels of humidity are the most detrimental to human comfort, productivity and health [7]. Sterling et al. [8] reviewed a number of studies regarding the effects of humidity on human health and comfort from various viewpoints and proposed that an RH range between 40 and 60% provides the best conditions for human occupancy. Therefore, it can be concluded that the humidity in the living hall in CSH 2 is more tolerable than that in CSH 1.

37.6 Recommendations for Modern Houses

Basically, as a large amount of cross-ventilation cannot be expected in an elongated row house like the traditional CSHs, a high thermal mass structure can be a good option even in a hot-humid climate for achieving indoor thermal comfort by lowering daytime air temperatures. Internal courtyards are effective in securing airflow as well as daylighting particularly in deep-plan house. It was found that a deep (in height), closed internal courtyard is better able to maintain lower indoor air temperatures in a hot-humid climate, particularly during the day, even though the reduction in air temperature will increase the indoor RH throughout the day. As a living space, it is important to utilize airflow to improve sweat evaporation of occupants while also avoiding excessive humidity, even though cross-ventilation would increase indoor air temperature to a certain extent. Therefore, to achieve indoor thermal comfort in and around the courtyards, closed, cross-ventilated courtyards with a low sky view factor (SVF) are recommended in hot-humid climates. One way to implement the closed but cross-ventilated courtyard in a deep-plan house is to design multiple closed courtyards that are connected with each other, as presented in CSH 2. A viable alternative, however, is to connect the courtyard with the street via a hallway, as suggested in previous publications [9, 10].

As evidenced in the previous sections, internal courtyards would serve as cooling sources to the surrounding spaces especially during the night. The cooling effect was increased when a courtyard was situated between buildings/walls of different heights (i.e., staggered form), which was often located at the end of the building. Furthermore, it was found by the previous study [3] that V-shaped roofs of a courtyard would contribute to increase the inflow of cooled air from the roofs during the night as illustrated in Fig. 37.3a. Hence, it is recommended that a staggered form courtyard with V-shaped roofs should be designed at the corner of a deep-plan house as a nocturnal cooling source, while designing above-mentioned closed but cross-ventilated courtyards within the same building.

Meanwhile, plants, particularly shade trees, would be effective at lowering air temperatures in and around the courtyards, though they would increase humidity levels simultaneously. The solar orientation should not be of primary importance because the solar exposure is basically limited in a narrow, deep internal courtyard, which is the main interest of this study. Rather, the wind orientation is more important for the narrow, deep courtyards to improve the cross-ventilation for the building. In summary, an elongated courtyard house should be oriented towards the prevailing wind directions as presented in most of the study CSHs in Malacca.

37.7 Conclusions

- The different types of courtyards performed different functions with respect to improving the indoor thermal comfort in the adjacent living halls. The living hall of CSH 1 was located next to the deep, closed courtyard (SVF: 0.2%). Although the indoor airflow was almost absent even during the day (<0.2 m/s), the air temperatures in the courtyard and the adjacent living hall maintained relatively low values that were approximately 4.0 °C lower than the corresponding outdoor temperature.
- The other living hall (CSH 2) was situated between the two courtyards of different types. The indoor wind speeds of approximately 0.2–0.4 m/s were generated due to the circulating airflow between the two courtyards during the day. Though this inflow increased indoor wind speeds, it simultaneously increased indoor air temperatures. The daytime air temperature in the living hall was only 1.5 °C lower, on average, than the outdoor temperature.
- When OT was used for the indoor thermal comfort evaluation, the living hall of CSH 1 was considered superior to that of CSH 2 because of its (CSH 1) lowered air temperatures. Nevertheless, when the effect of sweat evaporation was taken into account by using SET*, the condition of CSH 2 was similar to that of CSH 1 because of its higher indoor wind speeds. Meanwhile, as for RH, it was concluded that the humidity in the living hall in CSH 2 was more tolerable than that in CSH 1.
- For elongated row houses of high thermal mass structure in hot-humid climates, it is suggested that closed (with a low SVF), cross-ventilated courtyards be embedded to achieve indoor thermal comfort and avoid excessive humidity. Meanwhile, it is also recommended that a staggered form courtyard with V-shaped roofs should be designed at the corner of the house as a nocturnal cooling source.

Acknowledgements This research was supported by LIXIL Foundation and Nichias Corporation. The field measurements were conducted by the students of Hiroshima University in collaboration with *Universiti Teknologi Malaysia*. Particularly, we would like to express our sincere gratitude to Mr. Seiji Abe.

References

1. Kubota T, Toe DHC, Ahmad S (2009) The effects of night ventilation technique on indoor thermal environment for residential buildings in hot-humid climate of Malaysia. Energ Build 41 (8):829–839
2. Kubota T, Zakaria MA, Abe S, Toe DHC (2017) Thermal functions of internal courtyards of traditional Chinese shophouses in the hot-humid climate of Malaysia. Build Environ 112:115–131
3. Toe DHC, Kubota T (2015) Comparative assessment of vernacular passive cooling techniques for improving indoor thermal comfort of modern terraced houses in hot-humid climate of Malaysia. Sol Energy 114:229–258
4. Toe DHC, Kubota T (2013) Development of an adaptive thermal comfort equation for naturally ventilated buildings in hot-humid climates using ASHRAE RP-884 database. Front Archit Res 2(3):278–291
5. Tsutsumi H, Tanabe S, Harigaya J, Iguchi Y, Nakamura G (2007) Effect of humidity on human comfort and productivity after step changes from warm and humid environment. Build Environ 42:4034–4042
6. Rijal HB, Humphreys M, Nicol F (2015) Adaptive thermal comfort in Japanese houses during the summer season: Behavioural adaptation and the effect of humidity. Buildings 5:1037–1054
7. ASHRAE (2016), Humidifiers, Chapter 22, ASHRAE handbook, heating, ventilating and air-conditioning systems and equipment, ASHRAE, Atlanta
8. Sterling EM, Arundel A, Sterling TD (1985) Criteria for human exposure to humidity in occupied buildings. ASHRAE Trans 91(1B):611–622
9. Rojas JM, Galán-Marín C, Fernández-Nieto ED (2012) Parametric study of thermodynamics in the Mediterranean courtyard as a tool for design of eco-efficient buildings. Energies 5:2381–2403
10. Ok V, Yasa E, Özgunler M (2011) An experimental study of the effects of surface openings on air flow caused by wind in courtyard buildings. Archit Sci Rev 51(3):263–268

Chapter 38
Passive Cooling of the Traditional Houses of Nepal

Hom Bahadur Rijal

Abstract For the purpose of evaluating passive cooling effects in summer, a thermal investigation of traditional houses was conducted in the sub-tropical region of Nepal. The results were as follows: (1) daytime indoor air temperature is 4.6 K less than outdoor air temperature. (2) The daytime surface temperature of the thatched roof is lower than the clay tile and cement tile roof. The results indicate that passive cooling and moisture control effects are found in traditional houses with earthen floors, thatched roof, mud and brick walls and mud vessels which are effective in producing thermal comfort for residents.

Keywords Nepal · Traditional house · Mud vessel · Passive cooling · Thermal environment

38.1 Introduction

Sometimes extremely hot climatic conditions necessitate specially in adapted types of housing, which incorporate a variety of passive cooling methods designed to keep indoor environmental conditions thermally comfortable. In the sub-tropical region, traditional houses are designed to exploit building elements such as earthen floors, eaves, huge mud vessels and brick walls to produce high thermal mass. However, in these same structures, defects can be found. Large gaps in doors and between the top of walls and the roof severely detract from the overall cooling effect, instead creating excessive heat. Firewood combustion in kitchens also contributes to high indoor air temperatures. In addition, the gradual increase in the use of cement roofs further intensifies indoor heat.

There is no known, substantial research on thermal mitigation in traditional Nepalese houses. However, such research is imperative if this low-energy consumption lifestyle is

H. B. Rijal (✉)
Department of Restoration Ecology and Built Environment, Tokyo City University,
3-3-1 Ushikubo-nishi, Tsuzuki-ku, Yokohama, 224-8551 Japan
e-mail: rijal@tcu.ac.jp

to continue. By introducing new techniques into existing houses in sub-tropical Nepal, it would be possible not only to support a comfortable thermal environment but also to build a sustainable society based on energy conservation. The main aim of the present research is to determine the passive cooling techniques now in use in traditional houses in Nepal and to suggest improvements.

In the present research, a thermal environment survey of traditional houses was conducted in the sub-tropical region of Nepal. The main purpose of this research is to evaluate passive cooling methods in summer through the use of semiopen spaces, earthen floors, mud and brick walls, mud vessels and thatched roofs of Banke District [1].

38.2 Method

38.2.1 Investigated Area and Houses

The investigation was carried out in the Banke District (See Chap. 6), which is located in the sub-tropical plain region (latitude, longitude and altitude, 28°4' north, 81°37' east and 144 m, respectively). Generally, the area is hottest in May and coldest in January. Because of the landlocked nature of the country, the climate is dry and hot in summer. The mean outdoor air temperature (relative humidity) is 31.4 °C (53%) in May and 15.2 °C (80%) in January [2]. In summer, the relative humidity is within the comfort zone; thus, it is cool in shaded areas. Agriculture is the main occupation, and firewood is burned in open hearths for cooking year-round (Fig. 38.1).

The investigated houses are shown in Fig. 38.2. In sub-tropical climates, the mud walls are made using a mixture of mud, cattle or buffalo dung, rice husks (2:1:0.5 in ratio) and water, which is then plastered to the internal and external surfaces of the framework of the wooden structure. This wall cannot withstand rain, and consequently residents repair their homes every year in the dry season. The houses are

Fig. 38.1 Overall view of the investigated area (Photo by Kurihara Hiromitsu) [1]

Fig. 38.2 Plan and sections of the measured houses [1] (**a**) Plan (Mud house) (**b**) Section A-A (**c**) Plan (Brick house) (**d**) Section B-B ★: Measurement point, Unit: mm

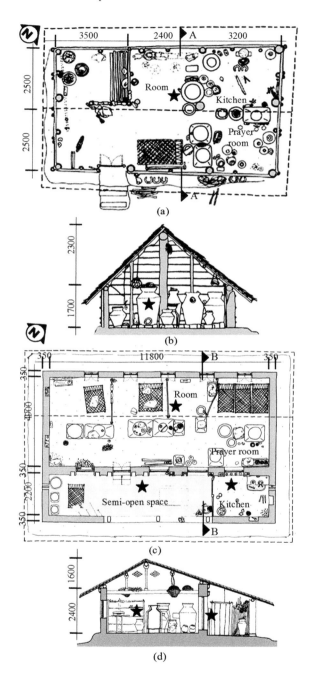

distinguished by a large floor plan, high ceilings and open spaces. The indoor space is divided by mud vessels (Fig. 38.3) and mud walls. Mud vessels are made by fashioning a mixture of rice husks, mud (1:1.5 in ratio) and water into different sizes and shapes. These vessels have the dual function of grain storage and wall partition. The height of this indoor wall is approximately 1.8 m, which maintains an open floor plan, creating the sense of a single room. For the purpose of admitting daylight, the external walls have two to three holes at the height of 1 m above the floor. Because of the weight of the mud wall, its thickness decreases gradually, and a gap of approximately 1 m exists between the top of the wall and roof (Fig. 38.3). This gap provides lighting, ventilation and smoke elimination. In addition, the houses have a working space and cattle shed, which are constructed of wooden pillars (without walls) and a thatched roof. Recently, because of modernization, the brick houses (350 mm thick external dry brick walls) with windows (Fig. 38.2c and d) and cement tile roofs are seen.

38.2.2 Measurements

The investigations were conducted during the summer (May 6–10, 2000) [3, 4]. Air temperature, relative humidity, globe temperature, surface temperatures, solar radiation and wind velocity and direction were measured. The main outdoor and indoor measurement points were at 1.5 m above-ground level and 0.6 m above floor level, respectively. Surface temperatures were measured at the centre of the floor, wall and

Fig. 38.3 Big mud vessel (left) and big gap between the top of the wall and roof (right) [1]

roof. One mud vessel was empty in the brick house; thus, we were able to measure the internal and external surface temperature in the centre of the vessel. All data were measured at 5 min intervals and recorded using a data logger. Measured data were calibrated.

38.3   Results

38.3.1   Semiopen Space

To evaluate the passive cooling effect of the semiopen space, air temperatures outdoors and in the semiopen spaces were compared. In the daytime, the air temperature of the semiopen space was 33.7 °C, which is 2.9 K lower than outdoor (Table 38.1; Fig. 38.4a). This difference in temperature may be due to the radiative cooling of the earthen floors and brick walls, which are not exposed to direct sunlight. The cooler environment of the semiopen space is used for work and rest in the daytime. Furthermore, night-time air temperature and wind velocity in the semiopen space are close to those of the outdoors, and thus the space is used for sleeping. Because of this region's low humidity, semiopen spaces could be developed and used by residents in their everyday life. In passive cooling design, semiopen spaces serve the critical function of providing a thermally comfortable area in which residents can conduct their daily lives.

38.3.2   Indoor Spaces

To evaluate the passive cooling of the indoor spaces, outdoor and indoor air temperatures were compared. In the daytime, the indoor temperature of the bedroom was 4.6 K lower than outdoor (Table 38.1; Fig. 38.4a). The results show that traditional houses are highly effective in mitigating the indoor thermal environment.

Table 38.1 Mean air temperatures indoors and outdoors [1]

House type	Room type	T_{in} & T_{out} [°C]			$T_{in}-T_{out}$ [K]		
		Daily	Daytime	Night-time	Daily	Daytime	Night-time
	T_{out}	32.4	36.6	28.8			
Mud	Bed	31.5	32.1	30.8	−0.9	−4.5	2.0
Brick	Semiopen	31.9	33.7	30.1	−0.5	−2.9	1.3
	Kitchen	35.3	37.7	33.0	3.8	5.6	2.2
	Bed	31.1	32.0	30.1	−1.3	−4.6	1.3

T_{in} & T_{out} indoor and outdoor air temperature, *Daytime* 6:00~18:00, *Night-time* 0:00~6:00 and 18:00~24:00

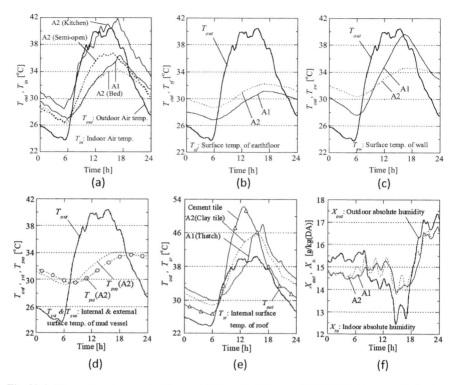

Fig. 38.4 Thermal environment of the mud (A1) and brick house (A2) [1] (**a**) Air temp. (**b**) Surface temp. of earth floor (**c**) Surface temp. of wall (**d**) Surface temp. of mud vessel (**e**) Surface temp. of roof (**f**) Absolute humidity

38.3.3 Position of the Kitchen

The kitchens in mud houses are strategically positioned in the corner of the structure so that heat accumulation due to firewood combustion may be efficiently expelled. However, in the brick house, a modern development of the more traditional mud house, the kitchen is closed off and located adjacent to the semiopen space. This disadvantageous arrangement encourages heat build-up in the kitchen area and diminishes the passive cooling effect of the semiopen space. The air temperature of the kitchen was 37.7 °C in the daytime and rose to a maximum of 41.9 °C (Table 38.1; Fig. 38.4a). This kind of brick house subjects residents to unnecessary discomfort and excessive heat, especially during cooking times, and is highly energy inefficient. Improvements in cooling brick houses are possible if the passive cooling design techniques of traditional houses are adapted into the planning of their brick counterparts.

Table 38.2 Mean surface temperature of floor, wall and mud vessel [1]

House type	Room type	$T_{sf}-T_{out}$ [K]			$T_{sw}-T_{out}$ [K]			$T_{svi}-T_{out}/T_{svo}-T_{out}$ [K]		
		Daily	DT	NT	Daily	DT	NT	Daily	DT	NT
	T_{out}	32.4	36.6	28.8						
Mud	Bed	−3.3	−7.4	0.3	0.8	−2.2	3.3			
Brick	Bed	−1.8	−5.9	1.8	0.4	−3.8	3.9	−0.8/ −0.5	−5.6/ −4.8	3.4/ 3.1

T_{out} outdoor air temperature, T_{sf} & T_{sw} surface temperature of earth floor and south wall, T_{svi} & T_{svo} internal and external surface temperature of the mud vessel, DT (daytime) 6:00~18:00, NT (nighttime) 0:00~6:00 and 18:00~24:00

38.3.4 Earthen Floors

To evaluate the passive cooling effect of floors, the surface temperature of the earthen floor was investigated. In the daytime, the floor temperatures of the mud and brick house were 7.4 K and 5.9 K lower, respectively, than the outdoor temperature (Table 38.2; Fig. 38.4b). Throughout the day, the floor temperature of the mud house is lower than that of the brick house. The difference between the two can be attributed to the placement of ventilation. The mud house is favoured with gaps located aloft between the tops of walls and roof, an arrangement which minimizes the effect of outdoor air on the earthen floor. In contrast, the brick house features windows that are positioned closer to the floor and are typically left open for the entire day. At this height, the outdoor air has a more direct effect on the floor of the brick house. To decrease the floor temperature of the brick house, the position of the windows should be raised to a height similar to the gaps seen in the mud houses. Unfortunately, two-storey houses have started to appear recently, and there is concern that earthen floors will eventually cease to be included in future home design.

38.3.5 Mud Walls and Brick Walls

To evaluate the passive cooling effects of internal walls, the surface temperatures of mud walls and brick walls were compared. In the daytime, the wall temperature was 2.2 K (mud wall) and 3.8 K (brick wall) lower than outdoor temperature (Table 38.2; Fig. 38.4c). The results show that the brick wall is more effective than the mud wall in maintaining the effects of passive cooling. The thermal capacity of mud walls (which tend to be thin, approximately 100 mm) is small, cooling quickly at night. In contrast, the thicker brick walls (350 mm) have a greater thermal capacity and retain heat longer. As mentioned above, mud walls are insubstantial when exposed to rain, forcing residents to repair them every year in the dry season. Technically, if mud walls were built with sufficient thickness and strength, they could produce a thermal effect similar to that of brick walls.

Table 38.3 Internal surface temperature of the roofs

Roof type	Internal surface temperature [°C]		
	Daily mean	Max	Diurnal range
Thatched	37.9	48.1	18.1
Clay tile	39.4	45.7	16.8
Cement	42.2	52.2	25.6

38.3.6 Mud Vessels

To evaluate the passive cooling effects of mud vessels, internal and external surface temperatures of mud vessels were compared in the brick house. In daytime, the internal and external surface temperatures of mud vessels were 5.6 K and 4.8 K lower than the outdoor temperature (Table 38.2; Fig. 38.4d). In the daytime, the internal and external surface temperature was larger than at night. As such, mud vessels are more effective in the daytime. Mud houses are more closed off than brick houses; therefore mud vessels would seem to be more effective in mud houses. Although mud houses are being replaced by brick houses as housing modernization progresses, the utilization of mud vessels is not changing, which is desirable from the viewpoint of the thermal environment.

38.3.7 Roof Materials

To evaluate the thermal properties of the roof, the internal surface temperature of thatched, thick clay tile and cement tile roofs were compared. The daytime internal surface temperature of the thatched roof is lowest among the three surveyed roofs (Table 38.3; Fig. 38.4e). Thatched and clay tile roofs are characterized by a resistance to heat. The effect of heat on cement tile roofs is completely opposite to the others. Due to the thinness of the thatched roof (200 mm), the maximum and diurnal ranges are high. To remedy this problem, the thatched roof might be better insulated by increasing its thickness. The results suggest that modern building materials are not necessarily better for achieving thermal comfort in houses.

38.3.8 Moisture Control Effect

To evaluate the moisture control effect of natural building materials, the absolute humidity in the mud house and brick house was investigated. The indoor absolute humidity in the mud house was 0.5 g/kg(DA) (daytime) higher and 0.6 g/kg(DA) (night-time) lower than outdoor (Table 38.4; Fig. 38.4f). The results show that the moisture control effect is found in earthen floors, mud walls and mud vessels, which is to say that indoor moisture absorption occurs when outdoor humidity is high and indoor moisture discharge occurs when outdoor humidity is low.

Table 38.4 Absolute humidity indoors and outdoors [1]

House type	Room type	X_{in} & X_{out} [g/kg(DA)]			$X_{in}-X_{out}$ [g/kg(DA)]		
		Daily	Daytime	Night-time	Daily	Daytime	Night-time
	X_{out}	15.1	14.2	15.9			
Mud	Bed	15.0	14.7	15.3	−0.1	0.5	−0.6
Brick	Bed	15.2	14.8	15.7	0.1	0.6	−0.2

X_{in} & X_{out} indoor and outdoor absolute humidity, *Daytime* 6:00~18:00, *Night-time* 0:00~6:00 and 18:00~24:00

38.4   Conclusions

For the purpose of evaluating passive cooling effects in summer, a thermal investigation was conducted in the sub-tropical region of Nepal. The following findings were obtained.

1. Daytime indoor air temperature is 4.6 K less than the outdoor air temperature. Passive cooling effects are found in the earthen floors, mud and brick walls and mud vessels, all of which are effective for keeping cool in the summer.
2. Daytime mean air temperature in the kitchen, which burns large quantities of firewood in an open hearth, was 37.7 °C. The results suggest that the problem of excessive heat in the kitchen must be solved.
3. The daytime surface temperature of the thatched roof is lower than the clay tile and cement tile roof. The results show that thermal insulation provided by the thatched roof is most favourable among the three types of roofing.
4. The mean indoor absolute humidity of the mud house is 0.5 g/kg(DA) (daytime) higher and 0.6 g/kg(DA) (night-time) lower than outdoors. Moisture control effect is found in the house with earthen floors, mud and brick walls and mud vessels.

Acknowledgements We would like to give thanks to Prof. Harunori Yoshida and Prof. Noriko Umemiya for their research guidance and valuable advice, to the 'Hata laboratory' for their drawings, to the investigated households for their cooperation and to our families and friends for their support. This research is supported by the Japan Society for the Promotion of Science.

References

1. Rijal HB, Yoshida H, Umemiya N (2005) Passive cooling effects of traditional vernacular houses in the sub-tropical region of Nepal. Proceedings of the 22nd Conference on Passive and Low Energy Architecture (Beirut), pp 173–178
2. H.M.G. of Nepal (1995, 1997, 1999) Climatological records of Nepal 1987–1990, 1991–1994, 1995–1996. Ministry of Science and Technology Department of Hydrology and Meteorology Kathmandu, Nepal

3. Rijal HB, Yoshida H (2002) Comparison of summer and winter thermal environment in traditional vernacular houses in several areas of Nepal. Adv Build Technol 2:1359–1366
4. Rijal HB, Yoshida H, Umemiya N (2002) Summer thermal environment in traditional vernacular houses in several areas of Nepal. J Archit Plann Environ Eng AIJ 557:41–48 (in Japanese with English abstract)

Chapter 39
Passive Cooling Strategies to Reduce the Energy Consumption of Cooling in Hot and Humid Climates in Indonesia

Tomoko Uno, Shuichi Hokoi, and Sri Nastiti N. Ekasiwi

Abstract This chapter proposes strategies for reducing the energy consumption of cooling in residences in hot and humid climates. Based on the results of field measurements and questionnaire surveys in Surabaya, Indonesia, a simulation of indoor thermal environments – with consideration of air conditioner operation – was conducted to evaluate energy consumption by air conditioners. This simulation program considers both heat and moisture transfers in building materials. The combined effects of building airtightness and the opening times of doors and windows for ventilation were examined. The simulation is conducted for the rainy season. Making the whole building airtight results in a small reduction in energy consumption, because the size of the air-conditioned area increases. Meanwhile, making only the air-conditioned room airtight is more effective for reducing cooling energy consumption. Regardless, nighttime ventilation of non-air-conditioned spaces is quite effective in reducing the sensible cooling load. In conjunction with economic growth and requirements for improving indoor environmental quality, energy consumption for cooling is expected to increase rapidly in hot and humid areas. Therefore, the introduction not only of high levels of insulation but also of airtightness, in addition to well-controlled ventilation, is required to achieve energy savings.

Keywords Use of air conditioner · Ventilation operating time · Airtightness · Energy saving for cooling · Hot and humid climate · Field survey

This chapter was revised the paper [4] in references, based on the recent researches.

T. Uno (✉)
Department of Architecture, Mukogawa Women's University, Nishinomiya, Hyogo, Japan
e-mail: uno_tomo@mukogawa-u.ac.jp

S. Hokoi
Kyoto University, Kyoto, Japan

Southeast University, Nanjing, China

S. N. N. Ekasiwi
Department of Architecture, Institut Teknologi Sepuluh Nopember, Surabaya, Indonesia

© Springer Nature Singapore Pte Ltd. 2018
T. Kubota et al. (eds.), *Sustainable Houses and Living in the Hot-Humid Climates of Asia*, https://doi.org/10.1007/978-981-10-8465-2_39

39.1 Introduction

In countries with a tropical climate, such as Indonesia, air conditioning (henceforth AC) is now used not only in offices but also in residences to create a comfortable indoor thermal environment [1]. This change is causing increases in household energy consumption.

To reduce energy consumption by air conditioning in tropical regions, researchers have previously considered strategies such as solar shading, natural ventilation, and façade design. The effects and the usages of these passive strategies are discussed in other chapters of this book. Particularly, many researchers have studied the advantages of buildings that use natural ventilation. However, natural ventilation is dependent on outside wind and other conditions and is not always available. Instead of relying on the airflow of natural ventilation, people have started to use AC. Therefore, we should consider the combination of AC use and natural ventilation as a means of effectively saving energy.

According to a field survey on the thermal environment and AC use in Surabaya, Indonesia [2–4], it was determined that residents of houses without AC lived under relatively hot conditions both day and night. More specifically, living room temperatures reached the same level as outside temperatures during daytime because of the large air exchange rates and heat transfer through ceilings that were heated by solar radiation. Additionally, nighttime bedroom temperatures did not decrease, as a combination of reduced ventilation and the high heat capacity of the building materials kept the bedroom temperature 2–3 °C higher than the outside temperature at night.

According to the results of the questionnaire, in houses with AC, residents first installed an AC in their bedroom, and more than 80% of residents used their AC from when they went to sleep until the morning [3]. They began to use AC at temperatures of 28–30 °C; the temperature then dropped to 25 °C and continued to decrease until morning. Sometimes, the temperature during the night fell below 24 °C, which seems too low for sleeping. It is likely that the AC is set at this low temperature because the high thermal capacity of the building materials keeps the indoor surface temperature high, and the high rate of air exchange caused by the building's low airtightness easily increases the room temperature. As a result, the residents tend to use the low-temperature setting to lower the room temperature quickly, and they continue to use AC for a long period of time.

As this situation indicates, from the perspective of saving energy, the thermal performance of today's houses unfortunately makes them unsuitable for the effective use of AC. Because these houses are not airtight and are poorly insulated, residents may be forced to lower the temperature setting, resulting in very low room temperatures. So, it is necessary to use AC more effectively, as well as to design buildings that are suitable for cooling.

In hot and humid climates, both temperature and moisture affect the energy consumed by AC. Because porous (hygroscopic) building materials are often used, the ab-/desorption properties of wall surfaces should be considered when evaluating latent cooling load, as they can have a significant effect on the energy consumed by AC. In this chapter, by using a simulation program that takes into account AC operation and the ab-/desorption of moisture by building elements, room temperature and humidity were calculated from the viewpoints of energy consumption and indoor thermal comfort. The simulation program was examined through a comparison of the calculated and the measured room air temperatures and humidity ratios. The effects of airtightness and ventilation strategy were examined.

39.2 Background of Simulation of Thermal and Moisture Conditions

39.2.1 Area and Climate

The investigated area was in Surabaya City (7°S 113°E), which is located in the eastern part of Java island in Indonesia. This island has both dry and wet seasons; the dry season is from April to November, while the wet season is from December to March. The monthly average temperature ranges from 27.2 to 29 °C, with an annual mean temperature of 28 °C [1]. The hottest months are October and November. The monthly average relative humidity ranges from 65 to 80%. The investigations reported in this chapter are mainly based on results measured and surveyed in February, during the wet season.

39.2.2 Plan and Details of the Simulated House

The residences analyzed in this study are semidetached houses that are quite common in urban areas in Java, Indonesia; the plan of these houses is shown in Fig. 39.1. Bricks are the main construction material used in these houses. On the windows and doors, there are many openings and upper panes that are left permanently open for ventilation in the non-air-conditioned room. This study considers temperature and humidity levels in the five rooms and one attic space (R6) of the houses.

---

[1]The monthly average temperature and humidity are the data between 1991 and 2010 at Juanda Airport meteorological station, which are calculated by Meteonorm ver7.

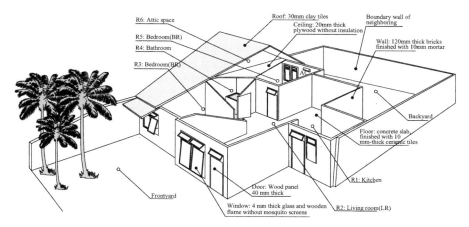

Fig. 39.1 Detail of the target houses

39.2.3 Outdoor Conditions Used in the Simulation

Outdoor temperature, humidity, and global solar radiation, measured between 17 and 23 February in 2001, are used as the inputs for outdoor conditions; these measurements were taken at an experimental house at the Institute of Technology Surabaya, close to the surveyed area. The measured solar radiation is separated into direct and diffuse components by using the Bouguer and Berlage equations [5]. The data are used repeatedly as inputs, with a cycle of 1 week.

The calculations are continued until the ground temperature becomes independent of the initial conditions. The outdoor temperature and humidity, averaged over the measured period, are used as initial conditions, and the ground temperature at a depth of 5 m is assumed to be the same as the average outdoor temperature.

39.3 Detail of the Simulation of Thermal and Moisture Conditions

39.3.1 Outline of the Simulation

The heat and moisture balances of the rooms (R1–R5 in Fig. 39.1) and an attic space (R6), as well as those rooms' building elements, are simulated. In calculating a room's temperature, the heat and moisture fluxes into the room, as shown in Fig. 39.2, are considered [4]. The volumes of inlet air into and outlet air from the room are balanced by the ventilation air volume when the air exchange rates are estimated and used as an input data. These values were estimated in order to obtain agreement between the measured and calculated temperatures [6]. The

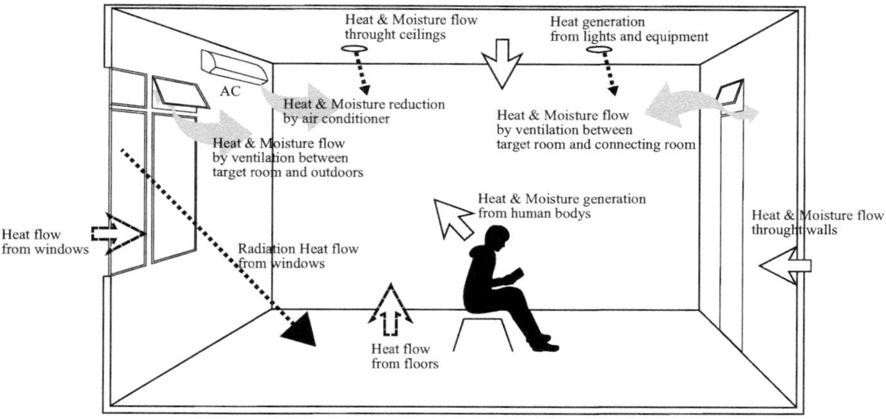

Fig. 39.2 Factors considered in the simulation of heat and moisture balances

simultaneous transfer of heat and moisture are taken into consideration in calculations of the temperature and humidity of the porous building materials used in the ceilings and wall [4, 8].

In this simulation program, the energy consumption of AC compressors is calculated. A simplified model of a heat pump is used, based on the heat and mass balances of the refrigeration cycle [7]. To evaluate the effects of various measures that could be used to reduce energy consumption, the energy consumption levels for cooling loads (total, sensible heat, and latent heat) are used.

39.3.2 Air Exchange Rates in Simulated Houses

Cross ventilation is generally considered important in Indonesian houses, which have many openings and low airtightness. In the houses being measured, the small windows above the door are always open. Thus, there was a certain amount of ventilation even if the external windows were closed [6].

The main factors affecting air exchange volume are the condition of the openings (open/closed) and the outdoor wind velocity and direction. The two patterns of air exchange, i.e., openings are opened or closed, make the analysis simpler and clearly indicate the impact that patterns of window opening and building airtightness have on the energy consumed by AC. The input values of the air exchange were estimated from 1 to 12 times/hour between the relevant room and outside and from 0 to 40 times/hour between the relevant room and connected rooms. These values are considered in the simple simulations, where the air exchange rates between the room and the outside are 23 times/hour in the living room (R2), 29 times/hour in the

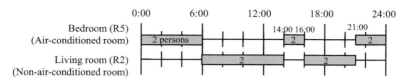

Fig. 39.3 The number of persons remaining in the rooms [4]

kitchen (R1), and 14 and 19 times/hour in the bedrooms (R3 and R5, respectively) [6]. The optimum values are determined as the input air exchange rate, which is the best fit for the measurement results of temperature and humidity [6].

39.3.3 Use of AC

In calculations performed to evaluate measures intended to reduce energy consumption by AC, it is assumed that one AC is installed in the bedroom (R5) in House 2 and that it is used from 14:00 to 16:00 and from 21:00 to 6:00, which is decided depending on the field measurement. The set-point temperature is assumed to be 25.5 °C. The cooling and moisture loads of the AC are calculated taking into consideration the heat and moisture balance between the indoor/outdoor air and the heat pump [7]. Two residents spend their time in the living room during the day, except from 14:00 to 16:00 when they move to the bedroom for a nap (Fig. 39.3). When residents remain in the bedroom, it is assumed that the AC is operated continuously.

39.3.4 Consideration of Energy Reduction by Airtightness and Ventilation Control

The effects of airtightness and adjustments of the ventilation schedule were examined as alternative methods of reducing cooling energy consumption. When ACs are used in several rooms of the house, improving the airtightness of the entire house would be best. On the other hand, because in the current model, AC is used only in the bedroom, improvements only to the airtightness of the bedroom might be effective enough to reduce energy consumption.

Two cases of airtightness are considered: one in which airtightness between the air-conditioned room and the other rooms is high (Fig. 39.4, case 2) and another in which airtightness is increased between the outdoors and the entire house (Fig. 39.4; case 3). In total, three cases are examined, including a non-airtight case (Fig. 39.4; case 1). In the simulations, the air exchange rate when airtightness is improved is

Fig. 39.4 Image of
improvement in airtightness

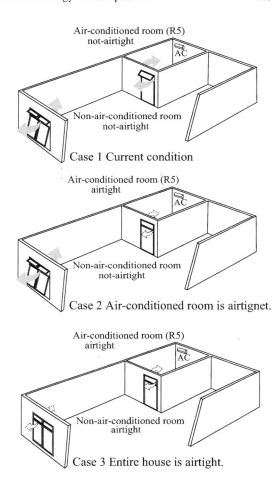

Air-conditioned room (R5)
not-airtight

Non-air-conditioned room
not-airtight

Case 1 Current condition

Air-conditioned room (R5)
airtight

Non-air-conditioned room
not-airtight

Case 2 Air-conditioned room is airtignet.

Air-conditioned room (R5)
airtight

Non-air-conditioned room
airtight

Case 3 Entire house is airtight.

assumed to be 0.5 times/hour as the minimum required air exchange ratio; the air exchange volumes are 38 m³/h in the living room and 4.5 m³/h in the bedrooms.

To determine the energy savings achievable by changes in the ventilation schedule, calculations are performed for four patterns of window opening schedules as shown in Fig. 39.5. Generally, residents open apertures during daytime to maintain airflow inside the house. Figure 39.5 shows the ventilation patterns used. Pattern A in Fig. 39.5 corresponds to the typical situation in Indonesia. The residents open windows during the daytime and close them at night. In pattern B in Fig. 39.5, the openings in the air-conditioned room are closed all day long, while all rooms are closed in pattern C. In pattern D, the windows are closed in the air-conditioned room, and the openings in the non-air-conditioned rooms are closed during daytime and open at night, providing nighttime ventilation.

Based on these three cases of airtightness and the four ventilation schedule patterns, simulations were carried out for the 12 resulting combinations.

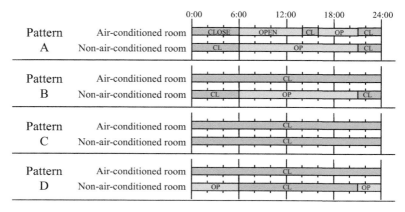

Fig. 39.5 Ventilation schedule [4]

39.3.5 Other Preconditions of the Houses Being Measured

Many factors need to be considered when attempting to save energy through the use of AC use. For example, reducing solar radiation into the air-conditioned room and insulating walls to reduce heat flow both appear to be effective. It is clear that internal insulation of the wall is better for the effective use of AC because it minimize the influence of the thermal capacity of the walls.

The houses studied here have been well designed to ensure shade from solar radiation. However, even now, the insulation of housing is not common in Indonesia. According to the simulation results for the houses studied here, a considerable amount of heat flows into rooms through the houses' thin wooden ceilings. To clarify the effects of the airtightness and ventilation patterns, the influence of other variables should be minimized. Therefore, the simulation presupposes 50 mm-thick insulation in the ceiling. Of course, insulation of the room's side of the exterior walls is effective at reducing the influence of the walls' thermal capacities and at quickly reducing the rooms' temperature.

39.4 Results and Discussion

39.4.1 Comparison Between Measured and Calculated Results

For the purposes of validating the simulation program, the calculated temperature and humidity ratios were compared with the measured results in Figs. 39.6 and 39.7. The air exchange volumes in this simulation were determined as a function of only two factors: the residents' window-opening pattern and whether the measurement was taken during the day or night. The calculated temperatures in the bedroom of

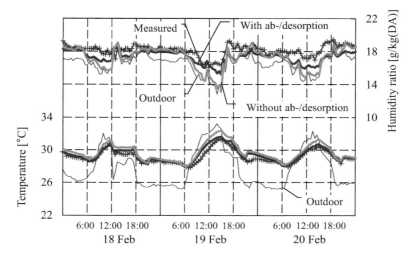

Fig. 39.6 Temperature and humidity ratios of BR in House 1 (Comparison between measured and calculated results) [4]

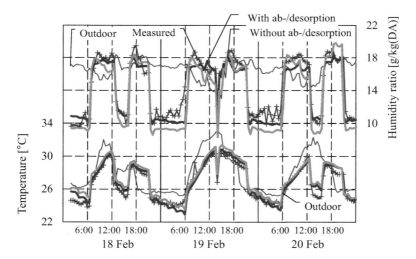

Fig. 39.7 Temperature and humidity ratios of BR in House 2 (Comparison between measured and calculated results) [4]

House 1 (Fig. 39.6) and in the bedroom of House 2 (Fig. 39.7) replicate the measured data very well. When the ab-/desorption of the walls are not considered, the humidity ratio is close to the outdoor value because ventilation is the dominant influence on changes in humidity. The calculated humidity ratio also closely matches the measured results when considering ab-/desorption, and ab-/desorption is quite important for predicting the humidity ratio.

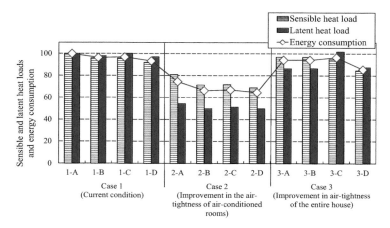

Fig. 39.8 Sensible and latent loads and energy consumption. The results are normalized to those for the standard case (Case 1-A), with a reference value of 100 [4]

39.4.2 Strategy for Reducing Energy Consumption

Figure 39.8 shows the calculated sensible and latent loads, along with energy consumption, for the 12 cases.

Effect of Airtightness (Cases 1, 2, and 3; Fig. 39.8)

The results in Case 2, in which airtightness is improved only in the air-conditioned rooms, show a 25% reduction in energy consumption and a 50% reduction in latent load. On the other hand, in Case 3, a maximum reduction in energy consumption of only 5–15% is recorded due to an increased cooling area. In other words, all rooms are air-conditioned by the AC in the bedroom. Because the humidity ratio of outdoors and in the non-air-conditioned rooms is high compared with that of the air-conditioned room, the latent load is larger. Therefore, an effective energy saving step is to reduce air exchange, not only between the air-conditioned room and the outdoors but also between the air-conditioned and non-air-conditioned rooms.

Effect of Ventilation Pattern (Patterns A, B, C, and D in Fig. 39.8)

In all cases of different degrees of airtightness (Cases 1, 2, and 3), energy consumption is decreased compared with pattern A, the standard ventilation pattern. It is clear that, from the perspective of the energy and AC load, ventilation during the daytime, when the room is not air-conditioned, introduces hot outdoor air into the rooms, which causes an increase in the cooling load. In Case 3, the energy consumption in pattern C is much lower. Because human bodies in the connected non-air-conditioned rooms generate heat and moisture loads (Fig. 39.3), both the temperature and humidity ratios become higher if the openings are closed all day long. On the other hand, both the temperature and humidity ratios are reduced when there is ventilation between the rooms and the outdoors during nighttime, as in Case 3-D.

In Cases 1-D, 2-D, and 3-D, which utilize nighttime ventilation, energy consumption is the lowest out of all the cases considered. Nighttime ventilation reduces the temperature in the non-air-conditioned rooms because the outdoor temperature is lower than in those rooms, thus reducing the sensible load. In one case (Case 2-D), the reduction in energy consumption is not large because the ventilation rate is restricted by partial airtightness.

These results indicate that nighttime ventilation in non-air-conditioned rooms is effective when there is a large amount of ventilation between the air-conditioned and non-air-conditioned rooms.

39.5   Conclusions

This chapter examined, based on the results of simulation, measures to reduce the energy consumption of cooling residences in hot and humid climates. The results of the numerical analysis of heat and moisture transfer are shown, taking the results of the field survey into consideration.

Improvements in the airtightness of air-conditioned rooms are effective in reducing energy consumption. This is because such improvements reduce the heat and moisture loads that are due to ventilation from the outdoors. On the other hand, because improvements in the airtightness of the entire house increase the total air-conditioned area, the net reduction in energy consumption for cooling is small.

Regardless of the degree of airtightness, nighttime ventilation of non-air-conditioned rooms is useful for reducing the sensible load.

References

1. Statistics Indonesia. http://www.bps.go.id/
2. Uno T, Hokoi S, Ekasiwi SNN, Funo S (2003) A survey on thermal environment in residential houses in Surabaya, Indonesia. J Archit Plan Environ Eng 564:9–15
3. Uno T, Hokoi S, Ekasiwi SNN (2003) Survey on thermal environment in residences in Surabaya, Indonesia: use of air conditioner. J Asian Archit Build Eng 2(2):15–21
4. Uno T, Hokoi S, Ekasiwi SNN, N Hanita AM (2012) Reduction of energy consumption by AC due to air tightness and ventilation strategy in residences in hot and humid climates. J Asian Archit Build Eng 11(2):407–414
5. Hokoi S, Ikeda T, Nitta K (2002) Ace architectural environmental engineering II – heat, moisture and ventilation. Asakura Publishing Co. Ltd., Tokyo, pp 4–8

6. Ekasiwi SNN (2007) Passive method for improving indoor thermal environment for residential buildings in hot-humid region (Indonesia). Ph.D thesis submitted to Kyoto University
7. Ito K, Ohkouchi K, Shibata K (1985) Multi objective optimal design of a heat pump system by considering partial load for air-conditioning. Trans of SHASE Jpn SHASE 29:51–61
8. Matsumoto M, Hokoi S, Ka E (1997) An analysis of coupled heat and moisture transfer in buildings considering the influence of radiant heat transfer. ASHRAE Trans 103(Part 1):573–583

Chapter 40
Indoor Thermal Environments in Apartments of Surabaya, Indonesia

Tetsu Kubota, Muhammad Nur Fajri Alfata, Meita Tristida Arethusa, Tomoko Uno, I Gusti Ngurah Antaryama, Sri Nastiti N. Ekasiwi, and Agung Murti Nugroho

Abstract This chapter presents the results of detailed field investigations of thermal conditions and occupants' window-opening behaviour in several apartments located in the city of Surabaya, Indonesia. In the public apartments, almost all of the respondents did not use air-conditioning, and approximately 70–80% opened their windows/doors on both front and rear sides, while 20–40% of them kept the rear opening opened at night. Meanwhile, most of the respondents in the high-rise private apartments depended on air-conditioning, and about 20% of them opened the rear window only during daytime. The results of field measurement showed that under the naturally ventilated conditions, the old public apartment unit provided better thermal conditions compared to those in the other types of apartments. It was difficult to achieve the thermal comfort without relying on air-conditioning in the high-rise private apartments.

Keywords Thermal comfort · Natural ventilation · Window-opening behaviour · Energy saving · Hot-humid climate

T. Kubota (✉)
Graduate School for International Development and Cooperation (IDEC), Hiroshima University, Hiroshima, Japan
e-mail: tetsu@hiroshima-u.ac.jp

M. N. F. Alfata
Research Institute for Human Settlement and Housing – Ministry of Public Works, Bandung, Indonesia

M. T. Arethusa
PT. Hadiprana Design Consultant, Jakarta, Indonesia

T. Uno
Department of Architecture, Mukogawa Women's University, Nishinomiya, Hyogo, Japan

I. G. N. Antaryama · S. N. N. Ekasiwi
Department of Architecture, Institut Teknologi Sepuluh Nopember, Surabaya, Indonesia

A. M. Nugroho
Department of Architecture, University of Brawijaya, Malang, Indonesia

© Springer Nature Singapore Pte Ltd. 2018
T. Kubota et al. (eds.), *Sustainable Houses and Living in the Hot-Humid Climates of Asia*, https://doi.org/10.1007/978-981-10-8465-2_40

40.1   Introduction

Indonesia has been experiencing high economic growth over the last few decades, and the middle class is now on the rise. According to the report [1], the proportion of the middle class, whose daily expenditure is US$ 2–20, was only 25.1% out of the total population in 1999, but it became 42.8% as of 2009. Meanwhile, the nation-wide final energy demand rose by 14 times over the last four decades. In order to accommodate the growing number of middle-class population, the number of high-rise apartments has been particularly increasing in major cities. In the city of Surabaya, for example, the number of apartments doubled over the last 5 years [2]. In these high-rise apartments, the lack of climatic consideration in building design results in severe indoor thermal conditions; thus, most of the households depend on the use of air-conditioning for thermal comfort. This chapter reports the results of detailed field investigations of thermal conditions and occupants' window-opening behaviour in several apartments located in the city of Surabaya, focusing especially on the high-rise apartments.

40.2   Apartments in Surabaya

Surabaya (7°9′21″S 112°36′57″E) is the second largest city, after Jakarta, in Indonesia with a population of more than 3 million as of 2012. The city has a hot-humid climate with monthly average temperature of 27–29 °C and the monthly RH of 66–81% [3]. Meanwhile, the monthly average wind speeds range from 2.1–3.1 m/s [3].

In Indonesia, apartments are broadly classified into the following three types: old public apartments, new public apartments and private apartments (Fig. 40.1). Air-conditioning is normally used only in the private apartments at the moment. In Surabaya, there are approximately 48 housing estates consisting of apartments as of 2013, and their number is rapidly rising. The public apartments have been constructed particularly from the 2000s. In the early stage, most of the public apartments were provided for low-income earners (e.g. for slum resettlements).

Fig. 40.1 Views of three types of apartments, (**a**) old public apartment, (**b**) new public apartment and (**c**) private apartment

Table 40.1 Brief profile of respondents

	Whole sample	Old public	New public	Private
Sample size	347	209	101	38
Household size (persons)	3.5	3.6	3.5	1.9
Ethnic group (%)				
Javanese	88.1	92.8	85.7	59.3
Maduranese	6	4.8	9.2	3.7
Others	6	2.4	5.1	37
Monthly income (%)				
<1 million (IDR)	13.3	9.8	10.4	–
1–2 million	39.8	50.2	25	20
2–3 million	22.8	22	29.2	10
3–4 million	13	11.7	15.6	30
>4 million	11.1	6.3	19.8	40
No. of apartments				
Built before 2008	7	5	–	2
Built after 2008	9	–	3	6
Floor area (m^2)	18–38	18–21	24–32	18–38

Since 2008, the Indonesian government extended the target to wider-income groups including low- to mid-income earners. These two apartment types, namely, 'old public apartments' and 'new public apartments', are clearly distinguishable as shown in Fig. 40.1a, b. In contrast, private apartments emerged from 2005 in the case of Surabaya. They normally target middle- to high-income earners.

40.3 Survey on Window-Opening Behaviour

The survey was conducted from September to October 2013 in Surabaya. A total of 347 households from eight public and eight private housing estates were selected through a proportional stratified sampling (Table 40.1). In the survey, the following information was obtained through face-to-face interview: (1) socio-demographic profile, (2) thermal sensation and preference, (3) duration and reasons for opening windows/doors and (4) usage of cooling appliances. The obtained answers were based on their responses to the average condition in the living room. The results are summarised in the following sections, but the details are presented in [4].

40.3.1 Profile of Respondents

As shown in Table 40.1, household sizes are averaged at 3.5–3.6 persons for the public apartments and 1.9 persons for the private apartments. As expected, the

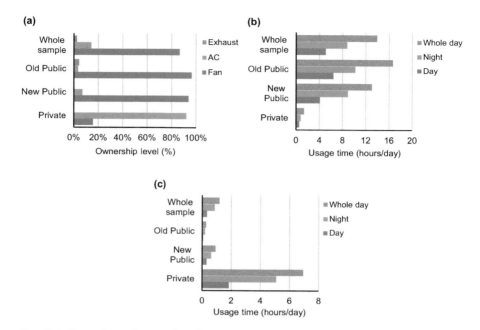

Fig. 40.2 Ownership and usage of cooling appliances. (**a**) Ownership levels, (**b**) usage time of fan (s) and (**c**) usage time of air-conditioning

monthly household income is the highest in the private apartments. Majority of the respondents in the old public apartments have lived in the apartments for more than 5 years (80%), whereas almost all in the new public and private apartments moved in less than 5 years ago (>95%).

40.3.2 Usage of Cooling Appliances

The results show that more than 90% of the respondents in both public apartments own one or more fans (most of them are standing fans), whereas more than 90% of those in the private apartments are equipped with air-conditioners (Fig. 40.2). During the dry season, the daily usage time of fan(s) is averaged at 16.7 h for the old public apartments, 13.1 h for the new public apartments and 1.4 h for the private apartments, respectively. The households in the public apartments tend to use fan (s) longer during night-time (10.2 and 8.8 h) than daytime (6.4 and 4.1 h). The usage of air-conditioning is the highest in the private apartments (5.1 h during night-time and 1.9 h during daytime).

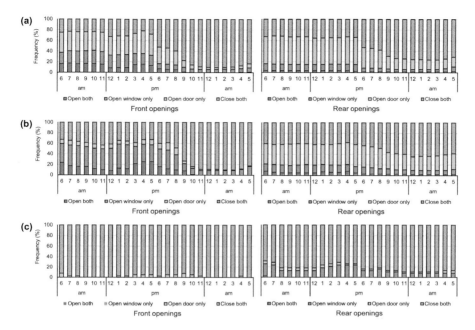

Fig. 40.3 Daily usage pattern of windows/doors (**a**) old public apartments, (**b**) new public apartments and (**c**) private apartments

40.3.3   Window-Opening Behaviour

Figure 40.3 shows the frequency of respondents who customarily open/close their doors/windows per day during the dry season. As shown, the respondents in the public apartments (both old and new) have relatively high usages of both front/rear doors and front window particularly during daytime (30–60%), while those in the private apartments rarely open doors/windows (0–30%). In general, most of the respondents in the public apartments tend to open only one opening for one side of units at one time. Less than 20% of the respondents open both windows and doors even during daytime. In the old public apartments, more respondents tend to operate the front door (40–60%) compared to the front window (30–40%). In contrast, the respondents in the new public apartments tend to open the front window (50–60%) more than the front door (20–40%). The usage patterns of rear openings are similar between old and new public apartments. About 40–50% of the respondents open the rear door only without opening the rear window at the same time. At night, the respondents in both apartments tend to close their windows, except that 20–30% continue to open the rear door throughout the night.

Figure 40.4 presents major reasons for respondents to open their windows/doors. As shown, 'obtaining fresh air (74%)', 'letting wind to enter (66%)' and 'providing cooling (45%)' are the highest reasons among the respondents. This implies that the respondents particularly expect ventilation and airflow through opening windows/

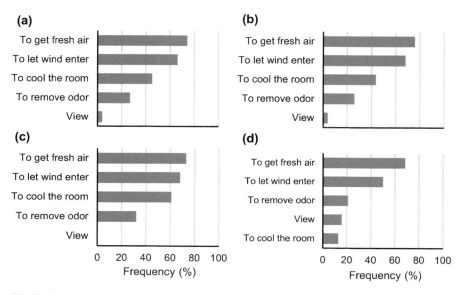

Fig. 40.4 Reasons for opening windows/doors. (**a**) Whole sample, (**b**) old public apartments, (**c**) new public apartments and (**d**) private apartments

doors. On the other hand, the reasons for not opening windows/doors are 'privacy (43%)', 'insects (41%)' and 'security (30%)' (Fig. 40.5). As shown, the major reasons are different for each apartment type: 'AC usage (53%)' for the private apartments, 'insects (53%)' for the old public apartments and 'privacy (57%)' for the new public apartments. As discussed previously herein, the pattern and average duration of opening front window and front door were different between old and new public apartments, unlike the rear door and window (see Fig. 40.3). It was seen that the respondents in both public apartments tend to open either front door or front windows, but not both. This means that when the respondents in the new public apartments cannot open their front door, they compensate it by opening front window instead. Figure 40.5 implies that the households in the new public apartments are more concerned about privacy and security than those in the old public apartments. In the old public apartments, the concerns of privacy and security should be less because most of the residents were relocated from the same areas. The importance of privacy issue for window-opening behaviour was also suggested in the previous studies in Singapore [5] and Hyderabad [6].

40.3.4 Thermal Sensation and Preference

Figure 40.6 presents the survey results of their average thermal sensations and preferences during daytime in the dry season. The sensation was measured in

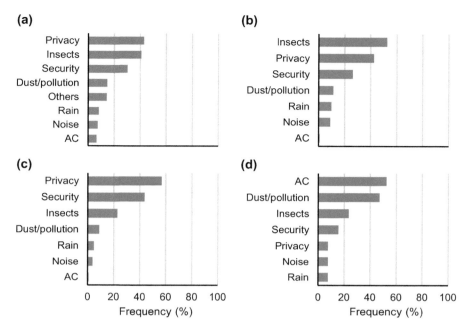

Fig. 40.5 Reasons for not opening windows/doors. (**a**) Whole sample, (**b**) old public apartments, (**c**) new public apartments and (**d**) private apartments

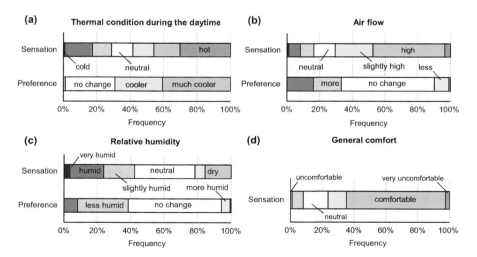

Fig. 40.6 Sensations and preferences for indoor environmental conditions

7-point scale (from cold to hot), while the preference was asked in 5-point scale (from much warmer to much cooler). As for the thermal sensation, more than 57% of the respondents answered 'warm' through 'hot'. Accordingly, the preference for cooler environment is evident (>68%). Despite the use of air-conditioning, more than 97% of the respondents in the private apartments prefer cooler indoor conditions. Meanwhile, as for the airflow, more than 70% of the respondents consider it to be 'slightly high' through 'very high'. More than half of the respondents do not prefer to change the current conditions, while about 30% prefer even higher airflow. This result also highlights the high preference and priority for sufficient indoor airflow. In general, however, more than 60% of the respondents regard their thermal conditions as 'comfortable', even though they prefer cooler conditions.

40.4 Field Measurement on Indoor Thermal Conditions

The field measurements were conducted in the three types of apartments during the dry season, where the above-mentioned survey was conducted, from August to October 2014 [7, 8]. One of the typical apartments from these three types was selected, respectively, for the field measurements. The selected old public apartment is the four-storey building with 48 units (see Fig. 40.1a). The measurement was conducted in the south-facing unit located on the second floor. The unit has a standard size of 21 m^2 with a relatively wide well-ventilated corridor space of 3.1 m width. A multifunctional balcony is placed in between the indoor unit and outdoors (Fig. 40.7a). Meanwhile, the new public apartment is the five-storey building with 96 units (see Fig. 40.1b). The measurement was conducted in the west-facing unit located on the top floor. The unit size is slightly larger (24 m^2), but the corridor space is narrower (1.9 m width) than those of the old public apartment

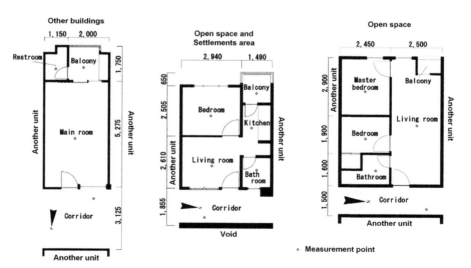

Fig. 40.7 Floor plans of measured apartment units, (**a**) old public apartment, (**b**) new public apartment, and (**c**) private apartment

(Fig. 40.7b). The corridor space is semi-open and adjacent to the void space. Basically, both of the public apartments are naturally ventilated. The structures of both apartments are of reinforced concrete block for walls with relatively high thermal capacities. Meanwhile, the private apartment is the 19-storey high-rise building comprising 762 units (see Fig. 40.1c). Almost all of the units are equipped with one or two air-conditioner(s). The measurement was conducted in the west-facing unit located on the eighth floor, with a total floor area of 31 m^2 (Fig. 40.7c). The small balcony is only used for placing the outdoor units of air-conditioners. Unlike the above public apartments, the long corridor space is not well-ventilated. Reinforced concrete construction is used for the main structure, while precast concrete and lightweight concrete are used for external wall and partition walls, respectively. The per-unit thermal mass is calculated at 1100 kg/m^2 which is 41–46% smaller than those for public apartments. Thermal insulation is not adopted to ceiling/roof and walls in all the three apartments.

Major thermal parameters including air temperature, RH, air velocity and globe temperature were measured at the centres of main rooms at 1.1 m above floor in the selected units from the three apartments, respectively (see Fig. 40.7). Air temperatures in the other spaces including corridor spaces were also measured. These indoor measurements were conducted in unoccupied units under different natural ventilation conditions (i.e. daytime ventilation, night ventilation, full-day ventilation and no ventilation). For the daytime ventilation, windows and doors were opened from 6:00 to 18:00. In contrast, all of them were opened from 18:00 to 6:00 during the night ventilation condition. The measurement for each ventilation condition was conducted for 6–7 days.

40.4.1   Indoor Thermal Environments

The outdoor air temperature during the field measurements ranges from 22.2 to 34.6 °C with the average of 28.3 °C, while the outdoor RH ranges from 22% to 86%. Figure 40.8 shows the temporal variations of air temperature, relative humidity and wind speed in different spaces during the daytime ventilation in all types of apartments, respectively. The daytime ventilation represents the current practice of most of the occupants. As shown, indoor air temperature in the old public apartment is 3.1–3.8 °C lower than the corresponding outdoor air temperature during the peak hours, whereas the nocturnal indoor temperature is approximately 5.0 °C higher than the outdoors. In the old public apartment, indoor air temperature is well stabilised due to the relatively high thermal mass (i.e. 2031 kg/m^2). Although the large corridor space is well ventilated during the daytime, the corresponding air temperature maintains lower than the outdoors. It can be seen that the corridor space and balcony play a role as a thermal buffer in reducing the heat gains.

In contrast, indoor air temperatures in the master bedrooms in both new public and private apartments are higher than the corresponding outdoor air temperature during most of the hours (Fig. 40.8). Unlike the old public apartment, the master

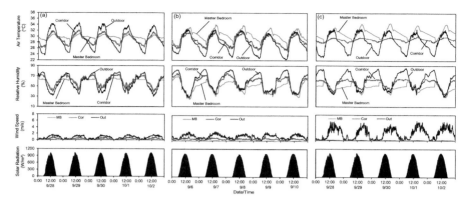

Fig. 40.8 Temporal variations of indoor thermal parameters during daytime ventilation in (**a**) an old public apartment, (**b**) new public apartment and (**c**) private apartment

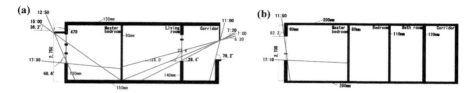

Fig. 40.9 Sun paths through windows in (**a**) a new public apartment and (**b**) private apartment

bedrooms of both apartments do not have balconies. During the peak hours, indoor air temperature in the master bedrooms of both apartments are up to 2.7 °C and 1.8 °C higher than the outdoors, respectively. The sharp peaks of indoor air temperatures observed during the afternoon are mainly due to the building orientation (both units face west). The new public apartment units received direct solar radiation in both living and master bedroom (Fig. 40.9).

40.4.2 Evaluation of Thermal Comfort

Evaluation of thermal comfort in the master bedrooms was conducted by using the ACE, which is an adaptive comfort standard developed for the hot-humid climates [9]. As shown in Fig. 40.10, the indoor operative temperatures in all the three apartments exceed the 80% upper limits for the whole daytime periods. As indicated in Table 40.2, full-day ventilation gives relatively shorter exceeding periods than the others in the cases of the old public and new public apartments, while night ventilation obtains the shortest exceeding period in the private apartment. It should be noted that thermal comfort cannot be achieved throughout the day in all the apartments if all of the windows and doors are closed (i.e. no ventilation condition).

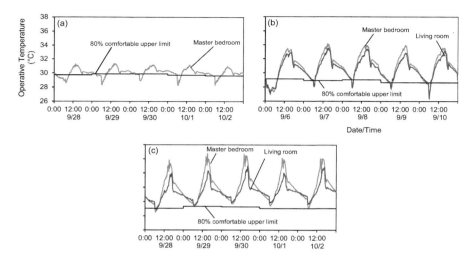

Fig. 40.10 Thermal comfort evaluation using adaptive comfort equation during daytime ventilation in (**a**) an old public apartment, (**b**) new public apartment and (**c**) private apartment

Table 40.2 Results of adaptive thermal comfort evaluation in the master bedrooms

Type of apartments	Ventilation conditions	Deviation of indoor operative temperature and 80% comfortable limit		Exceeding period
		Daily maximum (°C)	Daily minimum (°C)	(%)
Old public	Full-day	0.7 to 1.5	−1.4 to −2.3	34
	Daytime	1.3 to 1.5	−1.0 to −1.6	76
	Night	0.7 to 0.9	−1.0 to −2.4	48
	No	0.9 to 1.1	−0.2 to 0.2	89
New public	Full-day	12.5 to 16.0	−1.4 to −3.4	70
	Daytime	13.6 to 17.4	−0.7 to −1.9	88
	Night	15.5 to 18.8	−1.7 to −2.9	72
	No	9.4 to 17.5	−0.4 to 0.6	96
Private	Full-day	5.3 to 5.6	−0.1 to −2.2	79
	Daytime	6.7 to 7.8	−0.6 to 0.4	99
	Night	4.0 to 5.8	−2.8 to 0.3	72
	No	3.5 to 5.6	0.1 to 1.2	100

The above results clearly indicate that it is very difficult to achieve thermal comfort particularly in both new public and private apartments by means of natural ventilation alone, especially when the units are oriented towards west.

40.5   Conclusions

In Indonesia, the number of high-rise apartments has been particularly increasing in major cities, in order to accommodate the growing number of middle-class population. This chapter reported the results of detailed field investigations of thermal conditions and occupants' window-opening behaviour in several apartments located in the city of Surabaya, Indonesia. The key findings are summarised as follows:

- The respondents in the high-rise apartments rarely open windows/doors (0–30%). More than 90% of them are equipped with air-conditioner(s), and they use them for approximately 5.1 h during night-time and 1.9 h during daytime on average.
- The results of field measurements showed that indoor air temperatures in the master bedrooms in both new public and high-rise private apartments were higher than the corresponding outdoor air temperature during most of the hours per day.
- It was seen that it is very difficult to achieve thermal comfort without relying on air-conditioning particularly in both new public and high-rise private apartments.

Acknowledgements This research was supported by the JSPS KAKENHI (Grant No. JP 15KK0210), YKK AP corporation and Indonesia Endowment Fund for Education (LPDP). The field measurements were conducted by the students of Hiroshima University in collaboration with the Sepuluh Nopember Institute of Technology, Surabaya. Particularly, we would like to express our sincere gratitude to Mr. Naoto Hirata and Mr. Takashi Hirose.

References

1. Asian Development Bank (2010) Key indicators for Asia and the Pacific 2010, 41st edn. The rise of Asia's middle class, Asian Development Bank
2. Surabaya City Development Planning Agency (2013) Land use planning 2013. Surabaya City Development Planning Agency, Surabaya
3. NCDC. Climate data online. Global surface summary of day data. https://gis.ncdc.noaa.gov/map/viewer/#app=cdo. Accessed 18 May 2014
4. Arethusa MT, Kubota T, Nugroho AM, Ekasiwi SN, Antaryama IGN, Uno T (2015) Factors influencing window opening behavior in apartments of Surabaya: a structural equation modelling. Proceedings of 2015 TAU conference: mitigating and adapting built environments for climate change in the Tropics, Jakarta, Indonesia, 30–31 Mar
5. Wong NH et al (2002) Thermal comfort evaluation of naturally ventilated public housing in Singapore. Build Environ 37:1267–1277
6. Indraganti M (2010) Adaptive use of natural ventilation for thermal comfort in Indian apartments. Build Environ 45:1940–1507
7. Alfata MNF, Hirata N, Kubota T, Nugroho A, Uno T, Ekasiwi SN, Antaryama IGN (2015) Field investigation of indoor thermal environments in apartments of Surabaya, Indonesia: Potential passive cooling strategies for middle-class apartments. Energy Procedia 78:2947–2952
8. Alfata MNF, Hirata N, Kubota T, Nugroho A, Uno T, Antaryama IGN, Ekasiwi SN (2015) Thermal comfort in naturally ventilated apartments in Surabaya, Indonesia. Procedia Eng 121:459–467
9. Toe DHC, Kubota T (2013) Development of an adaptive thermal comfort equation for naturally ventilated buildings in hot-humid climates using ASHRAE RP-884 database. Front Archit Res 2 (3):278–291

Chapter 41
Indoor Thermal Environments
of Cambodian Houses

Hiroto Takaguchi

Abstract Currently, there are a variety of types of house from traditional stilt houses to contemporary town houses in Cambodia. Peoples have renovated their own houses based on their own needs or wants according to economic growth. Usually, they built a shading first for cutting direct sunlight and extend their space to outside for improving the quality of life. These actions for renovating their own houses, the quality of the indoor and outdoor environment, and occupant behavior influence each other. I report here about the difference in indoor temperature between several houses include the renovated house.

Keywords Cambodia · House · Energy consumption · Indoor temperature

41.1 Introduction

In rural areas of Cambodia, we can still find a very traditional detached house which is shown in Fig. 41.1. On the other hand, peoples have renovated their own houses based on their own needs or wants according to economic growth. Usually, they built a shading first for cutting direct sunlight and extend their space to outside for improving the quality of life. Next would be building walls for making the underfloor space of the stilt house to inside or adding a new kitchen. These actions for renovating their own houses, the quality of the indoor and outdoor environment, and occupant behavior influence each other.

We report here about the difference in indoor temperature between several houses included the renovated house, very traditional house in rural areas and urban areas.

H. Takaguchi (✉)
Department of Architecture, Waseda University, Tokyo, Japan
e-mail: takaguchi@waseda.jp

© Springer Nature Singapore Pte Ltd. 2018
T. Kubota et al. (eds.), *Sustainable Houses and Living in the Hot-Humid Climates of Asia*, https://doi.org/10.1007/978-981-10-8465-2_41

41.2 Indoor Thermal Environments in a Rural Area House

At the beginning, I would like to introduce the result of the survey conducted at Takeo on September in 2014. We measured indoor thermal environment of the traditional wooden stilt house which had a renovated big eave of zinc and expanded room for the kitchen (Fig. 41.2). The main part of the house was very traditional with Khmer tile roofing. Proximately 25% of underfloor space was changed to indoor space with a simple wall of bricks and mortar (Fig. 41.3). The upper floor of stilt was mainly used for the bedroom and closet of women and children. Other members of the family slept in under space of the floor or eave.

We measured the indoor temperature of each room and outside (Fig. 41.4). The thermometer was installed at a height of about 1.1 m from the floor.

Figure 41.5 shows the daily temperature change and solar radiation. The highest temperature among all was 32 °C that was measured in the kitchen, which is an expanded room. And the highest at night was the indoor space of underfloor. The kitchen had a zinc roof, and this roof was comparatively smaller than the building so that it was not able to shield solar radiation toward the walls. It is thought that the nighttime temperature did not decrease by storing this solar heat in the wall. On the other hand, the indoor space of underfloor did not have direct sunlight at all.

Traditional house   Expanded traditional house   Modern house

Fig. 41.1 Housing types in rural areas

Big eave

Expanded room

Main building

Fig. 41.2 Exterior of the measured house

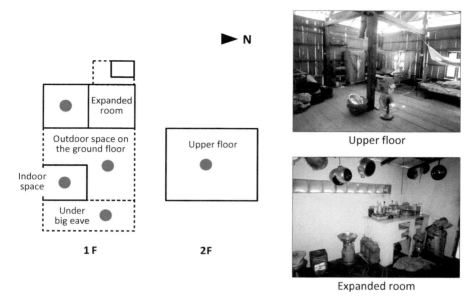

Fig. 41.3   Measurement points (● :thermometer) and interior

Fig. 41.4   Survey of thermal environment

Therefore the indoor temperature during the day was the lowest. As the heat transfer from the air warms the wall, the thermal storage effect of the wall prevented night temperature drop, and the nighttime temperature was the highest, even higher than that in the kitchen. This difference is due to radiation cooling. Therefore, most of the families used this type of expanded space as a kitchen, and very few used it as a bedroom, and inside space of under floor as a storeroom. The lowest temperature at night was in the space under the eave. The temperature of the outside under the floor was almost lowest among all day. The difference between the upper floor and

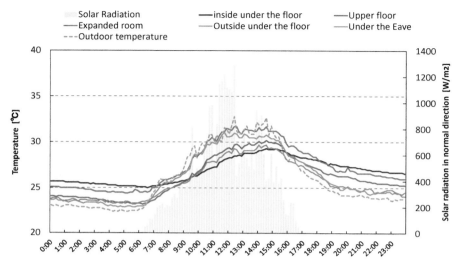

Fig. 41.5 Change of temperature during a whole day in a traditional house

underfloor was the temperature during the night, which the upper floor was a little warmer than underfloor. That is because of the heat storage of the roof tile.

41.3 Indoor Thermal Environments in an Urban Area House

This house is located at the center of Phnom Penh, which is a typical town house with a shop space on the ground floor (Fig. 41.6). We conducted the indoor measurement survey of this house in 季節 of 2009. It is constructed with concrete and bricks. The finish coat of the wall is mortar and plaster for its ceiling, and the floor is covered by tiles. In this case, the shop space was not used for shop, which was for an open living room. There were three bedrooms, and two of them had air conditioners, and one was for a housemaid without an air conditioner.

Figure 41.7 shows the change of indoor temperature of measured rooms. Two air conditioners were used for sleeping. Usually, the residents used the timer and did not use the air conditioners all the night. This day, one air conditioner was used for a very short time when the resident waked up. Basically, the air conditioners were not used during the day. The temperature of the rooms not air conditioned was stable at approximately 30 °C in the morning and around 32 °C in the daytime. It was higher than the outside at night and lower than the outside at daytime. This is considered to be due to the small amount of ventilation, the small inflow of outside air, and the heat storage in the block and concrete of walls.

Fig. 41.6 Exterior (left) and interior (right) of the measured house

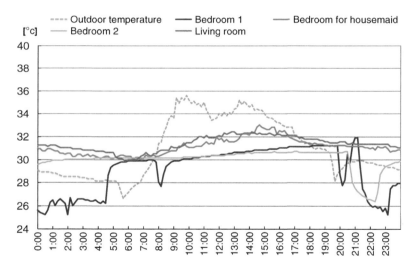

Fig. 41.7 Change of temperature during a whole day in a town house

41.4   Comparison of Indoor Thermal Environments of Five Housing Types

We conducted a measurement survey of the temperature at the bedrooms in Phnom Penh and Siem Reap in 2009 of October. It was for comparing a difference between the types of house. We measured five different types of house shown in Table 41.1, which were a traditional house, a detached house (mixed construction), a detached house (RC construction), a detached house (passive design), and a town house.

Table 41.1 Measured houses

Exterior					
Housing type	Traditional house	Detached house (mixed)	Detached house (RC)	Detached house (passive)	Town house
Construction	Wood	RC and wood	RC		RC
Roof	Zinc/palm	Tile	Khmer tile		Tile
Floor	Wood	Tile/wood	Tile	Flooring	Tile
Air conditioner	×		○	×	○

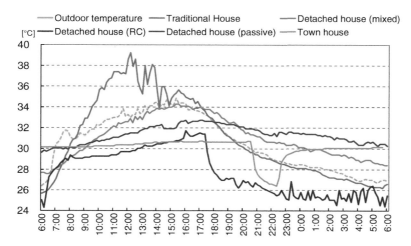

Fig. 41.8 Comparison of indoor thermal environments of five housing types

Figure 41.8 shows the hourly change of temperature in a representative bedroom in each type of house. The ambient outdoor temperature is from the traditional house. On the day of data collection, the outdoor temperature was 26 °C in the morning and 35 °C in the afternoon. Surprisingly, the temperature inside the traditional house surpassed the outdoor temperature and reached 39 °C at noon. The traditional house has recently upgraded the roof material from palm leaves to tin, and as such, the temperature of the living room increased because of radiation from the tin roof which became very hot in the direct sunlight. In the RC house that had an air conditioner, the effects of the cooling were observed. The indoor temperature remained relatively stable and was lower than the detached house which had an RC ground floor and wooden second floor.

The temperature was different between the traditional house shown in Figs. 41.5 and 41.8. In the house shown in Fig. 41.5, the room temperature was cooler than the outdoor temperature, but in the house shown in Fig. 41.8, the room temperature was hotter than the outdoor temperature. The biggest difference between these two houses was a roof material. The house shown in Fig. 41.5 had a Khmer tile roof, and other one had a tin roof. And another difference was the location of the kitchen. The kitchen of the house shown in Fig. 41.5 was independent, but the kitchen of the house shown in Fig. 41.8 was in the house. We are considering that the mixture of these differences affects the temperature.

41.5 Conclusion

Currently, there are a variety of types of house from traditional stilt house to contemporary town houses in Cambodia. Those cannot be classified neatly. The houses are renovated by the thought of occupant to improve the quality of life, and the materials and construction methods are overwrapped.

We have believed that the traditional house has good ventilation and easy to spend, but when the roof is renovated to tin, the radiation environment during the day is extremely poor. The indoor temperature was much higher than the ambient outdoor temperature. On the other hand, the temperature of the modern town and detached house is comparatively stable. But that means the room temperature is cooler than the outside air temperature during the day, and it is hotter than the outside air at night when the occupant wants to sleep. Therefore, many people come to use the air conditioners to fall asleep. It can be said that the heat storage performance of concrete and block does not fit the today's lifestyle and not work effectively. The materials of wall and roof, heat storage performance, and radiation environment have a great influence on the indoor comfort as well. Ventilation is also important; in addition, understanding of the characteristics of such materials and the natural environment is indispensable for creating a comfortable environment. For example, it might be useful to cool down the wall and floor by using night purge and to suppress solar gain from the windows and roofing to prevent the temperature rise or to add the insulation inside the bedroom to cut off the influence of heat storage.

Chapter 42
Actual Condition of the Indoor Environment of Houses in Nine Chinese Cities

Hiroshi Yoshino

Abstract Currently, China is undergoing a rapid economic development, and as a result, the standard of living and energy consumption are subsequently increased. Residential energy consumption is deduced to be 11% of the total; however, due to the improvement of indoor environment and the prevalence of adoption of electronic products, the energy consumption is expected to be increased significantly in the future. Given the enormous size of the Chinese population, the effect of increase in residential energy consumption on the global warming is estimated to be very large. Therefore, it is necessary to improve the standard of living without increasing the energy consumption. For this purpose, it is important to apply a variety of energy-saving measures to form a sustainable living environment, but the data of indoor environment and energy consumption is still scarcely available at the moment.

Over the years authors have carried out various investigation on the way of living, indoor thermal environment, energy consumption, and so on of apartments in different Chinese cities. The focus is placed on obtaining basic data for the proposal of design methods for energy conservation.

This chapter shows the actual situation of the indoor environment, based on the results of questionnaire and measurement survey in nine cities.

Keywords China · Indoor environment · Houses · Questionnaire · Measurement · Occupant behavior

Yoshino H (2012) Research on Thermal Environment and Indoor Air Quality of Residential Buildings –For developing comfort, healthy and low energy houses-, Tohoku University Press, Sendai pp.447–452 (In Japanese)

H. Yoshino (✉)
Institute of Liberal Arts and Science, Tohoku University, Sendai, Japan
e-mail: yoshino@sabine.pln.archi.tohoku.ac.jp

Fig. 42.1 Location of the cities for investigation

42.1 Outline of the Questionnaire and the Measurement Survey

Subjects for the investigation are apartments sited in urban area of nine cities, i.e., Harbin, Urumqi, Beijing, Xi'an, Shanghai, Changsha, Chongqing, Kunming, and Hong Kong, as shown in Fig. 42.1. Harbin and Urumqi belong to extremely cold region; Beijing and Xi'an cold region; Shanghai, Changsha, and Chongqing hot-summer-cold-winter region; and Kunming warm region. Hong Kong belongs to hot-summer-winter-warm region. Questionnaire and measurement survey in each city were performed twice in the winter (January) and summer (August) from 1998 to 2004. In the questionnaire survey, the characteristics of housing, housing equipment, way of living, and room temperature were asked, and liquid crystal thermometers were used to measure room temperature within the survey.

In the measurement, temperature and humidity in the living room, the bedroom, and the outside were measured at 30 min intervals for 5 days by using compact data loggers with temperature and humidity sensors. For the winter measurement, the number of apartments for questionnaire survey and the measurement survey was 810 and 76, respectively. For the summer measurement, it was 942 and 91, respectively.

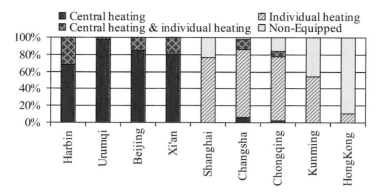

Fig. 42.2 Ratio of apartments with different space heating system or apparatus

42.2   The Results of Questionnaire Survey in the Winter [1, 2]

42.2.1   The Prevalence of Space Heating System and Appliances

Figure 42.2 shows the prevalence of heating system and appliances in each city. In Urumqi, Harbin, Beijing, and Xi'an, considered as extremely cold and cold regions, district heating (including the central heating system for an apartment building) has become widespread with nearly 100%. In Shanghai, Changsha, and Chongqing, hot-summer-cold-winter region, the ratio of usage of space heating appliances was 75–80%. But in Kunming belonging to the warm region, the ratio was 50%. In Hong Kong, it was as small as 11%.

42.2.2   Thermal Sensation During Evening Time in Winter

Figure 42.3 shows the thermal sensation during evening time (19:00–21:00) in winter in each city. In the apartments of Changsha, Chongqing, and Kunming with individual space heater, 60% of occupants felt "cold" or "rather cold." On the other hand, in the apartments of Urumqi and in Xi'an with the district heating, no occupant felt "cold," and 20% of occupants felt "hot" or "rather hot." Room temperature of apartment with district heating could not be controlled in most cases. When the room temperature rose, occupants opened the windows to adjust the room temperature.

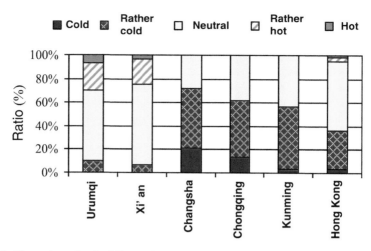

Fig. 42.3 Thermal comfort in different cities

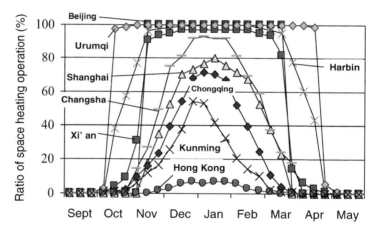

Fig. 42.4 Change of usage ratio of space heating in a year

42.2.3   Seasonal Change of Usage of Space Heating

Figure 42.4 shows the seasonal change in the usage ratio of space heating in each
city. In Urumqi, Beijing, and Xi'an (severe cold and cold region), which are installed
with district heating, apartments are heated from October (Urumqi) or November
(Harbin, Beijing, and Xi'an) to April (Urumqi) or March (Harbin, Beijing, and
Xi'an). In Shanghai, Changsha, and Chongqing (hot-summer-cold-winter region),
the pattern of seasonal change in usage ratio of space heating was found totally

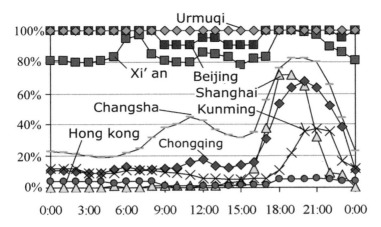

Fig. 42.5 Change of usage ratio of space heating usage in a day

different from the cold region. In this figure, there are different heights of bell curves depending on the level of cold climate. The period with more than 50% usage ratio in Changsha, Shanghai, and Chongqing was roughly from December to February. In Kunming, the period was very short in the end of the year. In the case of Hong Kong, the maximum ratio was less than 10%.

42.2.4   Daily Change of Usage of Space Heating

Figure 42.5 shows the daily change in the usage rate of space heating in each city. In the apartments of Urumqi, Beijing, and Xian, all rooms were heated all day long in many homes. In Shanghai, Changsha, and Chongqing, the pattern of usage ratio of space heating was completely different from the cold region. There was a peak of usage ratio during the evening time (supper hours), varied from 68% to 82%. In Kunming, this peak of usage rate was less than 40%.

42.2.5   Average Temperature and Standard Deviation of Living Room and Bedroom

Figure 42.6 shows the average temperature and the standard deviation of the living room, the bedroom, and the outdoor during the evening time (19:00 to 21:00). The average temperature of both the living room and the bedroom in Harbin, Urumqi, Beijing, and Xi'an with district heating was 18~22 °C. In Hong Kong with mild climate even in winter, temperatures of both rooms were about 20 °C. On the other

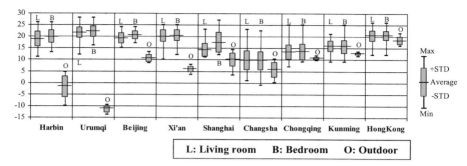

Fig. 42.6 Winter indoor temperature of the living room and the bedroom with the outdoor temperature in different cities

hand, in Shanghai, Changsha, and Chongqing with cold winter, the average room temperature was 10~17 °C. It can be seen that they lived in a low temperature environment comparing to other cities.

Variation in the room temperature was small in Harbin, Beijing, Xi'an, and Hong Kong, but this was found larger in the cities without district heating. It was because the usage of local space heaters was different between the apartments in these cities. Among all of the cities, the bedroom temperature was higher than that of living room temperature.

42.3   Examples of the Change of Indoor Temperature and Humidity in Winter

A typical example of 5-day change of indoor temperature and humidity in each city of Urumqi, Chongqing, and Changsha is shown as follows.

42.3.1   Urumqi (Fig. 42.7)

All rooms were heated by the district heating system. Although the outdoor temperature was around minus 10 °C, the temperature of both the living room and the bedroom during the measurement period was stable at around 20 °C. Temperature difference between the upper and lower in the living room was not found. The relative humidity was low and that of the living room was around 20%. In some houses, occupants put a shallow container with water on top of the storage furniture for the humidification.

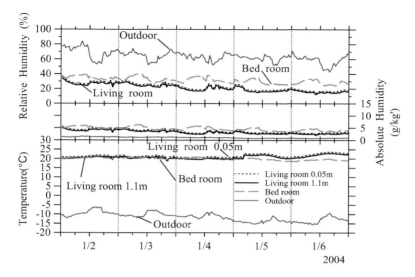

Fig. 42.7 Indoor thermal environment in an apartment of Urumqi

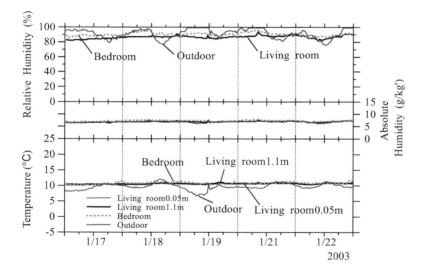

Fig. 42.8 Indoor thermal environment in an apartment of Chongqing

42.3.2   Chongqing (Fig. 42.8)

Heating equipment was not used. When the outside air temperature was varying between 5 and 10 °C, the room was stable at approximately 10 °C. Although the temperature was stable as measured in Urumqi, it was low and far below the range

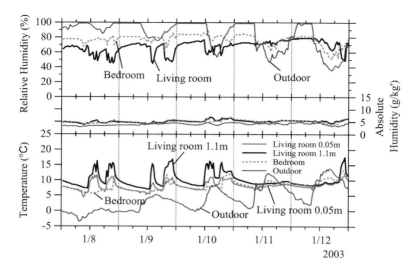

Fig. 42.9 Indoor thermal environment in an apartment of Changsha

that has been said to be thermally comfortable. Further, the humidity was more than 80%, and the problem of vapor condensation was easy to occur.

42.3.3 Changsha (Fig. 42.9)

Due to intermittent usage of space heating apparatus, the room temperature changed greatly high and low. When the outdoor temperature was varying between 0 and 5 °C, the temperature of the living room moved high and low from 7 to 16 °C with intermittently heated. The maximum temperature of the bedroom was 12 °C. During daytime, the temperature of the room without space heating was around 10 °C when the outside air temperature rose up to 13 °C. Humidity was at least 70% except when it decreased slightly during the heating time.

42.4 Measured Living Room Environment Compared to Comfort Zone by ASHRAE and Chinese Standard in Winter

Based on the 5-day measured room temperature and humidity for winter, the daily average values during the evening time (around 18:00 to 22:00) are plotted in the psychrometric chart with the comfort zone provided by the American Society of

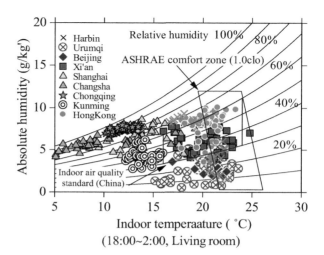

Fig. 42.10 Indoor thermal environment of apartments on psychrometric chart in winter

Indoor temperaature (˚C)
(18:00~2:00, Living room)

Heating, Refrigerating and Air-Conditioning Engineers (ASHRAE) [3] as well as China indoor environment standard [4] as shown in Fig. 42.10.

First of all, it is necessary to pay attention to the fact that the comfort zone in winter of the Chinese standard is placed in a lower-temperature part comparing to ASHRAE standard. The values decided by temperature and humidity of the living rooms of Harbin, Urumqi, Beijing, and Xi'an with district heating and Hong Kong with mild winter were distributed in and around the thermal comfort zones. Almost all of the rooms of Hong Kong and Xi'an fell into the zone of Chinese standard. Urumqi was shifted to the bottom region with the Chinese standard because relative humidity was low. However, the values of the rooms in the other cities were shifted to the left with the China standard. Especially room temperatures in Chongqing and Changsha were found to be very low.

42.5   The Results of Questionnaire Survey in the Summer [2]

42.5.1   The Prevalence of Air-Conditioning

Figure 42.11 shows the use of different types of air conditioners in each city. In Harbin and Urumqi with cool summer, the adoption of air-conditioning in living room and bedroom was below the 20%. In Beijing, Shanghai'03, Changsha, Chongqing, and Hong Kong, the adoption was from 60% to 100%. In Shanghai, the investigation had been done twice in 1998 and 2003. The prevalence of using air-conditioning in 2003 was higher than in 1998. In Xi'an, it was less than 60%. In Beijing, Shanghai'03, and Changsha, the prevalence in bedroom was higher than in living room. The major kind of air conditioner was mini-split unit type, but in

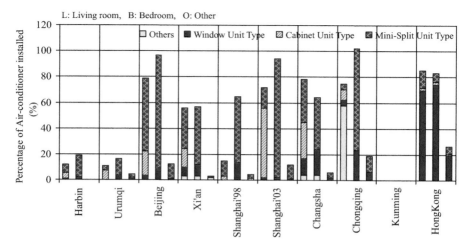

Fig. 42.11 Use of different types of air-conditioners

Fig. 42.12 Change of
usage ratio of
air-conditioning in a year

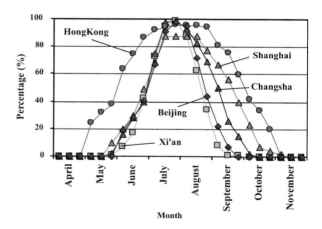

Shanghai'03 the large cabinet unit type which set on the floor shared 75%. In Hong
Kong, the window unit type was used in many apartments.

42.5.2 Seasonal Change of Usage of Air-Conditioning

Figure 42.12 shows seasonal change of usage of air-conditioning. Usage ratio was
over 90% from July to August in all cities. In Hong Kong, more than 50%
apartments used air-conditioning from June to September. In the other cities, occupants started to use it from July, but the end of usage period was different between all

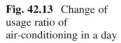

Fig. 42.13 Change of usage ratio of air-conditioning in a day

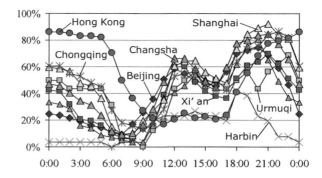

of the cities and was distributed from the end of August in Xi'an to the end of September in Shanghai.

42.5.3 Daily Change of Usage of Air-Conditioning

Figure 42.13 shows the daily change of usage of air-conditioning. The profile of Hong Kong was totally different from the other cities. More than 80% apartments in Hong Kong operated air-conditioning from midnight to 6:00 am. In the other cities, there were two peaks, from 12:00 to 14:00 and 19:00 to 22:00. In the nighttime, the usage ratio was 70–90%.

42.5.4 Ways to Cool the Rooms

Figure 42.14 illustrates the ways how to cool the rooms. In Kunming, most of the occupants opened windows for getting cool breeze. In Harbin and Urumqi, 60% of apartments opened windows, while the others opened windows together with electric fans on. In Beijing, Shanghai, Changsha, and Chongqing, which were different from those cities, 60–70% of the apartments used either air-conditioning or air-conditioning together with electric fans. This implied that climate conditions could affect the way to cool rooms.

42.5.5 Average Temperature and Standard Deviations of Living Room and Bedroom

Figure 42.15 shows the indoor temperatures of the living room and the bedroom as well as outdoor temperature. The cities where the indoor temperature was higher

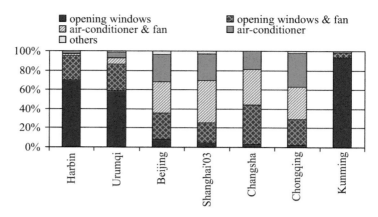

Fig. 42.14 Ratio of ways how to cool rooms in different cities

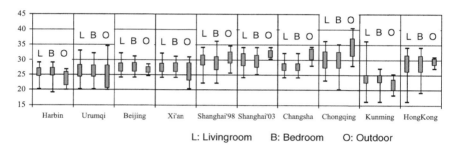

L: Livingroom   B: Bedroom   O: Outdoor

Fig. 42.15 Summer indoor temperature of the living room and the bedroom with outdoor temperature in different cities

than the outdoor temperature were Harbin, Urumqi, Beijing, Xi'an, and Kunming. The average indoor temperature was found from 24 to 28 °C. They do not use air-conditioning so much. In the other cities, the outdoor temperature was more than 30 °C, and the room temperature was kept lower than the outdoor temperature by using air-conditioning. In Chongqing and Shanghai, room temperature was around 30 °C. In Changsha, room temperature was around 28 °C. In these cities, the temperature of the bedroom was slightly lower than the living room.

42.6   Examples of the Change of Indoor Temperature and Humidity in Summer

A typical example of 5-day change of indoor temperature and humidity in Harbin and Shanghai is shown as follows.

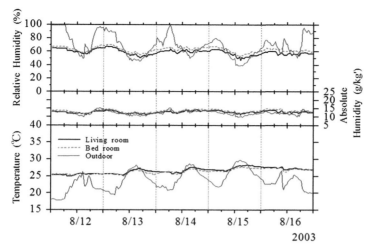

Fig. 42.16 Indoor thermal environment in an apartment of Harbin

42.6.1 Harbin

The measurement result of the indoor temperature and humidity is shown in Fig. 42.16. Although this house did not use the air-conditioning and the outside temperature during the day reached up to 30 °C, the room temperature was relatively stable at 25~28 °C. This is because a large heat capacity of the concrete wall. Relative humidity fluctuated slowly between 50% and 70%.

42.6.2 Shanghai

The measurement result of the indoor temperature and humidity is shown in Fig. 42.17. It was so hot during this period, and the outdoor temperature was varying between 28 and 38 °C. In this house, occupants used air-conditioning in the living room during evening time and in the bedroom during the sleeping hours. Indoor temperature was from 25 to 30 °C during the use of air-conditioning. In addition, room temperature even when the outside air temperature during the day is as high as more than 35 °C is suppressed to about 32 °C. The relative humidity during the day is from 60% to 80%, and it was reduced to 40% when the air-conditioning was operated during the night.

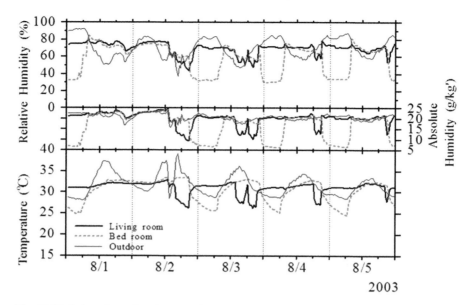

Fig. 42.17 Indoor thermal environment in an apartment of Shanghai

42.7 Measured Living Room Environment Compared to the Comfort Zone of ASHRAE and Chinese Standard in Summer

Based on the 5-day measured temperature and humidity for summer, the daily average values during the evening time (around 18:00 to 22:00) are plotted in the psychrometric chart with the comfort zone provided by American Society of Heating, Refrigerating and Air-Conditioning Engineers (ASHRAE) and Chinese indoor environment standard, as shown in Fig. 42.18. During the evening time, the living room temperature was recorded from 24 to 30 °C, and the relative humidity was obtained from 50% to 70% in Harbin, Beijing, and Xi'an. These values are plotted in and around the comfort zone. On the other hand, the living room temperature was from 24 to 34 °C, and the relative humidity was from 4% to 80% in Shanghai and Changsha. In this case, the values are distributed from the range of the comfortable zone to far from that. The living room temperature was plotted in the range of 28~32 °C, and the relative humidity was from 70% to 80% in Hong Kong. Indoor environment is hot and humid in this city. In comparison with the comfort range of China indoor environment standard in summer and ASHRAE, the range of relative humidity was higher, and the range of temperature was wider in the former one. The indoor environment of a part of Urumqi and Kunming is plotted in the ASHRAE comfort zone, but Harbin, Urumqi, a part of Shanghai, Beijing, Xi'an, and others are plotted in the zone of China indoor environment standard. Hong Kong,

Fig. 42.18 Indoor thermal environment of apartments on psychrometric chart in summer

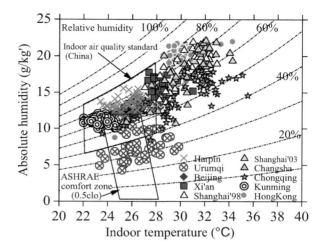

Chongqing, and a part of Shanghai were shown far from the comfort zone. The indoor environment of these locations is hot and humid.

For mitigating the hot and humid indoor environment, the sun shading is the most effective way. The second is the cross ventilation. The nighttime ventilation, while the outdoor temperature is relatively low, is effective for keeping the indoor temperature lower than the outdoor temperature during the daytime. The desiccant cooling is one of promising technologies in terms of energy conservation and thermal comfort.

42.8   Summary

1. Almost all of the houses in Urumqi, Harbin, Beijing, and Xi'an, with the severe cold weather conditions, were installed with district heating system.
2. In the houses with district heating system, 20% of the occupants replied "hot" or "rather hot" for thermal comfort.
3. Indoor temperature of the houses with or without local space heating was very low and far from the ASHRAE thermal comfort zone.
4. In more than 80% of the apartments in Beijing, Shanghai'03, Chongqing, and Hong Kong, the air-conditioning was used, but in the cold and severe cold regions, the prevalence was less than 20%. In Kunming, there was no apartment with air-conditioning.
5. For the indoor environment, some of the apartments in Urumqi and Kunming fell on the ASHRAE comfort zone, but the apartments of Harbin, Kunming, Shanghai, Beijing, Xi'an, and a part of Shanghai distributed to the comfort zone of Chinese standard.

References

1. Miyasaka H, Yoshino H et al (2004) A survey on indoor thermal environment of urban residential buildings in China, part 8, the results of winter investigation in Urumqi and Kunming, summaries of technical papers of Annual Meeting, Architectural Institute of Japan, Environmental Engineering, pp 145–146
2. Tanaka NH, Yoshino H et al (2005) A survey on indoor thermal environment of urban residential buildings in China, part 10, the results of field measurement in 9 cities of China summaries of technical papers of Annual Meeting, Architectural Institute of Japan, Environmental Engineering, pp 137–138
3. Addendum 55a ANSI/ASHRAE Standard 55-2004, ASHRAE, 2004
4. National Standard of People's Republic of China, GB/T18883-2002, Indoor Air Quality Standard

Part VI
Sustainable Houses in Asian Cities

Chapter 43
Energy-Saving Experimental House in Hot-Humid Climate of Malaysia

Tetsu Kubota, Mohd Azuan Zakaria, Mohd Hamdan Ahmad, and Doris Hooi Chyee Toe

Abstract In 2015, we constructed a full-scale experimental house in the city of Johor Bahru, Malaysia, in order to determine comprehensive energy-saving modification techniques through passive cooling for existing urban houses in the hot-humid climates. First, this chapter presents the results of numerical simulation to examine optimum combinations of modification techniques for the urban houses. Second, this chapter analyses the results of full-scale field experiments conducted in the above-mentioned experimental house to confirm the resulting effects of proposed modifications. The proposed modifications include (1) roof insulation, (2) external wall outer insulation, (3) external shading, and (4) forced ventilation (for the whole house and for the attic).

Keywords Energy saving · Modification · Thermal comfort · Natural ventilation · Hot-humid climate

43.1 Introduction

Most cities in Southeast Asia experienced dramatic rise of energy consumption in line with their economic growths over the last few decades. As these cities experience hot-humid climates throughout the year, the growing energy consumption for space cooling in buildings is becoming a major concern. In fact, the ownership levels of air-conditioners among households have been increasing rapidly in this region [1].

T. Kubota (✉)
Graduate School for International Development and Cooperation (IDEC), Hiroshima University, Hiroshima, Japan
e-mail: tetsu@hiroshima-u.ac.jp

M. A. Zakaria
Faculty of Civil and Environmental Engineering, Universiti Tun Hussein Onn Malaysia, Batu Pahat, Johor, Malaysia

M. H. Ahmad · D. H. C. Toe
Faculty of Built Environment, Universiti Teknologi Malaysia, Skudai, Johor, Malaysia

© Springer Nature Singapore Pte Ltd. 2018
T. Kubota et al. (eds.), *Sustainable Houses and Living in the Hot-Humid Climates of Asia*, https://doi.org/10.1007/978-981-10-8465-2_43

In Malaysia, the total population has almost doubled over the last three decades and reached about 30 million people at present. The stable economic growth over the same period induced the rapid increase in nationwide energy demand by more than five times [2, 3]. According to the report by ADB (2010) [4], the proportion of middle class was already more than 80% as of 1990 and reached around 90% at present. Meanwhile, the number of houses increased particularly from the 1990s to the 2010s, and the housing stock is now nearly 5 million [5].

Most of the urban houses in Malaysia are constructed of brick and concrete. The brick-walled houses account for approximately 91% of the urban houses [1]. These brick houses are becoming the norm across the cities in Southeast Asia. In the case of Malaysia, terraced houses are considered the most common housing type, which accounts for about 45% of all the urban houses [5].

It is not appropriate to discuss energy-saving techniques or policies for residential buildings in Southeast Asia uniformly because each of the cities or countries should have different conditions and development stages. Nevertheless, as far as Malaysia is concerned, one of the most important energy-saving strategies in the residential sector is to examine how to modify the millions of existing brick houses, particularly terraced houses, to become more energy efficient, especially in terms of space cooling. At the moment, Malaysia does not have specific energy-saving regulations for existing residential buildings.

In 2015, we have constructed a full-scale experimental house in the main campus of *Universiti Teknologi Malaysia* (UTM), which is located in the city of Johor Bahru, Malaysia. This study aims to develop comprehensive energy-saving modification techniques through passive cooling for existing urban houses in the hot-humid climate of Malaysia. First, this chapter examines the optimum combinations of modification techniques for the urban houses through numerical simulations using TRNSYS and COMIS. Second, we present the results of full-scale experiments conducted in the above-mentioned experimental house to confirm the resulting effects of the proposed modifications.

43.2   Energy-Saving Experimental House

The full-scale experimental house (two adjacent terraced houses) was constructed in UTM ($1°29'N$ $103°44'E$) in December 2015 (Fig. 43.1). The experimental houses were designed to represent typical floor plans of existing terraced houses. Each of the units measures approximately 6.7 m by 9.8 m with a total floor area of 127 m^2. As shown, the front façade of buildings is oriented towards north. It was constructed of brick and concrete and has single glazing windows. In addition to the original plans, small slit windows were added onto the upper and lower parts of the main windows, partition walls, and doors (Fig. 43.2). Moreover, exhaust fans were installed in the attic spaces, master bedrooms, and staircase halls. The houses were not installed with the insulation except for both end walls to eliminate the thermal influences from the outdoors.

Fig. 43.1 The full-scale experimental house in *Universiti Teknologi Malaysia* (UTM) campus

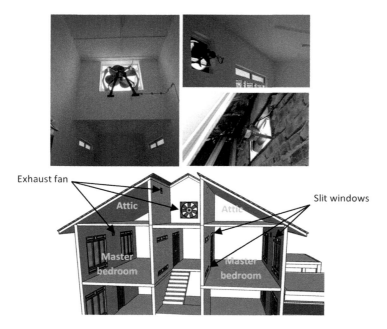

Fig. 43.2 Slit windows and exhaust fans employed in the experimental house

43.3   Numerical Simulation

43.3.1   Methods

The experimental house was modelled in this numerical simulation. Description of the construction layers of the house and their reference U-values in the numerical

Table 43.1 Construction layers and reference U-values of the numerical model

Building element	Constructional layers	Reference U-value [a] W/(m^2 K)
External and Internal walls	25 mm-thick cement plaster +100 mm-thick clay brick +25 mm-thick cement plaster	2.61
Party wall	13 mm-thick cement plaster +200 mm-thick clay brick +13 mm-thick cement plaster	2.19
First floor	15 mm-thick cement screed +100 mm-thick concrete slab +20 mm-thick cement plaster	3.49
Ground floor	150 mm-thick cement slab +600 mm-thick soil	1.40
Ceiling (master bedroom)	6 mm-thick ceiling board	4.55
Ceiling (other zones)	3.2 mm-thick ceiling board	5.54
Pitched roof	20 mm-thick concrete roof tile + aluminium foil	3.86
Flat roof	22 mm-thick cement screed +100 mm-thick concrete slab +20 mm-thick cement plaster	3.37
Window	6 mm-thick single layer float glass	5.61

[a]Includes convective and radiative heat transfer coefficients of 7.7 W/(m^2 K) for inside surface and 25 W/(m^2 K) for outside surface

model is given in Table 43.1. The model of each unit was composed of 12 thermal zones in TRNSYS with corresponding airflow zones in COMIS to represent each partitioned room or functional space including attic spaces.

Empirical validation of the numerical model was performed using the field measurement data, which was conducted in June 2016. This study focuses particularly on the results in the master bedrooms because existing households usually use air-conditioning mainly in the master bedrooms [6]. The results of the validation are detailed in [7].

Table 43.2 summarizes the simulation test cases of this study. The techniques were selected by considering their practicality to be applied to existing terraced houses through relatively simple building modification and/or behavioural adjustment. Some of the exceptions were mechanical exhaust fans as shown in Fig. 43.2. These exhaust fans were tested to examine whether they were able to strength the effects of nocturnal structural cooling.

Weather conditions for the simulation were taken from the typical meteorological year (TMY) data of Kuala Lumpur (3°07′N 101°33′E). The simulation was run using the above weather file for the whole months of April and May. Subsequently, simulation results for a 10-day period of continuous typical fair weather days were analysed in this study. The outdoor air temperature ranges from 25 to 36 °C, while outdoor relative humidity ranges from 40% to 95% over the period.

Table 43.2 Simulation test cases

Technique	Test conditions
Building orientation	East; West; South; North
Natural ventilation[a]	Night ventilation (20:00–8:00); daytime ventilation (8:00–20:00); no ventilation (0 h); full-day ventilation (24 h)
Forced ventilation	Attic space (40 ACH); master bedroom (40 ACH); whole house (30 ACH) [for night (20:00–8:00)]
Thermal insulation	Roof (R = 3 (m² K)/W); ceiling (R = 3); external wall – Outer surface (R = 3); external wall – Inner surface (R = 1.5); internal wall (R = 1.5); party wall (R = 1.5); floor (R = 1.5)
Window shading	External shading; internal shading (shading factor, SF, 0.75)
High reflectivity roof coating	Solar reflectance, 0.8; long wave emissivity, 0.8
Window glazing	Low-E glass (U-value, 2.54 W/(m² K); G-value, 0.44%)

[a]Open window period is given in brackets

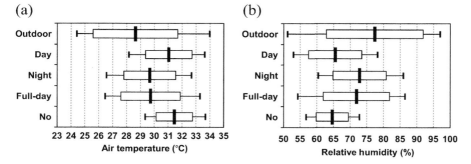

Fig. 43.3 Simulated indoor air temperature and relative humidity in the master bedroom (statistical summary of 5th and 95th percentiles, mean and ±one standard deviation)

43.3.2 Results and Discussion

Figure 43.3 presents the resulting daily averaged air temperature and relative humidity in the master bedroom under various natural ventilation conditions. As shown, night ventilation provides the lowest daily mean and maximum air temperature in the master bedroom, while full-day ventilation obtains slightly lower minimum air temperature at night. Meanwhile, it can be seen that the indoor RH exceeds 60% throughout the day when the night ventilation is adopted. As far as humidity condition is concerned, full-day ventilation is prior to the night ventilation. Hence, in the following analyses, the effects of respective modification techniques are examined under the night ventilation and full-day ventilation conditions, respectively.

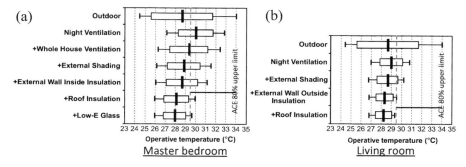

Fig. 43.4 Simulated indoor operative temperature in the master bedroom and living room under night ventilation condition (statistical summary of 5th and 95th percentiles, mean and ±one standard deviation)

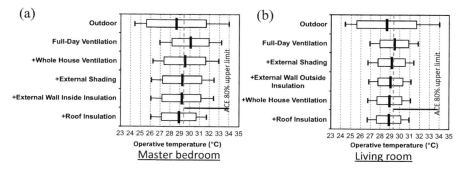

Fig. 43.5 Simulated indoor operative temperature in the master bedroom and living room under full-day ventilation condition (statistical summary of 5th and 95th percentiles, mean and ±one standard deviation)

In general, it was found from the previous study [6] that occupants in the Malaysian terraced houses tend to stay in the living room (ground floor) during daytime and stay at the bedrooms (first floor) during night-time. Then, we determined the optimum modifications based on the above assumptions by the means of design of experiment. The optimum combinations were determined to reduce the nocturnal operative temperature in the master bedroom as much as possible, while minimizing the exceeding period in the living room, which is the duration when the resulting operative temperature exceeds the required adaptive comfort limit given by the adaptive comfort equation (ACE) [8]. Figure 43.4 presents the resultant optimum combinations for the two rooms under the night ventilation condition, while Fig. 43.5 shows the corresponding results under the full-day ventilation condition. We adopted from more effective modifications to the less effective modifications accumulatively in addition to the respective natural ventilation techniques. Both results represent those at north orientation, for example.

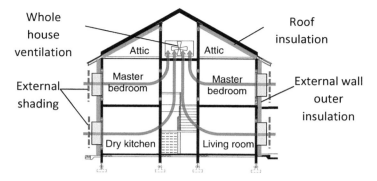

Fig. 43.6 Energy-saving modifications for the experimental house

As shown in Fig. 43.4a, the nocturnal operative temperature is reduced the most in the master bedroom by adding the whole house ventilation (using the exhaust fan) to the night ventilation with the temperature reduction of up to 0.8 °C at night and 0.4 °C during daytime. The next option is the external shading, which reduces the nocturnal operative temperature further by 0.2 °C as well as reduces the daytime peak operative temperature by 0.9 °C, followed by external wall inner insulation, roof insulation, and low-E glass. As for the living room, the external shading is better able to reduce the maximum operative temperature with a temperature reduction of about 0.6 °C (Fig. 43.4b). The external wall outer insulation was chosen as the next option, followed by the roof insulation.

When the full-day ventilation is applied, the whole house ventilation is found to be the most effective in reducing the nocturnal operative temperature in the master bedroom just like the case of night ventilation, followed by the external shading, external wall inner insulation and roof insulation (Fig. 43.5a). As for the living room, the external shading is found to be the most effective, followed by the external wall outer insulation, the whole house ventilation and roof insulation (Fig. 43.5b).

By combining the optimum combinations for the master bedroom and the living room, we determined the optimum combinations of energy-saving modifications for the experimental house (Fig. 43.6). As shown, they include roof insulation, external wall outer insulation, external shading, and whole house ventilation using the exhaust fan.

43.4   Full-Scale Measurement

43.4.1   Methods

The full-scale experiment was conducted in the two units of above-mentioned experimental house in UTM from June to September 2016 (see Fig. 43.1). In the field measurement, one of the units (House 2) was equipped with the proposed modifications (i.e. measurement unit), while the other unit (House 1) was remained unchanged as the control unit. The proposed modifications include (1) roof insulation, (2) external wall outer insulation, (3) external shading, and (4) forced ventilation (for the whole house and the attic). The roof insulation was installed underneath the roof board of the two bedrooms of House 2. Meanwhile, the thermal insulation was installed on the outer surfaces of the external walls for House 2. The insulations are of 100 mm thick rock wool form with a thermal resistance of 3 $(m^2 K)/W$. Moreover, the external shading devices were installed on all windows in House 2. The shading factor was set to be 0.75. Furthermore, the large exhaust fan was adopted at 30 ACH for the entire house, while small-scale fans were used for attic spaces at 40 ACH. These exhaust fans were operated only during night-time in order to reinforce the effect of night ventilation (i.e. structural cooling).

The detailed thermal conditions were investigated mainly in the master bedrooms (first floor) and the living rooms (ground floor) of the two units at 1.1 m height above the floor. The measured variables include air temperature, RH, wind speed, and globe temperature. In addition, air temperature and RH were also measured in the attic spaces and the other spaces.

43.4.2 Results and Discussion

Here we compare the results of measurement unit (House 2) with those of control unit (House 1) in terms of air temperatures at 1.1 m above floor in the master bedrooms (first floor). Figure 43.7 shows the temporal variations of measured air temperatures in Cases 1-3 that analyse the effects of proposed modifications under the night ventilation condition. Case 1 adopts (1) roof insulation, (2) external wall outer insulation, and (3) external shading for windows. Case 2 further adds (4) whole house ventilation using the large exhaust fan to Case 1, and Case 3 applies (5) attic ventilation using the small-scale exhaust fan in addition to Case 2.

As shown in Fig. 43.7a, as the techniques adopted in Case 1 are for reducing solar heat gain, the cooling effect can be seen only during daytime. The daytime peak air temperature is reduced by 0.9 °C with an average daytime reduction of 0.6 °C compared with the control unit (i.e. House 1). When the whole house ventilation is adopted in Case 2, the cooling effect becomes evident at night (Fig. 43.7b). In Case 2, the cooling effect is as almost the same as that of Case 1, but the nocturnal air temperature is reduced by 0.6 °C on average. After adding attic ventilation further, the cooling effects are increased during day and night (Fig. 43.7c). The daytime peak air temperature is reduced by 1.3 °C with an average reduction of 1.2 °C, whereas the nocturnal air temperature is reduced by 1.1 °C on average.

The detailed spatial distribution of air temperatures in Case 3 is illustrated in Fig. 43.8. The difference between the two units due to the roof insulation can be seen in the daytime peak air temperatures of their attic spaces. In House 1 (without roof

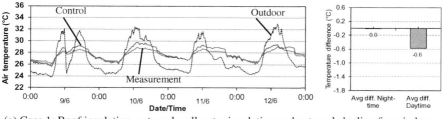

(a) Case 1: Roof insulation, external wall outer insulation and external shading for windows.

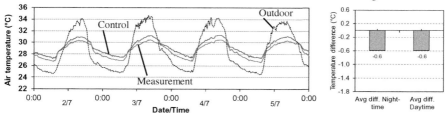

(b) Case 2: Roof insulation, external wall outer insulation, external shading for windows and whole house ventilation (at night)

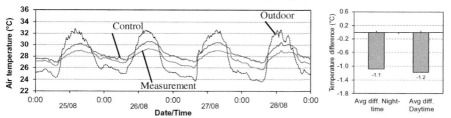

(c) Case 3: Roof insulation, external wall outer insulation, external shading for windows, whole house ventilation and attic ventilation (at night).

Fig. 43.7. Temporal variations of measured air temperatures in the master bedroom in Cases 1-3 under the night ventilation condition. (Note: control = House 1, measurement = House 2)

insulation), the air temperatures in the attic are increased to about 32 °C, which is almost the same level as the outdoors. In contrast, the corresponding air temperatures in House 2 (with roof insulation) are reduced to about 30 °C. The thermal conditions in these attic spaces directly affect the surface temperatures of the ceiling in the master bedrooms situated below, and in the case of House 1, the warmed ceiling surface temperatures increase the indoor air temperatures at the upper levels.

In Case 3, the forced ventilations are applied not only to the whole house but also to the attic spaces at night (20:00–8:00). Hence, the nocturnal air temperatures in the attic spaces in House 2 are relatively lower than those of House 1 (Fig. 43.8). Exceptionally, the slit windows located upper and lower parts of the main windows were opened instead of the main windows during night-time in Case 3, and thus the surface temperatures of ceiling and floors are reduced slightly more.

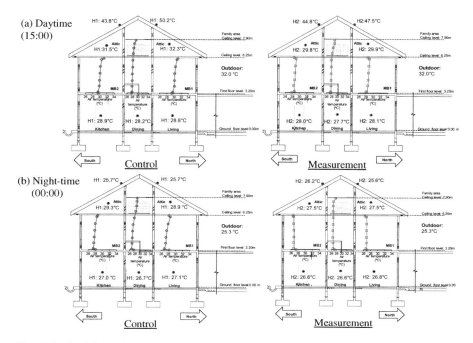

Fig. 43.8 Spatial distribution of indoor air temperature in the control (House 1) and measurement unit (House 2) in Case 3

Figure 43.9 shows the temporal variations of measured air temperatures in the master bedrooms in Cases 4-6 under the full-day ventilation condition. The modifications adopted in Case 4 are the same as those in Case 1, but the cooling effects are negligible in Case 4 not only during night-time but also during daytime mainly because windows/doors were open (Fig. 43.9a). When the whole house ventilation is applied additionally in Case 5, the cooling effect became evident at night as before (Fig. 43.9b). The nocturnal air temperature was reduced by 0.7 °C during night-time on average in Case 5, just like Case 2. After adding attic ventilation in Case 6, the cooling effect during night-time is slightly increased further (Fig. 43.9c).

The detailed spatial distribution of air temperatures in Case 6 is illustrated in Fig. 43.10. Just like Case 3 (night ventilation), it can be seen that the air temperatures in the attic spaces in House 2 are reduced to about 30 °C due to the roof insulation. Nevertheless, the cooling effects in other indoor spaces are limited because windows/doors were open except for the surface temperatures of ceilings. The surface temperatures of the ceilings in House 2 can be approximately 1.8 °C lower than those of the control unit. This means that although the proposed modifications are not able to reduce indoor air temperatures, they contribute to improve the radiant heat conditions in the rooms.

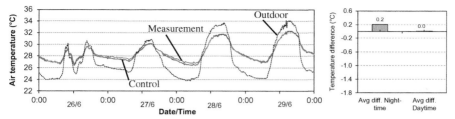

(a) Case 4: Roof insulation, external wall outer insulation and external shading for windows.

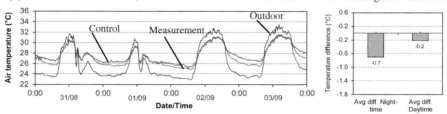

(b) Case 5: Roof insulation, external wall outer insulation, external shading for windows and whole house ventilation (at night)

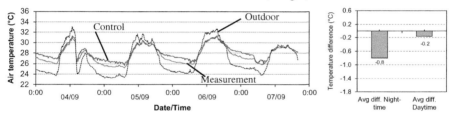

(c) Case 6: Roof insulation, external wall outer insulation, external shading for windows, whole house ventilation and attic ventilation (at night).

Fig. 43.9 Temporal variations of measured air temperatures in the master bedroom in Cases 4–6 under the full-day ventilation condition. (Note: control = House 1, measurement = House 2)

43.4.3   Thermal Comfort Evaluation

This section evaluates the resulting thermal comfort in House 2, which is equipped with the proposed modifications, under the two ventilation techniques, i.e. night ventilation and full-day ventilation. Figure 43.11 shows the daily average of indoor thermal environments in the master bedroom (first floor) and the living room (ground floor) under the two ventilation techniques, respectively.

As shown in Fig. 43.11, in the case of night ventilation, the daytime peak temperature in the master bedroom and the living room tend to be lower than the corresponding outdoor air temperature by up to 4.0 °C and 4.5 °C, respectively. Meanwhile, the air temperature in both spaces are about 2 °C higher than the corresponding outdoor air temperature during the night-time. In this case, the RH is always above 75% throughout the day.

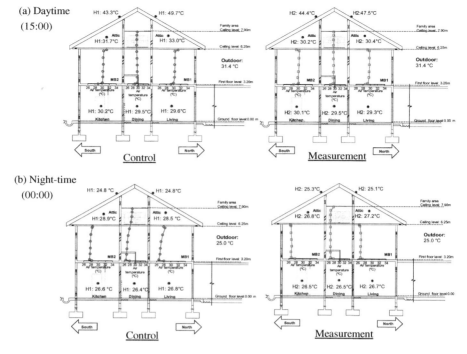

Fig. 43.10 Spatial distribution of indoor air temperature in the control (House 1) and measurement unit (House 2) in Case 6

In the case of full-day ventilation, the reduction of air temperature during daytime in the two rooms is smaller than those of the previous condition. The air temperatures in the master bedroom and the living room are about 2 °C and 3 °C lower than the corresponding outdoor air temperature. Nevertheless, it should be noted the RH of the rooms is lower than that of the previous condition as expected, which is as low as 65% during daytime. At night, the air temperature difference between indoor and outdoor is almost the same as that of the previous condition, which is about 2 °C higher than the corresponding outdoor air temperature.

Operative temperature (OT) and SET* are used as indices to evaluate thermal comfort in the master bedroom and the living room. As for the OT, the thermal comfort is evaluated based on ACE [8]. Meanwhile, a metabolic rate of 1 met and a clothing value of 0.4 clo are adopted in the calculation of SET*.

As presented in Fig. 43.11, if OT is used for evaluation, then the master bedroom and the living room in night ventilation condition record lower values (28.0–29.0 °C) than those in full-day ventilation condition (29.0–30.0 °C) during daytime. The calculated OT of the master bedroom and living room in night ventilation generally fell within the limits, while those in full-day ventilation exceed the limit over 30% of the measurement period. Nevertheless, when SET* is used for

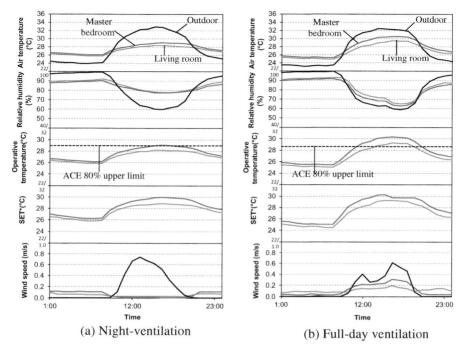

(a) Night-ventilation   (b) Full-day ventilation

Fig. 43.11 Indoor thermal environment of master bedroom and living room in the measurement unit (House 2) under night ventilation and full-day ventilation

evaluation, those in the two rooms during daytime under the two ventilation techniques show almost the same values, which is approximately 30.0 °C and 30.3 °C SET*. This indicates that if the effect of sweat evaporation is taken into account by using SET*, the resulting thermal comfort levels by the two ventilation techniques are considered almost equal under the given measurement conditions. On the other hand, it can be seen that relatively lower RH in the full-day ventilation condition is more tolerable than that in the night ventilation condition since the constant high humidity condition (>70%) may cause mould growth [9] and health-related problems [10]. Hence, it can be concluded that the cooling strategy based on full-day ventilation is probably preferable than that by night ventilation for the hot-humid climates such as in Malaysia.

43.5   Conclusions

In Malaysia, terraced houses are considered the most common housing type, which accounts for approximately 45% of all the urban housing stocks. One of the most important energy-saving strategies in the residential sector is to examine how to modify the millions of existing brick houses, particularly terraced houses, to become more energy efficient.

- We constructed a full-scale experimental house in the city of Johor Bahru to develop comprehensive energy-saving modification techniques through passive cooling for existing urban houses in Malaysia.
- The effects of respective energy-saving modification techniques were examined under the night ventilation and full-day ventilation conditions, respectively, through numerical simulations. Optimum combinations of modification techniques were determined. They include the roof insulation, the external wall outer insulation, the external shading, and the whole house ventilation.
- The effects of proposed modification techniques were analysed through full-scale experiments using the experimental house under the two different natural ventilation conditions, i.e. night ventilation and full-day ventilation. It was concluded that the cooling strategy based on full-day ventilation is probably preferable than that by night ventilation for the hot-humid climates such as in Malaysia.

Acknowledgements This research was supported by the grants from Nichias Corporation and Asahi Glass Foundation. The construction of the experimental houses, the prior simulations, and the following field measurements were conducted by the students of Hiroshima University in collaboration with *Universiti Teknologi Malaysia*. Many students were involved in this long-term project and particularly, we would like to express our sincere gratitude to Mr. Susumu Sugiyama, Mr. Satoshi Yasufuku, Mr. Masato Morishita, Mr. Toshichika Kusunoki, Mr. Makoto Ohashi, and Mr. Seiji Abe for their great contributions.

References

1. Mahlia TMI, Masjuki HH, Saidur R, Amalina MA (2004) Mitigation of emissions through energy efficiency standards for room air conditioners in Malaysia. Energ Policy 32 (16):1783–1787
2. Department of Statistics Malaysia, Population Distribution and Basic Demographic Characteristics, Population and Housing Census of Malaysia (2010, 2011) Department of Statistics Malaysia
3. Economic Planning Unit, Malaysia Plan (1986, 1991, 1996, 2001, 2006)
4. Chun N (2010) Middle class size in the past, present and future: a description of trends in Asia. ADB economics working paper series no.217, Asian Development Bank
5. Pusat Maklumat Harta Tanah Negara, Residential Property Stock Report (1990–2016)
6. Kubota T, Toe DHC, Ahmad S (2009) The effects of night ventilation technique on indoor thermal environment for residential buildings in hot-humid climate of Malaysia. Energ Build 41 (8):829–839
7. Kubota T, Ohashi M, Zakaria MA (2017) Full-scale experiments on energy-saving modification for existing urban houses in Malaysia Part 1. Numerical simulation on optimum combinations of passive cooling techniques. In: Summary of technical papers of annual meeting, AIJ, 2017, Aug 31 – Sep 3, Hiroshima, Japan

8. Toe DHC, Kubota T (2013) Development of an adaptive thermal comfort equation for naturally ventilated buildings in hot-humid climates using ASHRAE RP-884 database. Front Archit Res 2(3):278–291

9. Johansson P, Ekstrand-Tobin A, Svensson T, Bok G (2012) Laboratory study to determine the critical moisture level for mould growth on building materials. Int Biodeter Biodegr 73:23–32

10. Sterling EM, Arundel A, Sterling TD (1985) Criteria for human exposure to humidity in occupied buildings. ASHRAE Trans 91(1B):611–622

Chapter 44
S-PRH in Kitakyushu, Japan

Yusuke Nakajima

Abstract The S-PRH built in 2000 in Kitakyushu City, Fukuoka Prefecture, is the suburban model of the PRH (Perfectly Recycled House) Series of experimental houses. The house uses recyclable building materials and has a double-skin space. This double-skin space is designed as a tool for passive techniques, enabling regulation of the environment through operation of windows and sun shading fixtures by the residents. Result of experimentation in summer, it was confirmed that the double-skin space of the S-PRH could improve indoor comfort and reduce the air-conditioning load by allowing the resident to appropriately use different passive technique to suit outside weather conditions that change moment by moment.

Keywords Experimental house · Reuse/recycle · Double skin · Passive technique · Energy saving

44.1 Introduction

The S-PRH is the suburban model of the PRH (Perfectly Recycled House) Series of experimental houses. The components of these houses are more than 80% recycled, and the design aims to establish equipment systems and lifestyles to reduce the environmental impact of daily life in terms of energy, water, and wastes. The house was built in October 2000 in Wakamatsu-ku, Kitakyushu-shi, Fukuoka Prefecture, by the Ojima Laboratory (operating at that time) of the Department of Architecture, Waseda University. After finishing the construction of this new S-PRH, building performance measurements were taken for 1 month, and soon after that, a disassembly/rebuilding experiment was conducted, and the house was rebuilt in April 2001.

In terms of architecture, the house is designed for longer life by using a heavy steel frame and seismic isolation structure, and the reuse/recycling rate is improved

Y. Nakajima (✉)
Department of Urban Design and Planning, School of Architecture, Kogakuin University, Tokyo, Japan
e-mail: yusuke@cc.kogakuin.ac.jp

© Springer Nature Singapore Pte Ltd. 2018
T. Kubota et al. (eds.), *Sustainable Houses and Living in the Hot-Humid Climates of Asia*, https://doi.org/10.1007/978-981-10-8465-2_44

473

by using recyclable building materials, modularizing components, adopting panels for dry construction, and other techniques. A reuse rate of 98% was achieved in the disassembly/rebuilding experiment.

This chapter introduces this type of S-PRH, using a sustainable house in the hot region of Asia as an example. Special attention is given to energy-saving innovations using double skin. The contents are summarized on a series of researach and its results [1–5].

44.2 Overview of S-PRH

Elements seen in Japanese traditional *Minka* houses, such as inner courts, verandas, and outdoor corridor spaces were effective as buffer spaces for regulating the outside/inside boundary, and the indoor environment was kept comfortable by appropriately using the different spaces to suit the season and outside weather conditions. The plan for the S-PRH incorporates an atrium-type double-skin space on the south side to play the role of this buffering space. In addition to the systematized double-skin function used on the building's exterior, this feature is also designed as a tool for passive techniques, enabling regulation of the environment through operation of windows and sun shading fixtures by the residents. Table 44.1 shows an overview of the architecture, Fig. 44.1 photos of the outside and inside appearance, and Fig. 44.2 a floor plan.

44.3 Overview of Double-Skin Space

As described above, the double-skin space of the S-PRH combines the function of the buffer spaces of traditional *Minka* with a modern systematized double skin. Passive techniques are also proposed for effectively utilizing natural energy by making use of this space. Figure 44.3 shows the cross-sectional composition of the double-skin space, and Fig. 44.4 shows a photo of the inside appearance.

The entire southern side is made into an opening, and the double-skin depth is 1200 mm. The sliding glass windows on both sides can be opened in the same way,

Table 44.1 Overview of the S-PRH architecture

Location	Kitakyushu City, Fukuoka Prefecture	
Construction	Heavy steel construction	
Story	Two stories	
Total floor area	177.84 m^2	
Construction period	New construction	July 2000 ~ Oct. 2000
	Dismantlement	Nov. 2000 ~ Dec. 2000
	Reconstruction	Dec. 2000 ~ Mar. 2001

Fig. 44.1 S-PRH photos

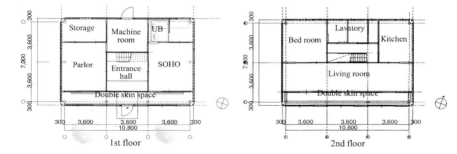

Fig. 44.2 S-PRH floor plan

Fig. 44.3 Cross-sectional composition

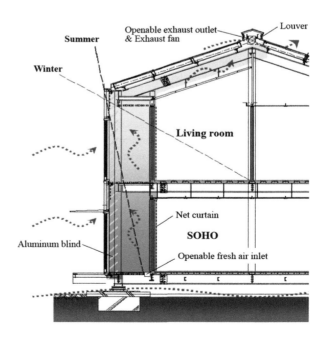

Fig. 44.4 Photo of the
double-skin space

so the design provides for plenty of ventilation and natural lighting. The floor has opening/closing air inlets, and the roof has opening/closing air outlets. The floor on the second story is a grating. As a result, in the summer air taken in from under the floor rises through the double-skin space, passes through the vent layer in the attic, and is discharged from the roof air outlets, thereby achieving efficient heat dissipation and ventilation. Forced air discharge fans are also installed at the roof air outlets, and the system keeps the indoor environment comfortable by opening and closing air inlets and outlets according to the season, indoor temperature, and other factors. The double-skin first-story floor is filled with gravel to achieve a heat storage effect in the winter.

In terms of sun shading fixtures, blinds are installed on the outside glass of the double skin and lace curtains on the inside glass. By operating these two types of shading according to the season and time slot, the house allows adjustment of sunlight, breezes, and lines of sight.

44.4 Summer Measurement Survey

44.4.1 Purpose of Measurement Survey

In the city of Kitakyushu, where the S-PRH is located, peak outside air temperature in the middle of summer reaches 35 °C or higher, and humidity is also high. Passive

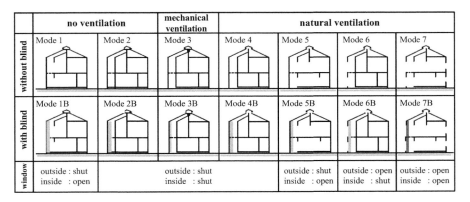

	no ventilation		mechanical ventilation	natural ventilation			
without blind	Mode 1	Mode 2	Mode 3	Mode 4	Mode 5	Mode 6	Mode 7
with blind	Mode 1B	Mode 2B	Mode 3B	Mode 4B	Mode 5B	Mode 6B	Mode 7B
window	outside : shut inside : open		outside : shut inside : shut	outside : shut inside : open		outside : open inside : shut	outside : open inside : open

Fig. 44.5 Overview of the usage modes of the double-skin space

techniques using a double-skin space also help reduce the load when using air-conditioning in the middle of summer, but beyond that, the aim is to allow residents to be comfortable in the early and late summer while using air-conditioning as little as possible. Therefore, a measurement survey was conducted to clarify the effectiveness of the various types of passive techniques using the double-skin space. The survey was conducted in June, which corresponds to the first half summer.

44.4.2   Setting the Usage Modes of the Double-Skin Space

Figure 44.5 shows an overview of the usage modes of the double-skin space. These usage modes are divided into 7 types based on window opening/closing conditions and the ventilation method, and if blind opening/closing is also included, then there are 14 types in total. If partitions and opening/closing of lace curtains are also included, the types are subdivided even further, and in actual daily life, a variety of usages are possible to suit the lifestyle of the residents, such as shutting the blinds on the first floor only or allowing a breeze through the second floor only. During the measurement period, the double-skin space was controlled with 14 types of modes to enable easy comparison of the effectiveness of passive techniques.

44.4.3   Temperature Changes by Type of Passive Technique

Five passive technique types (A–E) were established, employing the aforementioned usage modes in different time slots, and measurement was carried out while switching the types each day (Table 44.2).

Table 44.2 Five passive technique types

Passive technique types		Date	0:00–9:00	9:00–19:00	19:00–24:00
A	Sealed	June 1	Mode 4	Mode 2	Mode 2
B	Roof ventilation	June 9	Mode 5[*]	Mode 4	Mode 4
C	Roof ventilation+blinds	June 6	Mode 5	Mode 4B	Mode 4B
D	Roof ventilation+allowing a breeze	June 4	Mode 2B	Mode 7	Mode 5
E	Roof ventilation+allowing a breeze+blinds	June 3	Mode 4B	Mode 7B	Mode 2B

[*]North window open

Figure 44.6 shows a comparison of natural temperatures using passive techniques. The survey period was June 1–9, and the measurement was conducted while switching the usage mode on sunny days. Mode switching was done twice a day at 9:00 and 19:00.

With Type A, where the double-skin space was sealed throughout the day, peak temperature on the double-skin second floor exceeded 30 °C due to acquisition of solar heat, and SOHO temperature reached 28 °C. With Type B, where roof ventilation was carried out by providing air inlets under the floor of the double-skin and air outlets in the roof, temperature on the first floor of the double-skin dropped to about the outside air temperature. For this reason, with Type A the SOHO temperature rises higher than the outside air temperature at 10:00, but with Type B, the SOHO temperature is lower until 13:00. With Type C, a sun shading effect was added to the double-skin space by using blinds. As a result, there was a major drop in both double-skin space temperature and SOHO temperature. With Type D and E where a breeze is allowed throughout the day, the room temperature dropped to almost the same as the outside air temperature.

In this survey, windows were closed at night (19:00–9:00 the following day) as would be done if a person actually lived in the house. With Type E, when windows were closed at 19:00 and the double-skin space was sealed, the SOHO temperature rose by 1.5 °C. In this case, it is likely that heat was released into the room after accumulating in the joint parts of the outer wall panel located on the SOHO east side. With Type D, the SOHO temperature rise was suppressed by opening inside windows of the double skin and continuing roof ventilation even at night. Thus one advantage of the double-skin space is that the interior can be naturally ventilated even in cases where windows cannot be opened, due to the need to prevent crime at night or when out of the house.

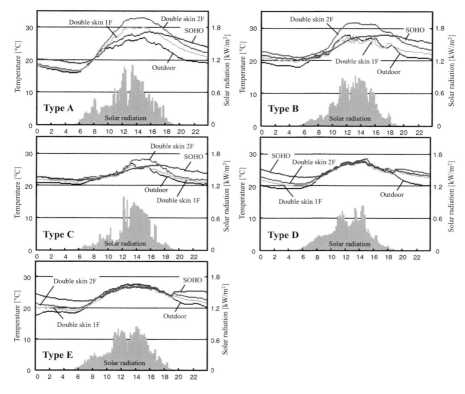

Fig. 44.6 Comparison of natural temperatures using passive techniques

44.4.4   *Effectiveness of Appropriately Using Different Passive Techniques*

Effectiveness in improving indoor comfort was verified by appropriately using different passive techniques in an even more complex regime throughout the day. Verification was carried out on June 11 and 12, sunny days with similar amount of insolation and changes in outside air temperature. Figure 44.7 shows the temperature variation of each living space, Fig. 44.8 variation in PMV, and Fig. 44.9 the schedule for differential use of passive techniques on the 12th.

On the 11th, roof ventilation was performed from 9:00 to 19:00, and a breeze was allowed by opening windows at other times. During the day, a vertical temperature difference arises in the double-skin space, and this promotes natural ventilation. However, due to the effects of sunlight, temperature in the living room rose to 30 °C, and the PMV value reached about 2. On the other hand, by appropriately

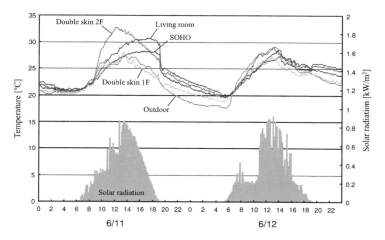

Fig. 44.7 Temperature variation of each living space (June 11–12)

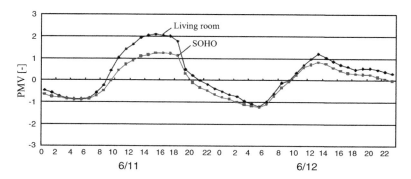

Fig. 44.8 Variation in PMV (June 11–12)

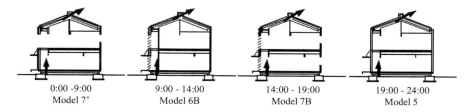

Fig. 44.9 Schedule for differential use of passive techniques on the 12th

using different passive techniques for the double-skin space in each time slot, it was possible to hold down temperature in the living room throughout the day to the same level as outside air temperature and keep PMV below a maximum of about 1.

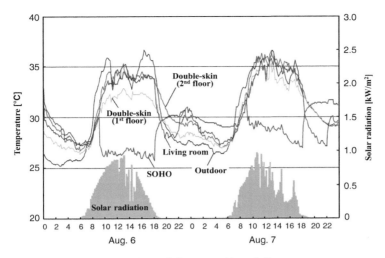

Fig. 44.10   Temperature variation of each living space (Aug. 6–7)

The control of the double-skin space shown in Fig. 44.9 is just one example, and the double-skin space of the S-PRH is designed to improve indoor comfort by allowing the resident to appropriately use different passive technique to suit outside weather conditions that change moment by moment.

44.4.5   Energy-Saving Effect When Air-Conditioning in the Summer

On August 6 and 7, sunny days with similar amount of insolation and changes in outside air temperature, the air-conditioning load was compared by operating air-conditioning in the first floor SOHO (9:00–18:00) and the second floor living room (18:00–22:00). As the passive technique type on the 6th, the outside window of the double-skin was opened, the blinds were lowered, and the floor air inlets and roof air outlets were both opened. On the 7th, a configuration close to an ordinary exterior (conventional type) was adopted by opening only the double-skin outside window.

Figure 44.10 shows the temperature variation of each living space; Fig. 44.11 shows the cooling load of living room and SOHO. As a result, it was confirmed that the approach using passive techniques had the effect of reducing the air-conditioning load by 17.2% compared to the conventional approach. SOHO temperature during air-conditioning on the 7th did not drop to the set temperature due partly to issues involving air-conditioning capacity, so it is presumed that an even greater reduction effect will be achieved in practice.

Fig. 44.11 Cooling load
(living room and SOHO)

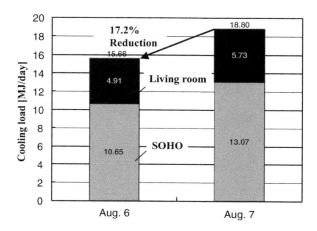

44.5 Conclusion

As indicated by the above measurement survey, the S-PRH is a house model designed to improve comfort and save energy by allowing the resident to appropriately employ the different double-skin space usage modes in each season and time slot. Here, discussion of the winter survey was omitted, but in winter too an improvement in comfort and reduction in heating load have been confirmed when the double-skin space is used as an insulating layer.

In the S-PRH, mode changes such as opening and closing blinds and windows have not been automated. Automation is one possible option, but this experimental house can help promote a lifestyle with a less environmental impact because the resident gains increased environmental awareness by mastering passive techniques.

References

1. Sugihara K, Yoshida K, Nakajima Y, Ojima T (2002) A investigation research on energy conservation by double-skin control in the S-PRH [part1]. Summaries of technical papers of annual meeting, Architectural Institute of Japan, D-1, pp 969–970
2. Yoshida K, Sugihara K, Nakajima Y, Ojima T (2002) A investigation research on energy conservation by double-skin control in the S-PRH [part2]. Summaries of technical papers of annual meeting, Architectural Institute of Japan, D-1, pp 971–972
3. Ikeda K, Sugihara K, Yoshida K, Nakajima Y, Ojima T (2002) A study on indoor thermal environment by using buffer space in the S-PRH. In: Proceeding of architectural research meetings, Kanto chapter, Architectural Institute of Japan, pp 469–472
4. Yoshida K, Nakajima Y, Ojima T (2001) Recycling construction method and environmental performances of the S-PRH [part2]. Summaries of technical papers of annual meeting, Architectural Institute of Japan, D-1, pp 961–962
5. Sugihara K, Yoshida K, Nakajima Y, Ojima T (2001) A study on comfortability by using buffer space in the S-PRH and its energy-saving effect. In: Proceeding of architectural research meetings, Kanto chapter, Architectural Institute of Japan, pp 597–600

Chapter 45
Eco-House in Kitakyushu, Japan

Weijun Gao and Didit Novianto

Abstract Due to its citizen's struggle to overcome the city environmental problems, Kitakyushu City of Fukuoka, Japan, was selected as an environmental model city in 2008. Being one of the global most famous cities in the environment conservation, Kitakyushu City has proposed a model of smart community in the Higashida District. The ecological house or more popular with eco-house is part of the smart community project. This model house was built with a passive design concept which utilized to maximize the natural light and the airflow into the house. In addition, some advanced energy-saving technologies such as hydrogen fuel cells and heat transfer solar panels are also used.

This study aims to conduct an investigation on people lifestyle and home energy consumption in eco-house model of Kitakyushu (north area of Kyushu). In this study, the energy use behavior based on field measurement in different seasons will be presented.

Keywords Eco-house · Energy saving · New technology

45.1 Kitakyushu Eco-House

The environment has been significantly improved to solve the pollution problem in Kitakyushu from the 1960s. In addition, Kyoto Protocol was initiated for the purpose of energy conservation and reduction of world carbon emissions. Japan's ambitious eco-house project has already completed 22 unique sustainable concept homes throughout the country with an aim to set a new national standard for building and environmental design [1]. At 34th G8 Summit in Hokkaido, Japan and the other member nations agreed to reduce carbon emissions by 50% before 2050. Now, Japan is working to surpass this goal and achieve a truly low-carbon society by reducing

W. Gao (✉) · D. Novianto
Faculty of Environmental Engineering, Department of Architecture, The University of Kitakyushu, Fukuoka, Japan
e-mail: weijun@kitakyu-u.ac.jp

emissions by as much as 80% within the next 40 years. A critical component of the plan is to reduce household's energy consumption which has risen by an alarming 40% since 1990s [2]. Therefore, the government has committed itself to improving the overall quality and performance of new and existing homes. Correspondingly, the Ministry of Environment established the program of an experimental eco-house model to build about 20 sustainable houses from Okinawa in the south to Hokkaido in the north (20 municipalities), of which each design and construction were adapting the local climate and site conditions. Each house model has been designed by a different local architect in conjunction with the city local governments and the Ministry of the Environment. Most are designed to utilize passive solar heating and use natural ventilation techniques to circulate warm and cool air throughout the houses. All but one of the projects is constructed from timber. The designs are combining the inherent sustainability of many traditional Japanese elements with a modern layout (Fig. 45.1).

Among them, Kitakyushu eco-house is a pioneer of various passive techniques which have been introduced in Japanese eco-houses. Fuel cells, photovoltaic solar cells, are equipped with power generation system together with hydrogen automobile. Kitakyushu eco-house is part of and located in the area of "Kitakyushu Environment Museum," which is becoming a center for environment-related learning and exchange of information that can be used by any citizens. This model house was designed as a place to experience an eco-friendly house for the public under the management of the museum. The eco-house was also proposed to promote "eco-friendly lifestyle and way of living." The environmental friendly principal and

Fig. 45.1 Physical appearance of Kitakyushu eco-house

technology performance of the eco-house enforce the public to minimize the energy for living as well as to enjoy a comfortable daily life. The eco-house has many features and characteristics, such as the simple wooden structure with high durability, a layout that can be easily changed in accordance with the changes of lifestyle or life stage of the family, an improved thermal conditioning system, and the use of various environmentally friendly materials. This house model also uses the natural energy as much as possible so that it can reduce the dependence on fossil fuels. For that reason, many efforts to maximize the use of natural sunlight, solar heat, winds, underground heat, and rain water by utilizations according to the local seasons and climate of Kitakyushu. In addition, Kitakyushu eco-house has been incorporating the traditional style of Japanese architecture with the adoption of conventional technologies and materials, such as the use of *doma* (earthen floor) at the entrance, *genkan* (verandah), bamboo lattice, and plastered walls. Kitakyushu eco-house also adopted the newest technologies and products including *eco-premium* local products approved by the local government.

Currently, the model house is to be functioned as an exhibition, education, and a tourist purpose without occupants or residents live in. For that reason, some experimental studies of this hybrid combined solar power system house were performed. This time, the study was conducted by setting the lifestyle of the household related with the use of home appliances, air conditioning, and ventilation method.

The basic information and features of Kitakyushu eco-house are shown in Table 45.1.

The eco-house energy utilization is a combination between active and passive design technologies. It was not only designed to supply the energy for its own but also for the surrounding neighborhood. The system will automatically distribute the electricity collected from solar PV and fuel cell to the commercial electric power company when oversupply. Contrary, when the energy demand of this house is higher, the electricity will be supplied automatically from the electric power company. The eco-house energy utilization system is shown in Figs. 45.2, 45.3 and 45.4.

The Kitakyushu eco-house layout is representing the Japanese detached house with the combination of the traditional and modern style of room layout. Equipped with various special technologies (such as solar thermal utilization system in the roof which absorb the heat from sunlight then circulate the warm air to the building in the winter and control the indoor temperature by sending the hot air outside), there are more utilization rooms in the layout compared to the common house. Due to the less of walls and partitions, the attached floor heating system in the living room area is planned to minimize the use of room air conditioner in the winter to keep the room comfort, especially Japanese do many activities in the floor. The house also adopted the cool tube under the ground which can distribute underground air to maintain the indoor comfort temperature in any seasons. Therefore, there are many pipes and tubes hidden.

Based on the government calculation prediction, the Kitakyushu eco-house can cut the electricity use to more than 50.12% compared to average homes electricity

Table 45.1 Kitakyushu eco-house basic information [3]

Location	Kitakyushu environment museum 33°52′05.5″N 130°48′28.8″E
House type	Two-story detached house
Total cost	95,000,000 JPY (2002)
Structure and floor	Conventional wooden construction method
Site area	4100.03 m^2
Construction area	130.83 m^2
Total floor area	183.43 m^2
Power generation devices	Solar power: capacity 3.29 kW, roof surface installation Fuel cell: using hydrogen gas with power generation capacity 700 W, 200 L hot water tank Roof air collector: thermal solar system (heat transfer solar panels over the entire roof surface, heat collection, heat pump, water heaters, and hot water supply for 460 L of hot water storage tank); Gas engine: cogeneration (using city gas, power generation capacity of 1000 W, hot water tank 200 L); Air conditioning: 2nd floor loft air conditioning 1: cooling capacity = 5.5 kW; COP = 3.3; heating capacity = 4.1 kW; COP = 3.7 2nd floor loft air conditioning 2: cooling capacity = 6.6 kW; COP = 3.1; heating capacity = 5.7 kW; COP = 4.1
Heating system	Air conditioning, heating, and gas engine cogeneration, air collection by roof thermal solar system, thermal solar heating by the second floor loft air conditioning, the combination of hot water and floor heating by fuel cell
Cooling system	The cooling system according to the 2nd floor loft air conditioning
Hot-water supply equipment	Heat collection by roof air collector thermal solar system
Hot-water supply system	Gas engine cogeneration (using city gas, power generation capacity 1000 W, hot water tank 200 L) Household fuel cell (using hydrogen gas, power generation capacity 700 W, hot water tank 200 L)

use in Japan [1]. The significant reduction is from air conditioning (heating-cooling), lightings, home appliances, and hot water supply, respectively. For that reason, we conducted the field measurement to investigate the performance of the house related to passive-active design, energy use behavior, and the efficiency of reducing heating and cooling loads (Figs. 45.5, 45.6 and 45.7).

The experiment and field measurement were conducted based on several patterns and some criteria [4]. Four members of research team each have a role as a nuclear family member, which are a couple (parent) and two children. In this study, some behavior cases were performed based on life schedule database collected from questionnaire through more than 3000 households in Kitakyushu City [5] before and supported by literature studies.

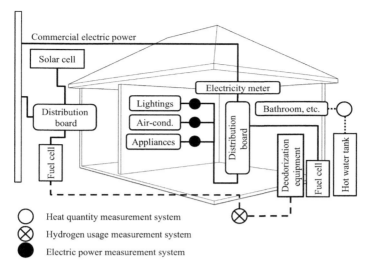

Fig. 45.2 Building energy system

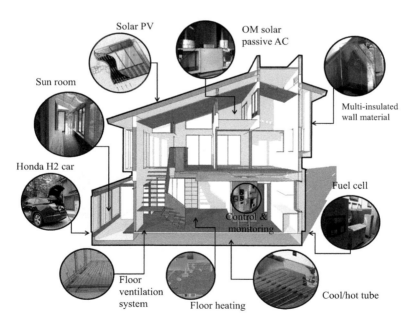

Fig. 45.3 Eco-house model building features

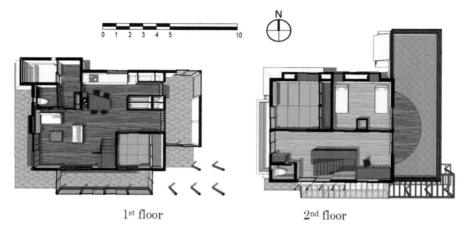

Fig. 45.4 Building floor plan

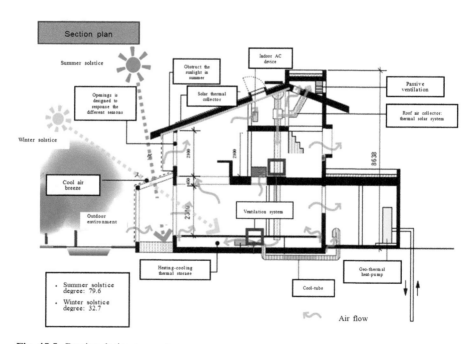

Fig. 45.5 Passive design concepts

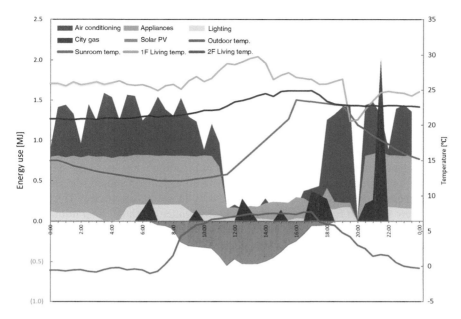

Fig. 45.6   Winter measurement (10~11 March 2014)

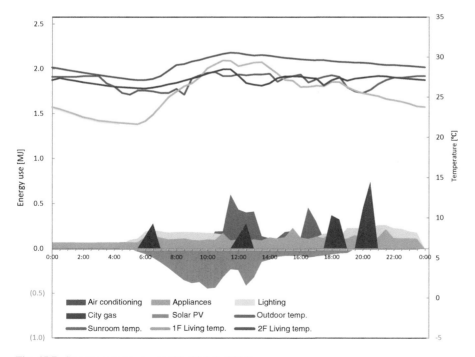

Fig. 45.7   Summer measurement (02~03 July 2014)

45.2 Summary

In this measurement, some of results and findings can be described as follows. Firstly, based on the field measurement results, the largest share of energy use in Kitakyushu eco-house model is the use of the electronic home appliances, especially in the winter case, which reached nearly 65% of total load in a day. Therefore lifestyle or behavior of the occupants plays important role in affecting the energy consumption. Secondly, as fuel cell was difficult to measure due to its distribution way, only the solar PV could significantly support the energy supply. In addition, in many cases of lifestyle, solar PV performs its maximum capacity in the day time, which is low of occupancy. Therefore, in the future, Kitakyushu eco-house could perform better if it was utilized with batteries to store the electricity. Thirdly, all rooms were set to be fully air conditioned during winter. With the total floor area of 183.4 m^2, Kitakyushu eco-house needs much more energy for the heating system in the winter. Contrary, in summer this model house successfully minimizes the energy use and performs the natural ventilation very well. Thus, the passive design concept in this model house has potential advantages for lowering energy consumption by not using the heating equipment if the similar model was built in hot and humid climate such as tropical region.

Further study will be needed in the way of finding the best practices and ideas for sustainable housing in Asian tropical cities by considering the local climates and cultures.

References

1. Kitakyushu Eco-house. http://www.env.go.jp/policy/ecohouse/challenge/challenge17.html
2. Statistical Handbook of Japan (2013) Chapter 7, Energy, Statistics Bureau, Japan, pp 77–82
3. Eco-House Follow-up Working Group, Japan Institute of Architects (2014) Introduction of Japan's Eco-House-performances from cold region to hot region. Japan (In Japanese)
4. Michisita T, Novianto D, Gao W, Kuroki S (2014) Study on lifestyle and energy consumption in Kitakyushu Eco-House, vol 2. Asia future conference. Bali, Indonesia
5. Novianto D, Gao W (2012) Analysis on residential lifestyle and energy consumption of Japanese family group in Kitakyushu City. 10th AIUE chino, Japan

Chapter 46
Sustainable Houses in Japan: An Overview

Miwako Nakamura

Abstract This chapter focuses on the sustainable housing in the tropical region of Japan.

It starts with the history of sustainable housing in Japan, followed by the specific look into techniques found in traditional housing in the tropical region of Japan where the climate is hot-humid and their adaptation to modern sustainable housing.

Keywords Japanese House · Tropical region · Hot-humid climate · Energy-saving · CO_2 emission

46.1 Introduction

During national disaster reconstruction and development under periods of economic boom, developers and institutions concentrate more on rapid development of the land, the supply of housing, as well as the generation of profit. These rapid developments cause environmental destruction by pollution to the land, water, and air. Subsequently traditional culture, lifestyle, and style of housing suitable for local climate are undermined, jeopardizing the life and health of the inhabitants. When we realized this and tried to recover these valuable assets, the process are often long and expensive.

After the Second World War, there has been a remarkable change in building materials, style, and construction method of the housing in Japan which has ignored the local features and traditional techniques. Furthermore, the popularization of home electronic appliances, especially air conditioning, has changed the way of living for people and has made their home living environment very different from prewar Japan. In architectural terms, we had been dealing with environmental problems and energy issues by installing even more mechanical facilities in the buildings.

M. Nakamura (✉)
MW Ecological Design, Tokyo, Japan

© Springer Nature Singapore Pte Ltd. 2018
T. Kubota et al. (eds.), *Sustainable Houses and Living in the Hot-Humid Climates of Asia*, https://doi.org/10.1007/978-981-10-8465-2_46

However, after the Great East Japan Earthquake, architects and engineers realized the importance of the source for primary energy and the use of passive techniques. We have come to a time where we need to reconsider a housing system using a technology based on passive methods. Many of these traditional passive techniques can be found in the housing system in the tropical region of Japan where the climate is hot-humid. These could be adopted into our modern lifestyle and houses and thus tackle global environmental problems at the same time.

46.2 The Road to Sustainable Housings in Japan

The word "sustainable" has been used in the Japanese architectural field since the end of the 1990s, though there have been various labeling of "environmentally friendly" housing. In the beginning, the word "energy saving" was used in Japan when the oil crisis occurred in 1972 as we faced the depletion of oil resources for the first time. At around the same time, Japanese housing policy had changed to improve the quality of housing, as the target housing to household ratio set for the restoration from the Second World War had already been fulfilled.

During this period, only certain architects and researchers of environmental engineering started to build or renovate experimentally in their own houses with solar heat utilization system and with passive techniques. These were known as "energy-saving houses." Their main aim was to reduce the heating energy in the winter season. The development of the alternative energy was another main issue for the energy saving, as it was important to reduce the use of oil.

During 1980–1987, passive solar housing as the "energy conscious house" continued to develop, and a variety of these housings, from low cost to expensive, have appeared by trial and error. House in Tsukuba by Yuichiro Kodama is one which later became a model for "the passive and low-energy detached house" (Fig. 46.1).

By the end of the 1980s, with the growth of the Japanese economy, there was a surge in land prices and development; people were generally more interested in improving the comfort of their home than to conserve energy. Along with the other home electric appliances, the use of heating and cooling equipment quickly spread to many homes in Japan. By using these appliances to control the indoor environment, homes became airtight and at the same time poorly ventilated. Architects concentrated on designing the indoor comfort and neglected the outdoor environment. The bubble economy led to a period where energy conscious housing was less favorable as people regarded home comfort to luxury and "energy saving" as stinginess.

In the 1990s, the problem of global warming was raised and quickly became a world issue. Japan accounted for a large percentage of carbon dioxide emissions which led to global warming and had an international obligation to reduce the emission rate. The Japanese government started the national architecture policy with the research and development of the "Environmentally Symbiotic Housing Project." The project created a framework for various environmental techniques,

Fig. 46.1 Tsukuba house

definition of the Environmentally Symbiotic Housing, and introduction of new techniques from European countries – such as insulation techniques and biotope. The resource circulation and symbiotic techniques toward nature and surrounding were also defined [2]. (Table 46.1).

As a result of the research, many multiple-dwelling complexes built by public organizations became national models, like the Fukasawa Environmentally Symbiotic Housing (Fig. 46.2). Energy and construction companies also took on the challenge to build their own environmentally symbiotic company houses.

Nevertheless, it has taken a long time for the typical detached housing and the multiple housing to become environmentally friendly in Japan. The term "sustainable housing" started to be used in the architectural field as it had meaning which covered a wider range of techniques from a much broader perspective.

During this time, the energy saving issue began to focus on techniques to reduce heat in the summer season. People started to be aware of the heat island phenomena as one of the problems in urban areas in the summer. The airtightness in houses, previously mentioned, was found to cause sick house syndrome; the use of natural ventilation and material was reconsidered, particularly in the case of children's health. With the consideration of the summer heat and energy saving, basic techniques such as shading and ventilation used in hot-humid area were more widely known and were adopted to other climatic regions. During this time, recycling and research into building material and the life cycle of housing were also introduced.

Sometime after 2000, the IPCC urged the world to adopt measures to counteract global warming: It was revealed that the total amount of CO_2 exhaust from the private household sector in Japan had increased when compared to levels in 1990.

Table 46.1 Eco-friendly techniques by "Environmentally Symbiotic Housing Project"

Category	Technique	Category	Technique
1. Energy saving	Reduce heat loss	2. Resource recycling	High durability
	Control of solar radiation		Flexible method of construction
	Passive solar-based utilization		Low emission
	Active solar-based utilization		The use of recycled material
	Unutilized energy		The use of water resources
	High-efficiency equipment		Architectural support of household waste
	Others		Others
3. Harmony with the local environment	Eco-friendly for local area	4. Health and amenity	Universal design/ handicapped accessible
	Low-impact to local water cycle		Cross ventilation
	Green and make green system		Natural material for the health
	Control of intermediate space		Protect from the noise
	Landscape conservation		Support of maintenance
	Local culture/local economy		Information services of the housing
	Others		Others

Fig. 46.2 Fukasawa Environmentally Symbiotic Housing

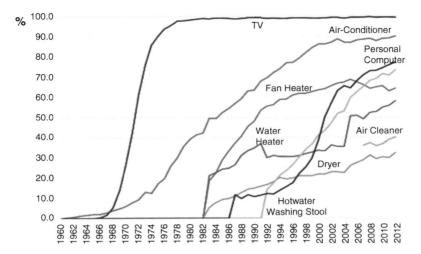

Fig. 46.3 Penetration rates of home appliances in Japan [5]

The reasons given were firstly, that each household tended to have more home electronic appliances than before (Fig. 46.3), and secondly, the number of person per household dropped, while the number of household was increased. The meaning of "energy saving" had also changed, targeting not only oil but electricity as well.

In the early 2000s, evaluation tools and building rating system were developed, such as CASBEE (Comprehensive Assessment System for Built Environment Efficiency) for detached houses [3] and "the housings of design method for low-energy housing" [4]. These assessment tools contributed to the spread of symbiotic housing and its technique, resulting in many kinds of environmentally symbiotic houses being constructed. The major house building companies have also started to engage in the development of environmentally friendly techniques as optional specifications. As a result, people tended to regard environmentally friendly houses as a luxury like the prius car, while the economy has been unstable since the collapse of the bubble economy.

The use of CO_2 emission figures as a new measure of eco-efficiency, began where previously only the number of environmentally friendly techniques was measured. It was revealed that the operation use of housing in Japan composed more than 70% of the life cycle CO_2 emission, and this indicated the remarkable change of peoples' lifestyle and usage of the heating and cooling devices. In the late 2000s, LCCM (life cycle carbon minus) housing [6] has been studied and constructed by BRI (Building Research Institute in Japan) forecasting the life cycle CO_2 emission. Thus, there were many kinds of experimental housing for the next generation. Another trend of this period was to utilize existing stocks effectively; especially the aging housing complexes provided during the high economic growth period were in need of maintenance or renovation as Japanese houses are reportedly rebuilt about every 25 years.

    In this period, people began to realize that designing sustainable housing had to be more suitable to their local climate. In 2010, the eco-house project was conducted by the Ministry of the Environment [7]; the 20 districts selected to exemplify Environmentally Symbiotic Housing showing the character of locality such as the climate, community, and regional economy. People realized it was easier and more effective to create sustainable housing for the housing complex or housing block rather than for a single detached house in terms of cost and techniques.

    In 2011, Japan entered a new era of energy saving as The Great East Japan Earthquake occurred. The nuclear power plants were damaged, and people living in the Tokyo metropolitan area had restrictions on their supply of electricity. They were aware of the importance of energy saving and the importance of the source of primary energy. It also has been noted that the alternative energy had changed, mostly from heat medium to solar power. Being eco-friendly became more a part of our daily life. After this incident, more architects and developers of housing have realized the importance of the environmental issue and designed houses more consciously using various kinds of renewable energy and techniques. Sustainable housings has now been widely recognized. Recently, a Japanese governmental project called "ZEH (net zero energy house)" [8] has started, a housing which had installed an energy creation system which is able to supply the entire energy consumed within the house. However, the existing houses and the lifestyle they offered still remain as a major problem that needs tackling. The amount of CO_2 emission from housing in Japan has been increasing since 1990, although people tend to notice environmental issues more than before, and Japanese population has been decreasing since 2006. We create more CO_2 emission per person. We need to consider these problems including the problem of economy and the effect on local communities.

46.3   Housing in the Tropical Region in Japan

The land of Japan stretches 3000 km long from north to south and east to west and surrounded by four oceans. There are roughly six energy saving reference regions [9] divided by heating degree day (Fig. 46.4). There are three cold regions: a temperate humid region, a hot region, and a tropical region. Okinawa prefecture is the only tropical region, and there are some areas in the Southern part of Kyushu and Pacific coastal area which also correspond to the tropical region. The temperate region which is the biggest area in Japan – includes Tokyo and Osaka – is where the summer is hot-humid (similar to that of the tropical region), yet their winter is cold and dry. Thus, there are many types of houses from ancient times which were designed specifically for the local climate, with most of the Japanese traditional houses built for hot-humid climate.

    In terms of the diffusion of energy saving housing and techniques, the development in the cold region has been more advanced than the tropical region. This is because less energy is used for heating, and no heating system is required in

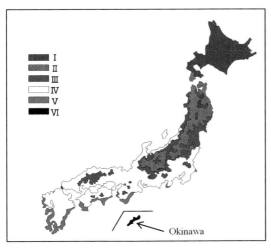

Fig. 46.4 Regional classification by heating degree day (left), and the typical traditional house in Okinawa (right above), and concrete housing after the war in the street in Ishigakijima, Okinawa (right down). (Photo by Shunsuke Sugimoto)

these houses. However, concrete surface of buildings and exhaust heat from air conditioning in the summer caused the heat island phenomena and in return wasted energy for additional air conditioning.

Originally, in tropical regions, there has been many traditional passive techniques for the protection against heat rather than for energy saving. For instance, the traditional houses in Okinawa are single-story wooden structure with deep eaves. They have a southern exposure with a stone wall in front of the house giving protection from typhoon. However, after the war, reinforced concrete housing were introduced with the influence from the American military base. The traditional environmentally friendly techniques were neglected.

Table 46.2 shows the representative sustainable housings in the area with hot tropical region from the 1980s to recent years. It revealed that most of the housing in this region uses the techniques of shading, air circulation, and natural ventilation. Subsequently, techniques of Resource Recycling, affinity with surrounding environment and eco materials followed. The techniques for affinity with surrounding environment include contributing townscape, local culture, and ecosystem. There were not many renewable energy examples such as solar energy because the focus has always been more on the protection against hot-humid climate rather on the generation of energy. The air tightness techniques are only seen in recent years because the popularization of air conditioning turned houses into a closed environment.

Sun shading is a passive technique that can be found in traditional Japanese housing in the tropical region. Roughly, there are two types of sun shading techniques; one is the holed or slit façade wall and deep eaves, and another is shading by greenery.

Table 46.2 The representative sustainable houses in hot-humid region in Japan from past to present

Year		D/M[a]										Description
1981	Rubble stone house	D	O				O		O	O	O	RC/2 stories/109.13m^2
1986	House in Kokuba	D	O	O	O				O			RC/3 stories/164.60m^2
1988	Sueyoshi riverside terrace	M	O				O		O		O	RC/2 stories/87–93m^2/16 houses
1990	Tozan house	D	O			O	O		O		O	RC/2 stories/205m^2
1994	Itoman Kaneshiro house	D	O				O	O	O	O	O	RCB/2 stories/136.58m^2
1997	House in Gishi	D	O			O	O	O		O	O	RC/2 stories/182m^2
1997	Taira Eimajo housing	M	O				O	O	O	O	O	RC wall/3 or 4 stories/13 buildings/ 177 units
1998	Mihama heights 2	D	O				O			O	O	RC/2 stories/53 units
2000	Yakushima symbiotic housing	DM	O	O		O	O		O	O	O	Wooden/2 stories/64–79m^2/50 houses
2000	Ozaki house	D	O		O		O		O		O	RC/ stories/220.69m^2
2000	Okinawa Hirara housing	M				O	O	O	O	O	O	RC/3–6 stories/180 units
2002	Shintoshinmekaru housing	M	O			O	O	O	O	O	O	RC/9 stories/135 units
2005	Tamanaha house	D			O	O	O		O	O	O	Wooden +RC/2 stories/147m^2
2006	Sustainable housing in Kiire	DM	O	O		O	O		O	O	O	RC/ 2 stories/53–79m^2/35 houses

Year	Name	DM[a]	1	2	3	4	5	6	7	8	9	10	11	12	13	14	15	16	Structure
2006	Sustainable housing in Kaseda	DM	O	O		O	O	O		O		O		O	O	O	O	O	Wooden/2 stories/78–89m²/120 houses
2010	Eco-house in Miyako-jima	D				O	O	O			O						O	O	RC/2stories/119.27m²
2011	Ma of wind	D	O	O	O	O	O			O					O	O	O		RC/2stories/109.13m²
2013	Shirasu	D	O	O	O	O		O	O						O	O			RC+ CB/2 stories/143.94m²
2013	Huis ten Bosch smart house	D	O	O					O						O	O	O	O	Wooden +S/1 story/83.5m²

1 insulation (Earth), 2 airtight, 3 mass, 4 shading/solar control, 5 daylight use, 6 ventilation/air circulation, 7 humidity control, 8 spraying water, 9 efficient energy devices, 10 solar energy, 11 renewable energy, 12 life extension planning, 13 resource recycling, 14 eco material, 15 ecosystem conservation, 16 surrounding environment

[a]Detached house or multiple housing

The holed or slit façade wall techniques mainly uses perforated concrete block wall; these can be seen on façade of House in Kokuba, House in Gishi, and Tozan House. The Sueyoshi riverside house has the same perforated concrete block but also made used of the leafy shade and bougainvillea, a local plant. The hole in the façade not only offers protection from typhoon and strong wind but also promotes natural ventilation (Fig. 46.5).

Ventilation and air circulation system can be incorporated when designing the plan and sectional structure of the house. A courtyard can serve as a buffer zone between the front wall and the housing. In order to prevent heat trapping inside of the houses, air circulation and natural or automated cross ventilation are very important. To avoid humidity, natural ventilation and water cooling systems can be used. However, water cooling systems are normally too expensive for general housing. On the other hand, if the indoor environment is controlled by air conditioning, there is no need to have humidity control system; however one pays the price with the problem of air tightness, loss of natural ventilation, and the merit of traditional housing design.

Since the late 1980s, the household air conditioning became widespread, and it has caused the heat island phenomena in urbanized area in this region and all over Japan. The passive techniques could help mitigate this problem. But the increase of home electrical appliances is one of the big reasons of the popularization of air conditioning, and the CO_2 emission has also been increasing.

Recently, there have been some experimental housing in area of the tropical region such as "Huis Ten Bosch Smart House." However, they are controlled by mechanical efficient energy devices and have not considered the locality such as the surrounding environment, ecosystem, etc. They also lose the passive techniques and traditional housing design, causing the wasting of resources and other incidences of environmental problems.

Fig. 46.5 Sueyoshi riverside house (Photo by Hirokazu Mitsuoka)

46.4 Summary

In conclusion, there were many passive and reducing energy techniques in our traditional housing, but nowadays we tend to rely on the mechanical equipment to control our indoor environment. We should learn from the original passive techniques in terms of not only energy saving but also the life cycle of housing and saving resources. With further research, we could apply advanced versions of these traditional techniques and incorporate them into modern design, modern lifestyle, and mechanical equipment.

In the future issues, I shall discuss how to select and introduce the environmentally friendly techniques basing on the following:

1. To select the techniques according to LCC (life cycle cost)and LCA (life cycle assessment) including the maintenances and disposal
2. To select the techniques which give multiple advantageous effect. For example, green has multiple effect such as shading, ecosystem conservation, CO_2 absorption, and healing people
3. To design the housing with the environmentally friendly techniques by overlooking the surrounding area and each unit of housing blocks, and with a future vision (not focus on the present)

Acknowledgment This article is based on the survey documents of sustainable housings in Japan according to architectural magazines and the actual investigation of Japanese sustainable housings, built since 1970 to the present, by the author [1].

References

1. Nakamura M, Iwamura K (2006) Outcomes of post-design on environmentally symbiotic Housing – case analysis on recent environmentally symbiotic multiple dwelling houses in Japan. J Archit Plann AIJ 610:1–8
2. Definition of Environmentally Symbiotic Housing (as of 2005) Environmentally Symbiotic Housing Promotion Council. http://www.kkj.or.jp/contents/intro_sh/index.html
3. Tools for housing scale, CASBEE Family and tools, Comprehensive Assessment System for Built Environment Efficiency, Institute for Building Environment and Energy Conservation (IBEC) (since 2007) http://www.ibec.or.jp/CASBEE/english/index.htm
4. Guidline forThe Housings of Design Method for Low Energy Housing (since 2005) Institute for Building Environment and Energy Conservation. http://www.jjj-design.org/design/index.html
5. Consumer Confidence Survey (conducted on March 24th, 2012) The Cabinet office, Japan. http://www.esri.cao.go.jp/en/stat/shouhi/shouhi-e.html
6. LCCM (Life Cycle Carbon Minus) Houses as advanced energy-efficient housing for a low-carbon society. Building Research Institute, Japan. http://www.kenken.go.jp/english/con tents/lccm/index.html
7. Eco Housing Model Project, Ministry of Environment. http://www.env.go.jp/policy/ecohouse
8. Agency for Natural resources and Energy. http://www.enecho.meti.go.jp/category/saving_and_ new/saving/zeh/
9. Regional classification by Energy-saving region references, attached table (2013) Manual of criterion by client and design and construction guidelines

Box A: An Example in Thailand: Floating in the Sky School for Orphans

Kikuma Watanabe and Masaki Tajima
School of Systems Engineering, Kochi University of Technology, Kochi, Japan
Email: tajima.masaki@kochi-tech.ac.jp

Abstract The floating in the Sky School, which is partly an adopted earthbag method, was built in Sangkhla Buri village, a mountainous area in Thailand. The earthbag method was adopted because the schoolchildren could engage in the construction, the materials can be procured locally, and as a result it can be built cheaper. Additionally, the stable cooler thermal condition of the daytime without any cooling system, which requires high electricity fee for the owner, even during the hot period of the dry season was aimed by the architect. Inside and outside of the architecture, measurement of temperature, relative humidity and globe temperature were operated for 3 days in a dry season. From the measurement results, the cool thermal storage at night was satisfactory to keep the inside temperature of earthbag chamber much lower than outside during daytime. This resulted as the night temperature could be sufficiently cooled in this area. Moreover, from the viewpoint of a risk of heatstroke, inside environment of the earthbag chamber is considered as much safer with the thermal storage performance.

Keywords Mountainous · Earthbag · School · Thermal storage · WBGT

Outline of the Building

The floating in the Sky School (see Photo 1) is built as a school for orphans in Sangkhla Buri village, a mountainous area in Thailand. It was designed by one of the authors, Kikuma Watanabe, aiming at realising a kid's dream of a flying ship. It is mainly employed by two components: the round, earthbag volumes on the ground and, the other, a light steel structure finished with bamboo and a grass roof. (Fig. 1)

The round volumes create a warm interior, fostering a sense of comfort for the children in prayer dome and classroom of earth. The floating level functions as learning area and a Buddhist room where children pray in the morning and the evening. The upper floor connects to the lower earthbag domes through two openings.

The earthbag method was adopted because the schoolchildren could engage in the construction, the materials can be procured locally, and as a result it can be built

© Springer Nature Singapore Pte Ltd. 2018
T. Kubota et al. (eds.), *Sustainable Houses and Living in the Hot-Humid Climates of Asia*, https://doi.org/10.1007/978-981-10-8465-2

Photo 1 The floating school

Fig. 1 Site plan

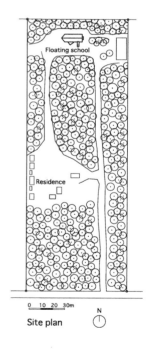

Site plan

cheaper. Moreover, it was considered that the temperature inside the earthbag construction was stable and children could spend comfortably even during the hot period of the dry season in the site without cooling system which required high electrical fee, because of its high thermal storage performance (Fig. 2).

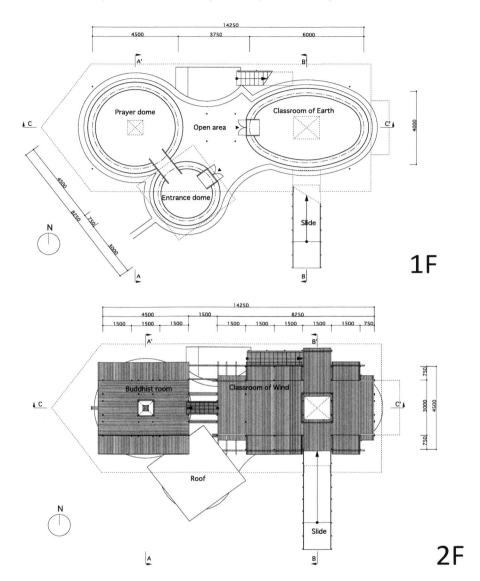

Fig. 2 Plan

The school is a semi-outdoor space and employs no air conditioning. Therefore, real measurements were conducted for the purpose of the present condition of thermal safety.

Measurement of Environment

To grasp the environment surrounding the children in dry season, which means the hottest season, real measurement was operated from 2nd to 4th of March 2015. Mainly, air temperature, air humidity and globe temperature were obtained in the target spaces such as outdoor, inside the dome, the classroom of Earth, (Photo 2) and the Buddhist room (Photo 3) using measurement instruments (examples are shown in Photos 4 and 5).

Photo 2   Inside the Dome

Photo 3   The Buddhist room

Photo 4 Thermometer

Photo 5 Globe temperature measurement

Measurement Results

Figures 3 and 4 represent outdoor temperature and outdoor relative humidity obtained at Bangkok and the school site. It is indicated that it is hotter during the daytime and cooler during the night-time in the site compared with in Bangkok which is very urban and locates approximately 200 km away from the school site. This condition in the site helps to keep Dome's indoor temperature much lower than outside as describes later.

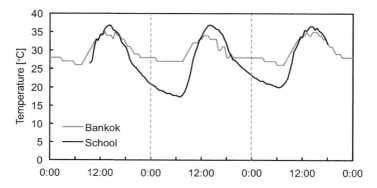

Fig. 3 Outside temperature (Bangkok and the school site)

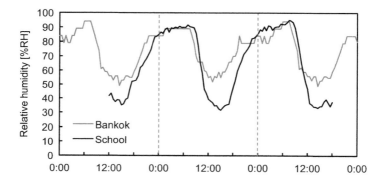

Fig. 4 Outside humidity (Bankok and the school site)

Figure 5 represents the air temperature of outside and inside the dome and the Buddhist room. During the daytime, air temperature of the Buddhist room is almost same as outside; however, inside the dome, its temperature is much lower than outside. The highest air temperature was just 30 degrees in Celsius whenever the outside temperature was over 37 degrees in Celsius. On the other hand, during the night-time, both spaces' air temperature were warmer than outside. It is suggested that this temperature differences are caused by the thermal storage of the earthbag.

Figures 6 and 7 represent boxplots of WBGT (WetBulb Globe Temperature), which is given by Eq. (1), of each time by using measured 3 days data obtained inside the dome and in the Buddhist room. These figures contain dot line on 25 degrees in Celsius, which means a heatstroke danger temperature. Inside the dome, most WBGT was lower than 25 degrees in Celsius even during the sunny daytime. In the Buddhist room, its WBGT was higher than the danger temperature

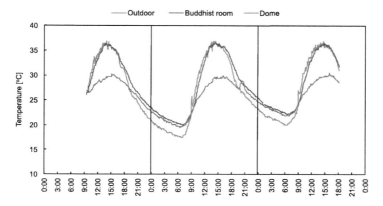

Fig. 5 Temperature(3 days)

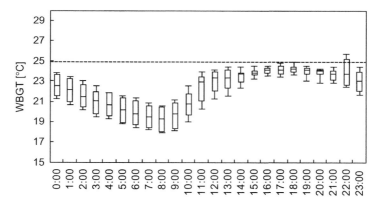

Fig. 6 WBGT inside the Dome (3 days)

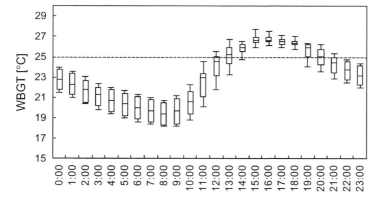

Fig. 7 WBGT in the Buddhist room (3 days)

through the sunset from noon. There is a need for frequent rest in the space during the time.

From the measurement results, inside the Dome made of earthbag, even in the dry season without any cooling system, its temperature and WBGT indicated that it is considered as safety for children from the viewpoint of a risk of heatstroke.

$$WBGT = 0.7WB + 0.3GT \qquad (1)$$

Where

WBGT: Wet bulb globe temperature [°C]
WB: Wet bulb temperature [°C]
GT: Globe temperature [°C]

Conclusions

In and around the floating in the Sky School which is partly an adopted earthbag aiming at stable thermal condition even during the hot period of the dry season, measurement of temperature, relative humidity and globe temperature was operated for 3 days in a dry season. From the measurement results, the cool thermal storage at night was satisfactory to keep the inside temperature of earthbag construction much lower than outside during daytime. This resulted as the night temperature could be sufficiently cooled in this area. Moreover, from the viewpoint of a risk of heatstroke, inside environment of the earthbag chamber is considered as much safer compared to the environment in a construction without thermal storage performance.

Part VII
Climate Change and Urbanization

Chapter 47
Climate Vulnerability in Tropical Asia

Takao Yamashita and Han Soo Lee

Abstract The tropical zone encompasses some of the wettest areas on Earth, as well as some of the world's driest deserts. This zone also includes some of the countries that are the most vulnerable to natural disasters, due to population pressures that drive settlement in flood- or drought-prone areas. Some of the less-developed countries also lack the resources to build structures resilient to climatic extremes. The limitations of physical and human infrastructure also contribute to the limited capacity to warn of or respond to major disasters, although this is an area where large improvements have been made in many countries in recent decades.

Tropical Asia's near future climate vulnerability in terms of extreme temperatures, precipitation and typhoons was analysed using the output of the general circulation model (GCM) MIROC4h in the Coupled Model Intercomparison Project Phase 5 (CMIP5). The analysis of the long-term sea level rise in the Pacific Ocean, using the dataset of the Reconstructed Sea Level Version 1 (RSLV1), indicates that sea level rise is occurring at an alarming rate.

Keywords ITCZ · Trade winds · ENSO · MIROC4h in CMIP5 · Precipitation · Extreme temperature · Typhoon · Sea level rise · RSLV1

47.1 Tropical Atmospheric Circulation

The most fundamental feature of the tropical atmospheric circulation is Hadley cells (see Fig. 47.1), wherein the tropical warm air rises from the surface to the upper atmosphere (convection) and moves towards the pole before descending in the subtropics, creating a zone of high surface pressures (the subtropical high). To

T. Yamashita
Japan Port Consultant Co., Ltd., Kobe, Japan

H. S. Lee (✉)
Graduate School for International Development and Cooperation (IDEC), Hiroshima University, Hiroshima, Japan
e-mail: leehs@hiroshima-u.ac.jp

© Springer Nature Singapore Pte Ltd. 2018   513
T. Kubota et al. (eds.), *Sustainable Houses and Living in the Hot-Humid Climates of Asia*, https://doi.org/10.1007/978-981-10-8465-2_47

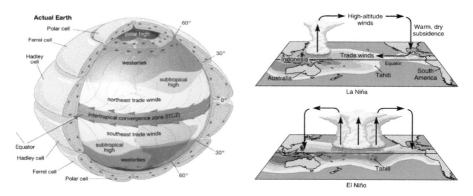

Fig. 47.1 (Left) Global circulation and intertropical convergence zone (ITCZ) [2] and (right) a diagram of El Niño and La Niña [3]

complete the vertical circulation cell, air then flows at the surface level from the subtropical high towards the equator.

The zone of near-equatorial low surface pressure is known as the intertropical convergence zone (ITCZ), which moves seasonally, going from being centred north of the equator during the Northern Hemisphere (NH) summer to south of the equator during the Southern Hemisphere (SH) summer. The dominant easterly winds between the subtropical ridge and the ITCZ are called the "trade winds", the direction of which is southeasterly in the NH and northeasterly in the SH.

The climate has several internal modes of variability caused by atmosphere and ocean interactions. The El Niño-Southern Oscillation (ENSO) is the most dominant oscillation in the tropical zone and is characterised by changes in ocean temperatures in the eastern and central equatorial Pacific that are caused by changes in the trade wind intensities. El Niño events are associated with a high risk of dry conditions over areas such as the western Pacific but are associated with high rainfall in east Africa and on the west coast of South America. El Niño events also have strong influences on the rainfall of many Pacific Islands. La Niña events are the near reverse of El Niño events – that is, waters are cooler than normal in the eastern and central equatorial Pacific – and have nearly the opposite climatic impacts [1].

47.2 Near Future Projection of Extreme Events in Asia

47.2.1 Model Simulation of Variability and Extremes

Based on the Intergovernmental Panel on Climate Change (IPCC) the Fifth Assessment Report (AR5) Working Group 1 (WG1), several general circulation model (GCM) outputs for the tropical circulation are summarised below [4].

(1) Earlier assessments of a weakening Walker Circulation from models and reanalysis have been tempered by the subsequent observational evidence that actual tropical Pacific trade winds have strengthened since the early 1990s. (2) Model outputs suggest that the widths of the Hadley cells should increase. (3) The tendency in a warming climate for wet areas to receive more precipitation and subtropical dry areas to receive less is called the "rich-get richer" mechanism.

The ability to simulate climate variability, both due to the unforced internal variability and the forced diurnal and seasonal variabilities, is important. The low-frequency climate variability must be estimated, at least in part, from long control integrations of climate models. This process also has implications for the ability of models to make quantitative projections of changes in climate variability and the statistics of extreme events given a warming climate. In many cases, the impacts of climate change will be experienced more profoundly via the frequency, intensity or duration of extreme events, such as heat waves, droughts, extreme precipitation and tropical cyclone. The accuracies of climate variability simulations are also central to achieving better skills in climate prediction by initialising the models from observed climate states. Model evaluation for climate variability prediction skill should be done to explore the representations of certain processes, such as the coupled processes underlying ENSO and other important modes of variability.

Extreme events are realisations of the tail of the probability distribution of weather and climate variability. These occurrences are higher-order statistics and thus are generally more difficult to reproduce in climate models. Shorter time scale extreme events are often associated with smaller scale spatial structures, which may be better represented as the model resolutions increase. For near future projections of extreme events in Asia, the outputs of MIROC4h, which is a new high-resolution atmosphere-ocean coupled general circulation model, were analysed in terms of the maximum/minimum temperatures, precipitation and tropical cyclones.

47.2.2 Near-Term Climate Prediction by MIROC4h in CMIP5

A new high-resolution atmosphere-ocean coupled general circulation model named MIROC4h has been developed. The atmospheric components of MIROC4h have a resolution of 0.5625 deg., and the oceanic components are eddy-permitting with a horizontal resolution of 0.28125 deg. (zonal) by 0.1875 deg. (meridional). The model performance in a 120-year control experiment (including a 50-year spin-up) under the present conditions was examined before the short-term prediction of 2006–2035. Compared with the previous versions MIROC3h, many improvements have been achieved. For example, the errors in the surface air temperatures and sea surface temperatures are smaller, there is less drift of the ocean temperature in the subsurface-deep ocean, and the frequency of the heavy rain is comparable to that observed. The fine horizontal resolution in the atmosphere makes orographic wind and its effects on the ocean more accurate than those of the previous models, and the treatment of the coastal upwelling motion in the ocean has been improved. The phenomena in the atmosphere and ocean related to ENSO are now closer to those observed than those of previous model outputs [5].

Using the model MIROC4h, a set of near-term climate prediction experiments was conducted under the protocol of CMIP5 [6] (RCP4.5 scenario) and was contributed to the IPCC AR5, in which four ensemble experiments were carried out. In the following analysis, the output of one experiment, named "r1i1p1", was used.

47.2.3 Maximum/Minimum Temperature

In terms of the historical climate changes, GCMs generally capture the observed trends in temperature extremes from the second half of the twentieth century. The left panel in Fig. 47.2 shows the time series of all the daily maximum and minimum temperature changes simulated from 2006 to 2035 at the representative cities of Hanoi, Singapore, Jakarta, Bangkok, Kathmandu and Dhaka. The middle and right sections of Fig. 47.2 show the changes in the annual means of the maximum and minimum temperatures, respectively. A linear regression equation in the figure indicates the annual rate of increase of the annual mean of the daily max/min temperatures of these cities. The annual rates of increase of the annual means of the daily max/min temperatures of 15 major cities are listed in Table 47.1

The table indicates that the annual mean of the daily maximum temperature of Beijing rises by 10.03 (deg) and the maximum temperature rises by 6.55 (deg) after 100 years. The rising rates in the subtropical cities are quite high compared with those of the tropics. In Yangon, the gradient of the linear regression of the annual mean of the daily maximum temperature is negative, dropping by -0.0101 (deg/year) instead of following the trend of the strongly increasing rate of the minimum temperature of 0.0503 (deg/year).

Figure 47.3 shows the distribution of the annual rates of increase (deg/year) of the annual means of the daily maximum/minimum temperatures (2006–2035) in Asia Oceania

The highest rate of increase of the maximum temperatures was observed in the north of the Bo Hai Sea, China, and the highest rate of increase of the minimum temperatures in southern Tibet. The rates of increase of the maximum temperatures in Northwest Australia and the Guangxi province of China are notably high. In Central Asia, the ratio of the maximum/minimum temperatures will increase in the northern regions and decrease in the south.

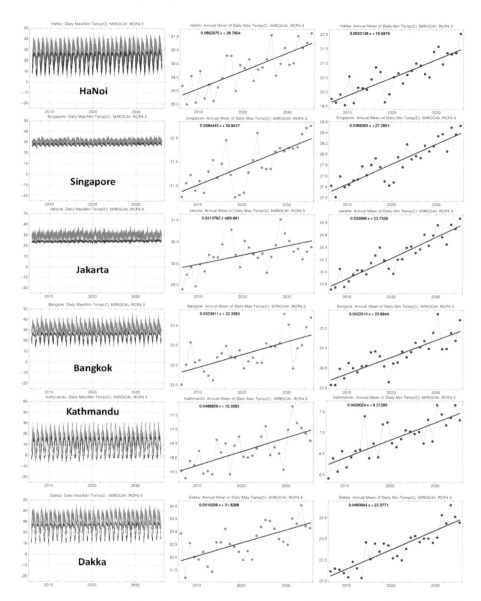

Fig. 47.2 Maximum (red) and minimum (blue) temperature changes simulated for 2006–2035 (left) and the changes in the annual means of the daily maximum temperaures (middle) and those of the minimum temperatures (right)

Table 47.1 Predicted annual rates of increase of the annual means of the daily minimum and maximum temperatures at 15 major cities in Asia Oceania (2006–2035)

	Min (deg/year)	Max (deg/year)
Sapporo	0.0599	0.0584
Tokyo	0.0551	0.0564
Beijing	0.0655	0.1003
Shanghai	0.062	0.0649
Hanoi	0.0632	0.0662
Ho Chi Minh	0.0444	0.0345
Singapore	0.0369	0.0384
Jakarta	0.034	0.0214
Bangkok	0.0423	0.0324
Yangon	0.0503	−0.0101
New Delhi	0.0436	0.0264
Kathmandu	0.042	0.019
Kabul	0.0352	0.0282
Perth	0.0162	0.0394
Sydney	0.0202	0.0537

47.2.4 Precipitation

For extreme precipitation (hourly and daily), the observational uncertainty is much larger than that for temperature, making model evaluation more challenging. It has been noted from comparisons to satellite-based data sets that the majority of models underestimate the sensitivity of extreme precipitation intensities in the tropics and globe.

Using GSMaP (Global Satellite Mapping of Precipitation), the Japan Aerospace Exploration Agency (JAXA) offers hourly global rainfall maps in near real time (about 4 h after observation) using a combined MW-IR algorithm with GPM-Core GMI, TRMM TMI, GCOM-W AMSR2, the DMSP series SSMIS, the NOAA series AMSU, the MetOp series AMSU and Geostationary IR data. Background cloud images are globally merged with IR data produced by the NOAA Climate Prediction Center (CPC), using IR data observed by JMA's MTSAT satellite, NOAA's GOES satellites and EUMETSAT's Meteosat satellites [7].

A distribution of the monthly precipitation (mm/month) in 2014 mapped by GSMaP is shown in Fig. 47.4. This distribution depicts the typical seasonal changes of the precipitation of Asia Oceania together with the characteristics of the ITCZ seasonal changes. From December to April, the Asian continent has extremely low rainfall. After May, the Indian summer monsoon starts and the ITCZ shifts to the north to bring rainfall to the land of the Asian continent, with especially heavy rainfall in the Bay of Bengal/South China Sea in June and July and considerable rain in the Himalayan region in May to August.

The geographical pattern of precipitation illustrates strong locality because the climate model confirmed that it is difficult to predict precipitation even within the next few years. The differences between the precipitation data measured by satellite (GSMaP) and the projections from MIROC4h in the overlapping period of

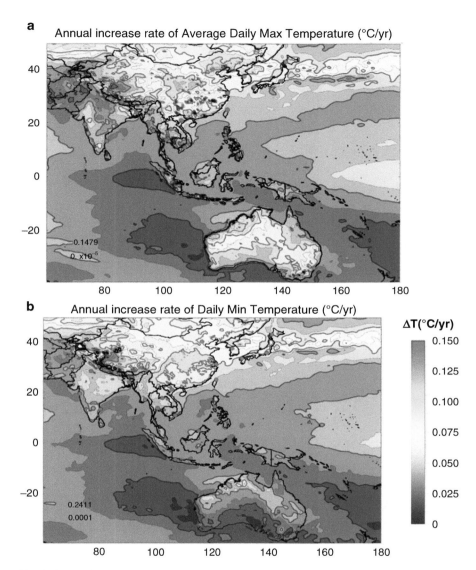

Fig. 47.3 Distribution of the annual rate of the increase of the annual means of the daily maximum/minimum temperatures (2006–2035) in Asia Oceania

2006–2014 were used as the learning data for the model projection error to create an algorithm to correct the MIROC4h output for predicting future precipitation. The correction method used is summarised below.

(1) Using the interpolation of the GSMaP data (0.1-degree resolution), convert the daily precipitation distribution of GSMaP to the 0.5-degree grid system of the MIROC4h projection and obtain the differences between the two results that can be

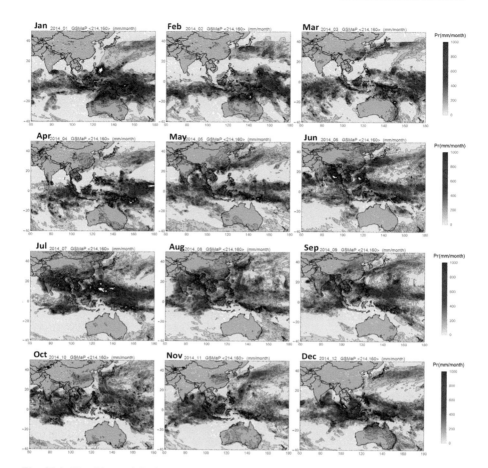

Fig. 47.4 Monthly precipitation observations mapped by GSMaP in 2014

defined as the MIROC4h projection error at each grid point. (2) Determine the time series of the MIROC4h projection error for an average monthly rainfall at each grid point for the period of 2006–2014. (3) Assuming SARMAP (Seasonal Auto Regressive Moving Average Process) is used for the monthly MIROC4h projection error, then estimate the processes and forecast the future error of the monthly rainfall from 2015 to 2035. (4) Although SARMAP corrects the future oscillation processes of the error, it does not correct the trend (long-term mean). The correction coefficients were computed for all the grid points to adjust for this gap of trends.

Multiplying by this correction coefficient and using the SARMAP corrections, we can get the corrected precipitation in which the GSMaP observations (blue) and the corrected MIROC4h projection (red) are continuous, as shown in Fig. 47.5. Rainfall

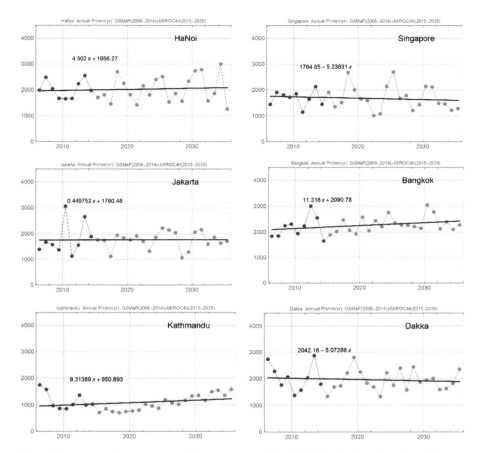

Fig. 47.5 Annual changes in precipitation observed by GSMaP (blue) and projected by the corrected MIROC4h (red)

in Hanoi, Bangkok and Kathmandu will increase, and there will be a slight decrease in Singapore. Strong rainfalls are projected in 2019 and 2026 for Singapore, in 2034 for Hanoi and in 2020 for Dhaka. The linear regression equations are shown in the figures.

The distribution of the annual rate of increase of the yearly precipitation (mm/year) estimated by the corrected MIROC4h for 2006–2035 was computed as shown in Fig. 47.6, in which the precipitation will increase in almost all the cities in tropical Asia. Drought is caused by long-term scale (months or longer) variability of both precipitation and evaporation. Because the temperature will increase and the precipitation will decrease in northern Central Asia, drought is a concern for this region.

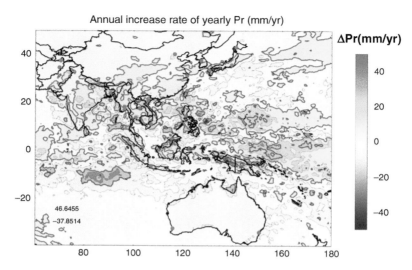

Fig. 47.6 Annual rate of increase of the yearly precipitation (mm/year) estimated using the corrected MIROC4h output for 2006–2035

47.2.5 Typhoon

Vortices of surface winds in the tropical cyclones can be detected and traced in the outputs of GCMs, but generally, their intensities are too weak, even in the MIROC4h results. Using the daily average surface wind speeds and sea level pressures of the MIROC4h outputs, vorticity was calculated, and the lowest pressure point, which has a vorticity of more than 0.000135 (1/s), was used to detect a typhoon centre in the Pacific Ocean. When a typhoon centre continued for 4 days or more, it was defined as a typhoon track in this analysis. This method detects typhoons in the period of 2006 to 2035, and the obtained annual typhoon track charts are shown in Fig. 47.7.

As their intensities are too weak due to the coarse resolution of the model, some correction of the central pressures should be done using the observations and projection data within 2006–2014. Fitting the frequency distribution of the lowest central pressure observed, and subsequently projecting the typhoon in the Weibull distribution, the same frequency of occurrence of the two cumulative frequency distributions can allow the corrections of the lowest central pressures of the observations and projections. Using this correction method of typhoon central pressures, the annual changes in the lowest typhoon pressures were obtained along with the number of typhoon occurrence from 2006 to 2035, as shown in Fig. 47.8.

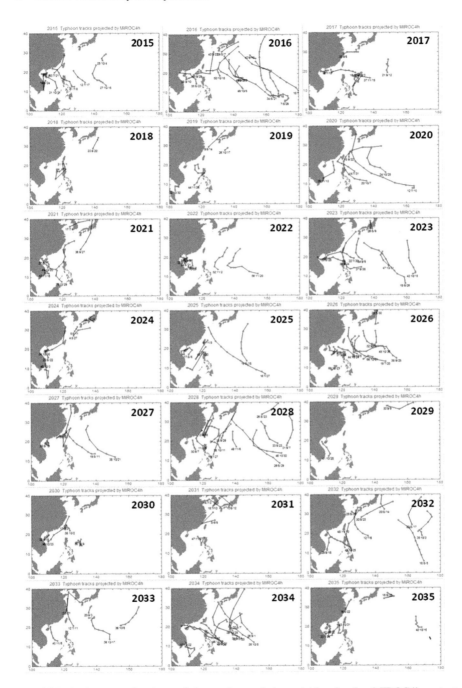

Fig. 47.7 Typhoon tracks detected by surface wind vorticities in the MIROC4h output (2015–2035)

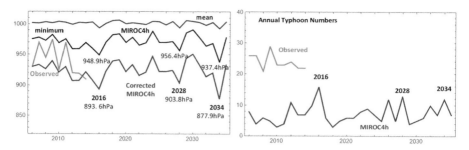

Fig. 47.8 Changes in the lowest typhoon pressures (left) and number of typhoon occurrences (right) in the period of 2006–2035

47.3 Sea Level Rise in the Pacific Ocean

47.3.1 Reconstructed Sea Level Version 1

A dataset named "Reconstructed Sea Level Version 1" (RSLV1) provided by NASA (National Aeronautics and Space Administration) contains sea level anomalies derived from satellite altimetry and tidal gauges. The satellite altimetric record provides accurate measurements of sea level with near-global coverage but has a relatively short time span, beginning in 1993. Tidal gauges have measured sea level over the past 200 years, with some records extending back to 1807, but they provide only regional coverage, not global. Combining the satellite altimetry with tidal gauges via a technique known as sea level reconstruction results in a dataset of record length comprising tidal gauges and the near-global coverage of satellite altimetry. The reconstructed sea level dataset has a weekly temporal resolution and a half-degree spatial resolution [8].

47.3.2 Sea Level Changes, ENSO and Trade Winds

Using the dataset of RSLV1; the ENSO index MEI (Multivariate ENSO Index) [9], in which positive values indicate El Niño and negative La Niña; and the WMO wind dataset; the relations of sea level changes in the Pacific Ocean, ENSO and trade winds were analysed for the period of 1950–2009. Figure 47.9 shows the typical relative sea levels (cm) together with the names of the output points (top left); the annual changes in the 1-year total of the hourly wind speeds (m/s) at Tarawa, Kiribati (top right); the changes in MEI (middle, left); and the relative sea level changes at the three output points of the Line Islands, Solomon, and Palau.

There are clear dominant periods of El Niño (1950–1977) and La Niña (1978–2007) in the time series of MEI and the relations with trade winds and

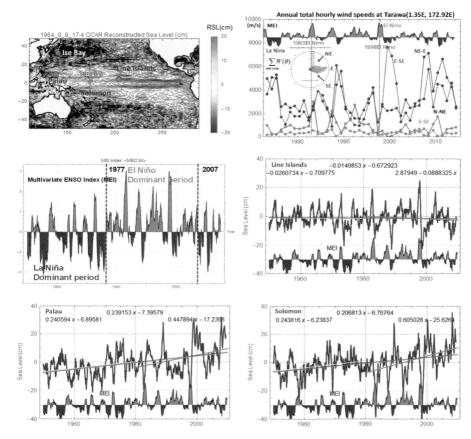

Fig. 47.9 The typical relative sea level of Aug 8, 1964, and the names of the output points (top left), the annual changes in the 1-year totals of the hourly wind speeds (m/s) at Tarawa and Kiribati (top right), the changes in MEI (middle, left) and the relative sea level changes at the three output points of the Line Islands (middle, right), Solomon (bottom, left), and Palau (bottom, right)

ENSO. Attenuation of the trade winds (NE-E-SE) causes El Niño, and their strengthening causes La Niña, with an overall decadal cycle. As a result, the sea level changes in the tropical Pacific are clearly under the influence of ENSO. The weekly averaged sea levels increase with the phase of El Niño at the Line Islands (east tropical Pacific) while decreasing at Palau and Solomon (west tropical Pacific) with ranges of plus or minus several tens of centimetres. The amplitudes of the ENSO-induced short-term sea level changes seem to have increased after 1980, which is consistent with the evidence that the tropical Pacific trade winds have strengthened since the early 1990s [10].

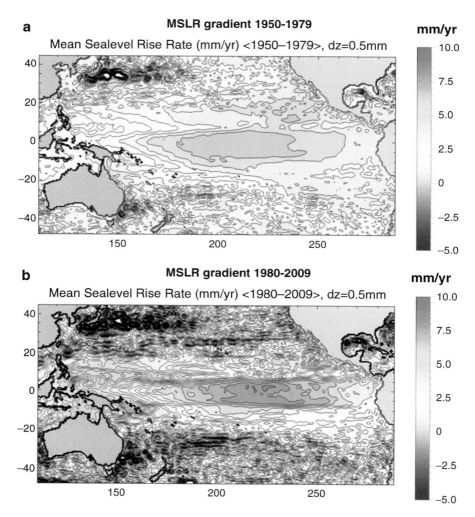

Fig. 47.10 Annual rates of the long-term sea level rise (mm/year) in the Pacific Ocean, using linear regression

47.3.3 Long-Term Sea Level Rise

As shown in Fig. 47.10, before and after 1980, the gradients of the linear regressions of long-term sea levels have clearly changed. Divided into two terms, 1950–1979 and 1980–2009, two maps of the annual rates of increase of the long-term sea levels (mm/year) of the Pacific Ocean were determined by applying the linear regressions for all the data points. Figure 47.10(a) shows the annual rates of increase of the long-term sea levels (mm/year) in the period of 1950–1979 and (b) shows that of 1980–2009. These figures depict a clear difference of the rates of sea level rise and

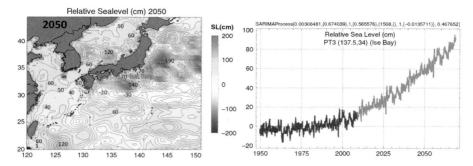

Fig. 47.11 Predicted sea level (cm) of Japanese waters in 2050 (left), and the sea level changes at the point of "Ise Bay" (right), wherein the blue line indicates the restructured data and the red depicts the predicted sea level changes from the SARIMAP model

the global increase of the long-term mean sea level rise, except over the eastern tropical Pacific Ocean, where the mean sea level decrease is in the range of several millimetres per year.

Figure 47.11 shows the predicted sea level (cm) around Japan in 2050, and the sea level is shown to change at the point of "Ise Bay", in which the blue line indicates the restructured data and the red depicts the predicted sea level changes from the SARIMAP (Seasonal Auto Regressive Integrated Moving Average Process) model, which can consider the effects of the increasing rate of sea level rise. The height of the sea level rise of Ise Bay will reach 100 cm by 2070.

47.4   Summary

Using the short-term projections of MIROC4h, the extreme temperature, precipitation and typhoons were analysed for 2006–2035, and the sea level changes in the Pacific Ocean were analysed with an observational dataset of the Reconstructed Sea Level Version 1 (RSLV1). The major outcomes are detailed below.

(1) The highest rate of increase of the maximum temperatures is observed in the north of the Bo Hai Sea, and the highest rate of increase of the minimum temperature is observed in southern Tibet. The rate of increase of the maximum temperatures in Northwest Australia and the Guangxi province of China are notably high. In Central Asia, the ratio of the maximum/minimum temperatures will increase in the northern regions and decrease in the south.

(2) Precipitation will increase in almost all cities in tropical Asia. Drought is a concern in northern Central Asia.

(3) Using the surface wind vorticities, the annual typhoon track charts are obtained for 2006–2035.

(4) The weekly averaged sea levels increase in the phase with El Niño in the eastern tropical Pacific and decrease in the western tropical Pacific, with ranges of plus or minus several tens of centimetres. The trend of increase in the rate of sea level rise clearly began changing in 1980 and continued onwards, except for in the eastern tropical Pacific, where the mean sea level decreased in the range of several millimetres per year. Using the Seasonal Auto Regressive Integrated Moving Average Process model analysis, it was observed that the sea level rise in Japanese waters will exceed 100 cm by 2070.

References

1. Trewin B (2014) Essay 1 – the climates of the tropics, and how they are changing, in state of the Tropics, pp 39–51
2. Online EB. Atmospheric circulation. https://www.britannica.com/science/atmospheric-circulation?oasmId=107938. Cited 15 July 2017
3. Cunningham W, Cunningham M (2016) Chapter 9: climate. In: The principle of environmental science. McGraw-Hill
4. IPCC (2013) Climate change 2013: the physical science basis. Contribution of working group I to the fifth assessment report of the intergovernmental panel on climate change. Stocker T, Qin D, Plattner G-K et al (eds). Cambridge University Press, Cambridge, UK/New York, p 1535
5. Sakamoto TT, Komuro Y, Nishimura T et al (2012) MIROC4h – a new high-resolution Atmosphere-Ocean coupled general circulation model. J Meteorol Soc Jpn Ser II 90 (3):325–359
6. Taylor KE, Stouffer RJ, Meehl GA (2011) An overview of CMIP5 and the experiment design. Bull Am Meteorol Soc 93(4):485–498
7. Okamoto K, Iguchi T, Takahashi N et al (2005) The Global Satellite Mapping of Precipitation (GSMaP) project. In: 25th IGARSS proceedings
8. Hamlington BD, Leben RR, Strassburg MW et al (2014) Cyclostationary empirical orthogonal function sea-level reconstruction. Geosci Data J 1:13–19
9. Wolter K, Timlin MS (1998) Measuring the strength of ENSO events: how does 1997/98 rank? Weather 53:315–324
10. Merrifield MA, Maltrud ME (2011) Regional sea level trends due to a Pacific trade wind intensification. Geophys Res Lett 38(21):L21605

Chapter 48
Urban Climate Challenges in Hanoi: Urban Heat Islands and Global Warming

Andhang Rakhmat Trihamdani, Han Soo Lee, Tetsu Kubota, Satoru Iizuka, and Tran Thi Thu Phuong

Abstract Urban climate of rapidly growing cities such as Hanoi will alter not only owing to land use changes but also global warming effect. This chapter investigates the contributions of land use changes and global warming to the future increases in urban temperature in Hanoi for 2030 through a numerical simulation using the Weather Research and Forecasting (WRF). The future climate data utilized the fifth phase of the Coupled Model Intercomparison Project (CMIP5) projected by the Model for Interdisciplinary Research on Climate Version 5 (MIROC5) through a direct dynamical downscaling method. In the 2030s, the average air temperature increase in the existing urban areas was projected to be up to $2.1\,°C$, of which up to 1.5 and $0.6\,°C$ are attributable to global warming and land use changes, respectively. The future increase in urban temperature will likely exceed the cooling effects of any urban heat island (UHI) mitigation measures.

Keywords Urban heat island · Developing cities · Master plan · Global warming · Dynamical downscaling

48.1 Introduction

Currently, major cities in the Southeast Asia tend to propose the large-scale urban development plans and increase their populations further. In Hanoi, Vietnam, a long-term urban development plan, namely, the Hanoi Master Plan 2030, was created and

A. R. Trihamdani (✉)
YKK AP R&D Center, PT. YKK AP Indonesia, Banten, Indonesia

H. S. Lee · T. Kubota
Graduate School for International Development and Cooperation (IDEC), Hiroshima University, Hiroshima, Japan
e-mail: leehs@hiroshima-u.ac.jp; tetsu@hiroshima-u.ac.jp

S. Iizuka
Graduate School of Environmental Studies, Department of Environmental Engineering and Architecture, Nagoya University, Nagoya, Japan
e-mail: s.iizuka@nagoya-u.jp

T. T. T. Phuong
Vietnam Institute of Urban and Rural Planning, Hanoi, Vietnam

© Springer Nature Singapore Pte Ltd. 2018
T. Kubota et al. (eds.), *Sustainable Houses and Living in the Hot-Humid Climates of Asia*, https://doi.org/10.1007/978-981-10-8465-2_48

529

its urban population is projected to increase from 6.7 million in 2010 to 9.2 million by 2030 [1]. The expansion of the city would result in a dramatic change in land use and therefore the urban climate. Meanwhile, the effects of global warming vary according to geographical locations and are highly anticipated in urban areas even in the future as near as 2030 [2]. Thus, the urban climate of rapidly growing cities such as Hanoi is predicted to change not only owing to land use changes but also global warming.

In this chapter, we analyse the contributions of land use changes and global warming to future increases in urban temperature in Hanoi for 2030. From the current land use condition in 2010, we adopt the land use condition proposed by the Hanoi Master Plan for the year 2030. Then, numerical simulations – specifically, regional climate modelling using Weather Research and Forecasting (WRF) – were conducted for the present land use as well as the land use condition of the master plan under the same weather conditions for the detection of only the impact of land use changes. Subsequently, WRF simulations were conducted for the land use condition of the master plan under the influence of global warming in the 2030s, utilizing the climate data from the fifth phase of the Coupled Model Intercomparison Project (CMIP5) projected by the Global Climate Models (GCMs). The GCM data were incorporated into WRF as initial and boundary conditions using a dynamical downscaling method [3].

48.2 Overview of Hanoi Master Plan 2030

The city of Hanoi includes 10 urban districts and 18 outskirts districts, which make up an area of approximately 3300 km^2. The city centre is located at 21°11′2″N 105°48′E, in the area along the Red River delta, which is approximately 90 km inland from the coastal line (Fig. 48.1). Hanoi has a humid subtropical climate,

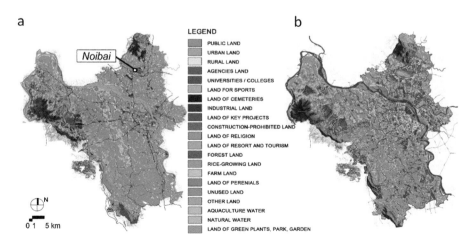

Fig. 48.1 Land use and land cover for (**a**) current condition and (**b**) the Hanoi Master Plan 2030 (Source: VIAP [1])

which experiences a hot-humid season (April to October) and a cool and relatively dry season (November to March). The maximum monthly average air temperature is observed in June, which is approximately 30.3 °C, whereas the minimum average value is recorded in February (approximately 17.5 °C).

To cope with rapid urbanization, the long-term urban development plan, the Hanoi Master Plan 2030, was established in 2011 (Fig. 48.1b). In the master plan, the population of Hanoi is projected to reach 7.3–7.9 million by 2020, 9–9.2 million by 2030 and approximately 10.8 million by 2050. The major target of the master plan is to develop Hanoi as a green-cultured and civilized modern city. To achieve the above target, the master plan proposes a series of spatial development strategies for the city, including the network of green spaces consisting of the established green corridor, green belts, green buffers and other green spaces. As a result, the green coverage would account for approximately 52% of the total land of the city, while the urban green spaces will increase remarkably, from only 745 ha (0.2%) in 2009 to 15,770 ha (5%).

48.3   Methodology

48.3.1   WRF

Numerical simulations were performed using WRF to obtain basic weather data, such as air temperature, humidity and surface winds. The present study used WRF version 3.6 [4]. Three simulation domains were configured so that the innermost domain covers the entire Hanoi City. The horizontal grid resolution of the three domains is 4.5 km, 1.5 km and 0.5 km for domains 1, 2 and 3, respectively (Fig. 48.2). The urban canopy model (UCM) was set up with default parameters as listed in [5] for simulation of the effect of urban geometry on urban climate [6]. The physics schemes used in numerical experiments are listed in Table 48.1. The validation of this method together with sensitivity tests for selecting the model parameterization is described in [7].

48.3.2   Experimental Design

Most of UHI studies under future climate conditions have relied on the outputs of GCMs. However, the spatial resolution of GCMs is too coarse (approximately 50–100 km) for urban-scale climate simulation. Downscaling methods have been developed to overcome this problem and obtain local-scale weather and climatic features, particularly at the surface level, from global- and regional-scale atmospheric GCM output variables. The dynamical downscaling uses a limited-area, high-resolution regional climate model (i.e. WRF in this study) driven by initial and boundary conditions from a GCM to derive smaller-scale features of the local

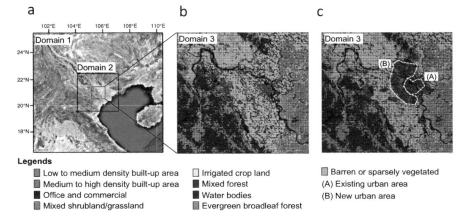

Fig. 48.2 (**a**) Three computational domains for WRF simulations. Domain 3 (indicated with red-line square) covers the whole Hanoi City. Land use map of domain 3 with 0.5 km horizontal grid resolution for (**b**) current condition and (**c**) master plan condition [8]

Table 48.1 WRF physics scheme

Physics scheme	Physics options
Cumulus scheme	Kain-Fritsch scheme
Microphysics	WRF single-moment (SM) 3-class scheme
Planetary boundary layer	Yonsei University (YSU) scheme
Longwave radiation	RRTM scheme
Shortwave radiation	Dudhia scheme
Surface layer	Monin-Obukhov similarity scheme
Land surface	NOAH land surface model (NOAH LSM) with UCM

climate. Among CMIP5 datasets from various GCMs, the Model for Interdisciplinary Research on Climate Version 5 (MIROC5) was used in this study [9].

The projected climatic data by CMIP5 are based on the assumption that the concentrations of greenhouse gases such as CO_2 will change in the future. The IPCC refers to it as the Representative Concentration Pathways (RCPs), comprising four pathways, known as RCP8.5, RCP6.0, RCP4.5 and RCP2.6 [10]. In this study, the dynamical downscaling processes with the WRF were performed for RCP4.5 and RCP8.5 for moderate and extreme cases, respectively. Furthermore, in order to consider climate variability and uncertainty from the interannual variability in climate projection, the WRF simulation was conducted for each June between 2026 and 2035. These 10-year repetitive simulations of each June will also provide the upper and lower bounds from the dominant interannual variability and further insight for a multi-model ensemble mean and variation over the Hanoi region. The full months of June were chosen as the simulation periods so that the maximum air temperatures during the summer can be observed.

In order to understand the contributions of land use change and global warming to the future increase in urban temperature in Hanoi, this study assessed four scenarios, namely, cases 1, 2-1, 2-2 and 2-3. Case 1 represents the current land use condition (Fig. 48.2b). The simulation for case 1 was conducted in June 2013 in which the monthly average air temperature (33.6 °C) is relatively close to the average air temperature in June over the 15-year period from 2000 to 2015. Accordingly, the initial and boundary conditions for WRF simulation were imposed by the NCEP Final (NCEP-FNL) Analysis 6-hourly data from June 2013.

The land use conditions in cases 2-1, 2-2 and 2-3 represent the proposed land use plan in the Hanoi Master Plan 2030 (Fig. 48.2c). In case 2-1, the WRF simulation was conducted under the same weather conditions as those in case 1. Thus, the impact of land use change on UHI can be observed by contrasting the results between cases 1 and 2-1. On the other hand, cases 2-2 and 2-3 were examined under the two scenarios utilizing RCP4.5 and RCP8.5 datasets of MIROC5, respectively, in order to analyse the influence from global warming. The simulations for both cases were conducted in each June, from 2026 to 2035. In addition, only the days with prevailing southeasterly winds (i.e. winds coming from directions between 90° and 180°) were taken into account in the analysis. Further explanations are included in [8].

48.3.3 Model Validation

The WRF simulation was performed for case 1 from 00:00 UTC on 12 June 2013 to 00:00 UTC on 24 June 2013. The simulation results were compared with the observation data from the weather station located in Noibai International Airport (see the location in Fig. 48.1a). The validation results are shown in Fig. 48.3. Overall, the simulation results agree with the observed values with the coefficient of determination (R^2) of 0.77 and 0.28 for the air temperature at 2 m and wind conditions at 10 m above the ground.

48.4 Results and Discussion

48.4.1 Impacts of Land Use Changes on Urban Climate

This section discusses the results from cases 1 and 2-1 to analyse solely the impacts of land use changes on UHIs under the same climatic conditions. Figures 48.4 and 48.5 illustrate the average air temperature distribution at 2 m above the surface and the wind directions at 10 m above the surface for all cases at 1:00 and 14:00, respectively. For a fair comparison of the results of all cases, the analysis in this section shows the temporally averaged results from the days with prevailing south-easterly winds (winds coming from directions between 90° and 180°).

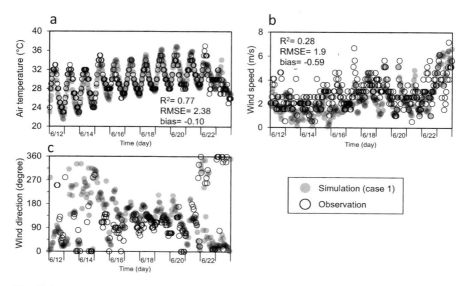

Fig. 48.3 Comparison between observed and simulated (**a**) air temperature, (**b**) wind speed and (**c**) wind direction at Noibai weather station, (21.22° N and 105.8072° E) [8]

As expected, the average air temperatures in urban areas are higher than those over the agricultural fields (i.e. rural areas). In general, the air temperature difference between urban and rural areas becomes significant at night, reaching up to 2–3 °C while it is up to 1 °C during the daytime in case 2-1 (Figs. 48.4b and 48.5b).

In the current condition (case 1), the maximum nocturnal average air temperatures are approximately 30–31 °C. The peak average air temperature of 31 °C is generally observed in urban areas (Fig. 48.4a). At night, the implementation of the master plan results in the expansion of the hotspots significantly both in existing and new urban areas (Fig. 48.4b). The average air temperature in these newly emerged hotspots increases by up to 2–3 °C over that of the same areas in case 1.

During the daytime, peak average air temperatures are observed in most parts of the urban areas in both cases (Fig. 48.5a and b). Although the peak average air temperatures remain almost at the same level even after the implementation of the master plan, the new hotspots with temperatures between 37 and 38 °C expand over the newly developed urban areas in case 2-1 (Fig. 48.5b). The average air temperature in these newly emerged hotspots increases by up to 1 °C over that of the same areas in case 1.

48.4.2 Impacts of Global Warming on Urban Climate

In this section, the impacts of global warming are discussed based on the comparisons among case 2-1, case 2-2 (RCP4.5) and case 2-3 (RCP8.5). The analysis in this

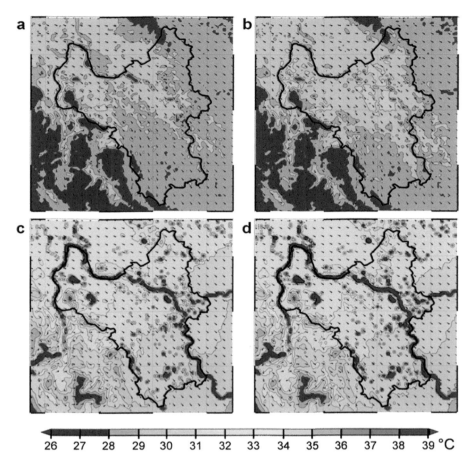

Fig. 48.4 Spatial distribution of average air temperature at 2 m and wind directions at 10 m above the surface at 1:00 for (**a**) case 1, (**b**) case 2-1, (**c**) case 2-2 and (**d**) case 2-3. The analysis is based on the temporal-averaged results calculated from the days with prevailing southeasterly winds. The results of cases 1-1 and 1-2 are the mean value from the days with southeasterly winds from 10 simulated months (every June from 2026 to 2035). The black line highlights the city's boundary [8]

section is generally divided into two types of urban areas: existing and new urban areas (see Fig. 55.2c). Although there are various land use types within each urban area, the results in this section are calculated only from the values taken from the built-up lands.

Figure 48.6 shows the statistical summary of air temperature for 10 simulated months (2026–2035) in cases 2-2 (RCP4.5) and 2-3 (RCP8.5), calculated for existing and new urban areas. For comparison, the results of cases 1 and 2-1 are also included. In general, the future climate conditions show a dynamic variation over the 10-year period for different RCP scenarios. For instance, the year 2027 is

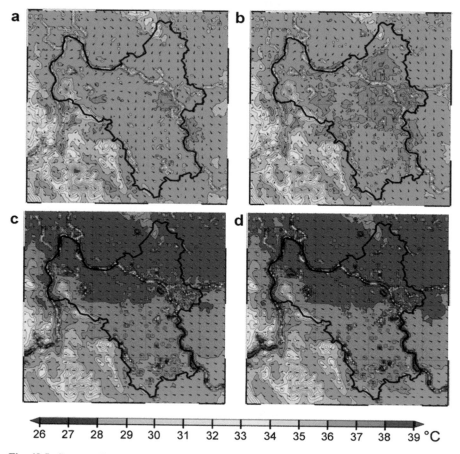

26 27 28 29 30 31 32 33 34 35 36 37 38 39 °C

Fig. 48.5 Same as Fig. 5 but at 14:00 [8]

the hottest year under RCP4.5, while 2035 is the hottest for RCP8.5. The future air temperature can even be cooler than present, as seen in 2032 and 2033 in case 2-2 (Fig. 48.6a and c). This may be due to interannual climate variations in existing urban and new urban areas for RCP4.5. However, the interannual variability in RCP8.5 results is not distinct, and the variations in air temperature are rather narrower than that of RCP4.5. Nevertheless, hotter years would occur more often in the near future than would cooler years.

Global warming would significantly alter the urban climate in Hanoi City. Under global warming conditions (i.e. cases 2-2 and 2-3), temperature increases are observed in the whole region (Figs. 48.4 and 48.5c, d). As shown in Fig. 48.6, the maximum average temperature in the global warming conditions is up to 1.7 °C greater than that of case 1 and 1.4 °C greater than that of case 2-1. The minimum

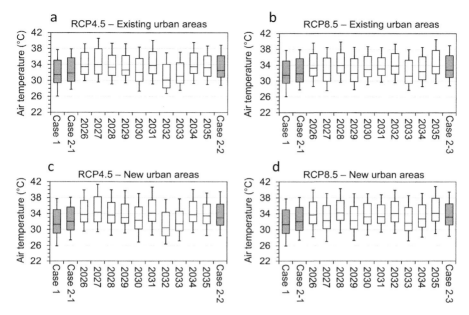

Fig. 48.6 Statistical summary (maximum, minimum, 25th percentiles, 75th percentiles, and median) of average air temperature in the existing urban area for (**a**) case 2-2 (RCP4.5), (**b**) case 2-3 (RCP8.5) and at the new urban area for (**c**) case 2-2 (RCP4.5) and (**d**) case 2-3 (RCP8.5). In order to see the climate variations over the 2030s, the results from 10 simulated months are shown. The box-plot at the far right of each graph is the mean value from 10 simulated months [8]

average air temperature that would be experienced by the urban areas in the future scenarios would also increase by up to 2.7 °C from case 1 and by up to 1 °C from case 2-1.

48.4.3   Contributions of Land Use Changes and Global Warming to the Future Temperature Increase

Figure 48.7 shows the diurnal variations of spatially and temporally averaged air temperatures in existing and new urban areas for all cases. As shown in Fig. 48.7a, the implementation of the master plan slightly increases the average air temperature in existing urban areas both in the daytime (6:00–17:00) and at night (0:00–5:00, 18:00–0:00). The increases in air temperature due to the land use changes in existing areas are up to 0.5 and 0.6 °C in the daytime and at night, respectively. In the new urban areas (Fig. 48.7b), the massive land use changes brought by the master plan result in a greater increase in average air temperatures than in the existing urban areas, by 0.9 and 1 °C in the daytime and at night, respectively.

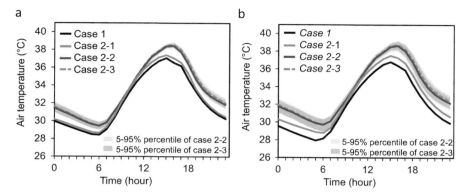

Fig. 48.7 Diurnal variation of spatial- and temporal-average air temperature in (**a**) existing urban area and (**b**) new urban area for all cases. The results are calculated from the days with prevailing southeasterly winds, and only the values from the built-up land use types are taken into account [8]

Meanwhile, under the global warming scenarios, the average air temperatures for the whole day are generally higher than in present condition. In the 2030s, the air temperatures in the existing urban areas increase by up to 1.3 and 1.5 °C for cases 2-2 and 2-3, respectively, over those in case 2-1 (Fig. 48.7a). A similar trend is also observed in the new urban areas (Fig. 48.7b), where the average air temperatures increase by up to 1.5 and 1.7 °C in case 2-2 and 2-3, respectively, over those in case 2-1. The average air temperatures under RCP8.5 scenario (case 2-3) are slightly warmer than those under RCP4.5 (case 2-2), by up to 0.2 °C.

In total, global warming and land use changes contribute to the temperature increases of up to 2.1 and 2.7 °C on average in existing and new urban areas, respectively. In the existing urban areas, out of the average increase of 2.1 °C, up to 1.5 °C (or 71%) and 0.6 °C (or 29%) are attributable to global warming and land use changes, respectively. Meanwhile, in the new urban areas, the corresponding contributions out of the average increase of 2.7 °C, up to 1.7 °C (or 63%) and 1 °C (or 37%) are attributable to global warming and land use changes, respectively.

48.5 Conclusions

The impacts of land use changes and global warming on the urban climate in Hanoi in the 2030s were investigated using numerical experiments with WRF modelling. For the near future climate conditions, direct dynamical downscaling with WRF was conducted with initial and boundary conditions from the MIROC5 datasets for RCP4.5 and RCP8.5.

The main findings are summarized as follows:

- Hotspots widely expanded following the pattern of new built-up areas in the master plan scenario. The implementation of the master plan increased the number of hotspots with peak average air temperatures in the existing urban areas, largely at night.

- The land use changes proposed in the master plan are expected to increase the average air temperature in the existing and new urban areas by up to 0.6 and 1 °C, respectively.
- The results illustrated that the urban air temperature was projected to increase along with global warming. In the 2030s, the average air temperature increase in the existing urban areas was projected to be up to 2.1 °C, of which up to 1.5 °C (or 71%) and 0.6 °C (or 29%) are attributable to global warming and land use changes, respectively.
- The increase in air temperature in the near future (2030s) will likely exceed the cooling effects from any UHI mitigation measures. Under this circumstance, adaptation measures to urban warming might be prioritized rather than the mitigation measures particularly in the growing cities of emerging economies.

References

1. Vietnam Institute of Architecture Urban and Rural Planning (2011) The Hanoi capital construction master plan 2030 and vision to 2050, Comprehensive report. VIAP, Hanoi
2. Hartmann DJ et al (2013) Observations: atmosphere and surface. In: Climate Change 2013: the physical science basis. Contribution of Working Group I to the Fifth Assessment Report of the Intergovernmental Panel on Climate Change, Cambridge, UK/New York
3. Bhuvandas N et al (2014) Review of downscaling methods in climate change and their role in hydrological studies. Int J Environ Chem Ecol Geol Geophys Eng 8(10):648–653
4. Skamarock WC, Klemp JB, Dudhia J, Gill DO, Barker DM, Duda MG et al (2008) A description of the advanced research WRF version 3 (NCAR technical note NCAR/TN-475 +STR no. TN-475+STR). Boulder
5. Chen F et al (2011) The integrated WRF/urban modelling system: development, evaluation, and applications to urban environmental problems. Int J Climatol 31(2):273–288
6. Kusaka H et al (2001) A simple single-layer urban canopy model for atmospheric models: comparison with multi-layer and slab models. Bound-Layer Meteor 101(3):329–358
7. Kubota T, Lee HS, Trihamdani AR, Phuong TTT, Tanaka T, Matsuo K (2017) Impacts of land use changes from the Hanoi master plan 2030 on urban islands: part 1. Cooling effects of proposed green strategies. Sustain Cities Soc 33:295–317
8. Lee HS, Trihamdani AR, Kubota T, Iizuka S, Phuong TTT (2017) Impacts of land use changes from the Hanoi master plan 2030 on urban islands: part 2. Influence of global warming. Sustain Cities Soc 31:95–108
9. Watanabe M et al (2010) Improved climate simulation by MIROC5: mean states, variability, and climate sensitivity. J Clim 23(23):6312–6335
10. van Vuuren DP, Edmonds J, Kainuma M, Riahi K, Thomson A, Hibbard K et al (2011) The representative concentration pathways: an overview. Clim Chang 109(1):5–31. https://doi.org/10.1007/s10584-011-0148-z

Chapter 49
Assessment of Future Urban Climate After Implementation of the City Master Plan in Vinh City, Vietnam

Satoru Iizuka, Masato Miyata, and Kaede Watanabe

Abstract An attempt to project the future urban climate in the summer (a month of June) of the 2030s in Vinh City, Vietnam, using a regional atmospheric model, the Weather Research and Forecasting (WRF) model, is introduced in this chapter. A city master plan for Vinh City proposed by a Japanese civil engineering consulting company, Nikken Sekkei Civil Engineering Ltd., is incorporated into the future projections. The proposed city master plan targets the year 2030, with a population of 900,000, and the total planning area covers approximately 250 km^2. Particularly, in the future projections, the effects of future global warming and introducing the proposed city master plan on the urban climate in Vinh City are quantitatively discussed. Moreover, the impacts of concentrating and decentralizing urban districts in the city master plan on the urban climate are numerically clarified.

Keywords Future urban climate · City master plan · Urban district structure · Vietnam · Weather Research and Forecasting (WRF) model

49.1 Introduction

Vietnam (Socialist Republic of Viet Nam) is a rapidly developing country, and the recent growth rates of the economy and population are about 6% and 1.2%, respectively. The expansion of urban areas is progressing in the country. In response to these situations, many city master plans have recently been proposed in Vietnam. Nghe An Province, located in the northern part of the country, asked a Japanese civil

S. Iizuka (✉)
Graduate School of Environmental Studies, Department of Environmental Engineering and Architecture, Nagoya University, Nagoya, Japan
e-mail: s.iizuka@nagoya-u.jp

M. Miyata
Mitsubishi UFJ Research and Consulting Co., Ltd., Nagoya, Japan

K. Watanabe
Nikken Sekkei Civil Engineering Ltd., Tokyo, Japan

© Springer Nature Singapore Pte Ltd. 2018
T. Kubota et al. (eds.), *Sustainable Houses and Living in the Hot-Humid Climates of Asia*, https://doi.org/10.1007/978-981-10-8465-2_49

541

Table 49.1 Physics options used in the WRF projections

Category	Option
1. Microphysics	WRF single-moment 3-class (WSM3) scheme
2. Surface layer	Eta surface layer scheme
3. Land surface	Noah land surface model (LSM) + urban canopy model [4]
4. Planetary boundary layer	Mellor-Yamada-Janjic (MYJ) model
5. Atmospheric radiation	Shortwave: MM5 (Dudhia) scheme
	Longwave: Rapid radiative transfer model (RRTM)

engineering consulting company, Nikken Sekkei Civil Engineering Ltd., to create a city master plan for Vinh City, the capital of Nghe An Province. The proposed city master plan targets the year 2030, with a population of 900,000, and the total planning area covers approximately 250 km^2.

In this chapter, an attempt to project the future urban climate in the summer (a month of June) of the 2030s in Vinh City is introduced. The future projections were performed by introducing the abovementioned city master plan and its modifications (northern or southern concentration of new urban districts) and by using a regional atmospheric model, the Weather Research and Forecasting (WRF) model [1], combined with a pseudo global warming method [2, 3]. By comparing the projected results with those obtained from the present climate case and the present land-use case, the effects of future global warming and introducing the proposed city master plan on the urban climate in Vinh City are quantitatively clarified. The impacts of the modified city master plans on the urban climate are also investigated.

49.2 Outline of Future Projections

49.2.1 Climate Projection Model

The WRF modeling system, Version 3.0.1.1, with the Advanced Research WRF (ARW) dynamic solver [1], was used to project the future urban climate in Vinh City, Vietnam. The target period was a month of June in the 2030s. June is a hot month in the target city under the present climate. Table 49.1 shows the physics options used in the WRF projections. As described in the next subsection, the pseudo global warming method proposed by Kimura and Kitoh [2] was introduced to perform the future projections.

49.2.2 Pseudo Global Warming Method

In order to consider the effect of future (2030s) global warming, the abovementioned pseudo global warming method [2] was combined with the WRF projections. In this

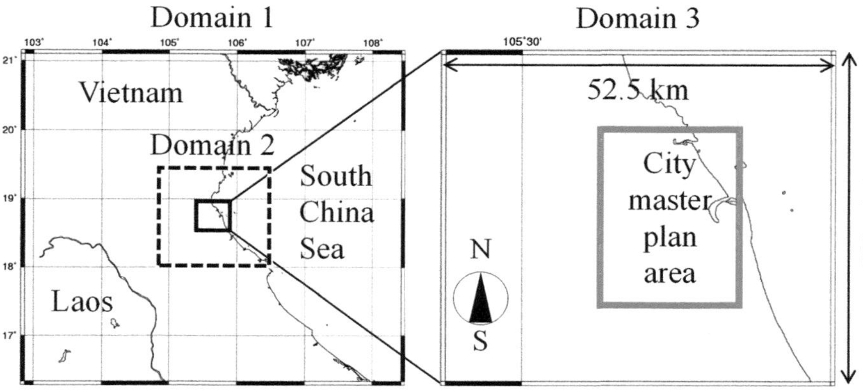

Fig. 49.1 Nested domains in the WRF projections

method, five components to formulate the pseudo global warming data, i.e., horizontal wind components, potential temperature, geopotential height, sea surface temperature, and ground surface temperature, were used as the initial conditions for all nested domains (cf. Subsection 49.2.3) and the boundary conditions for the largest domain in the WRF projections. The pseudo global warming data were generated using a general circulation model, GFDL-CM3, with the Representative Concentration Pathways (RCP) 8.5 scenario [5] adopted in the IPCC Fifth Assessment Report [6]. For further information, refer, for instance, to Iizuka et al. [3].

49.2.3   Projection Domains

Figure 49.1 shows three nested domains in the WRF projections. The sizes of Domains 1, 2, and 3 were 585 km × 540 km (4.5 km horizontal grid resolution), 177 km × 159 km (1.5 km horizontal grid resolution), and 52.5 km × 46.5 km (0.5 km horizontal grid resolution), respectively. In all of the domains, the vertical direction was divided into 34 layers (about 21 km). The region enclosed by a red line in Domain 3 indicates the area of the city master plan proposed by Nikken Sekkei Civil Engineering Ltd.

49.2.4   Projected Cases

Five cases, shown in Table 49.2, were tested. Four land-use models, i.e., the present land-use (Case 1; Fig. 49.2(1)), the city master plan proposed by Nikken Sekkei Civil Engineering Ltd. (Case 2; Fig. 49.2(2)), and two modifications of the city master plan (Cases 3 and 4; Figs. 49.2(3) and (4)), were introduced. As a

Table 49.2 Projected cases

	Period	Land-use
Case 0	June 2011 (present climate)	Present land-use
Case 1	June in the 2030s (future climate)	Present land-use
Case 2		City master plan
Case 3		Modified city master plan (1): Northern concentration of new urban districts
Case 4		Modified city master plan (2): Southern concentration of new urban districts

comparison, the present (June 2011) climate case with the present land-use model (Case 0) was also tested.

In the city master plan (Case 2) and its modifications (Cases 3 and 4), the urban area was classified into three categories, i.e., the central business district (CBD), existing urban district, and new urban district. For the two modifications of the city master plan, the new urban districts were concentrated in the northern (Case 3) or southern (Case 4) part of the planning area. The total area of new urban districts was the same in Cases 2, 3, and 4. The average height of buildings and the building (green) coverage ratio were 26 m and 50% (50%) in the CBD, 12 m and 70% (30%) in the existing urban district, and 12 m and 60% (40%) in the new urban district.

49.3 Examples of the Projected Results

49.3.1 Effect of Global Warming in the 2030s

Figure 49.3 shows the space-averaged (over all existing urban districts) and monthly-averaged (for a month of June) diurnal variation of the difference in air temperature at a height of 2 m between Case 1 (future climate) and Case 0 (present climate). The time-averaged temperature difference over 24 h depicted in the figure was 1.76 °C. This indicates a temperature increase by estimated global warming in the 2030s under the RCP 8.5 scenario, which is the pathway with the highest greenhouse gas emissions among all RCP scenarios.

49.3.2 Effect of Introducing the City Master Plan

Figure 49.4 shows the space-averaged (over all urban districts) and monthly-averaged (for a month of June in the 2030s) diurnal variation of the difference in air temperature at a height of 2 m between Case 2 (city master plan) and Case 1

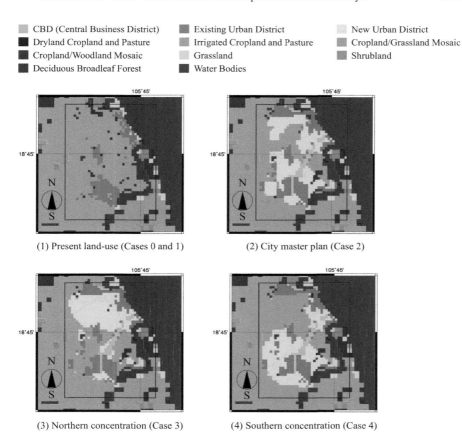

Fig. 49.2 Present and future land-use models

(present land-use). The temperature difference in the nighttime is larger than that in the daytime, and the maximum temperature difference is 0.33 °C (8 p.m.). This is considerably smaller than the temperature increase by global warming in the 2030s (1.76 °C; cf. Subsection 49.3.1). However, the small temperature difference is never negligible, considering the fact that the increase in the global average surface temperature over the period from 1880 to 2012 was 0.85 °C [6]. Therefore, actions to mitigate the temperature increase due to expansion of the urban area will be necessary in the near future.

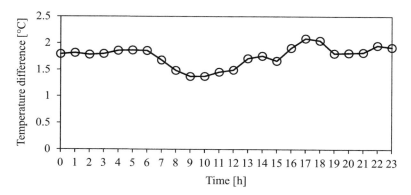

Fig. 49.3 Space-averaged and monthly-averaged diurnal variation of the difference in air temperature at a height of 2 m between Case 1 (future climate) and Case 0 (present climate)

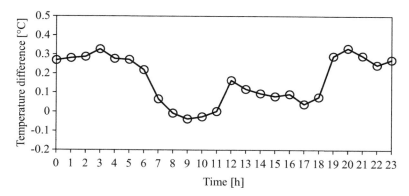

Fig. 49.4 Space-averaged and monthly-averaged diurnal variation of the difference in air temperature at a height of 2 m between Case 2 (city master plan) and Case 1 (present land-use)

49.3.3 Effect of Concentrating and Decentralizing the New Urban Districts

Figures 49.5 and 49.6 show the space-averaged and monthly-averaged (for a month of June in the 2030s) diurnal variations of the difference in air temperature at a height of 2 m between Case 3 (northern concentration of the new urban districts) and Case 2 (original city master plan) and those between Case 4 (southern concentration of the new urban districts) and Case 2, respectively. With regard to space averaging, averaging over each urban category (CBD, existing urban district, and northern and southern new urban districts) is introduced in the figures.

In Case 3 (northern concentration of the new urban districts), the temperature in the CBD is only a little lower than that in Case 2; however, the temperature in the northern new urban districts is higher than that in Case 2. The time-averaged

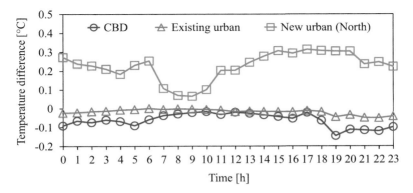

Fig. 49.5 Space-averaged and monthly-averaged diurnal variations of the difference in air temperature at a height of 2 m between Case 3 (northern concentration) and Case 2 (original plan)

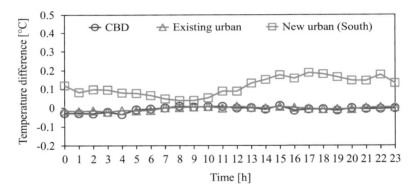

Fig. 49.6 Space-averaged and monthly-averaged diurnal variations of the difference in air temperature at a height of 2 m between Case 4 (southern concentration) and Case 2 (original plan)

temperature increase over 24 h is 0.23 °C. There is almost no difference in the temperature in the existing urban districts between Cases 2 and 3.

In Case 4 (southern concentration of the new urban districts), the temperatures in the CBD and existing urban districts are almost the same as the corresponding results in Case 2. The temperature in the southern new urban districts is a bit higher than that in Case 2. The time-averaged temperature increase over 24 h is 0.11 °C.

Compared to Case 2 (original city master plan), the further expansion of new urban districts in the northern part (Case 3) or in the southern part (Case 4) causes the abovementioned temperature increase in the corresponding area. Considering these projected results, the original city master plan, in which the new urban districts are decentralized in both northern and southern parts of the planning area, is better than the two modifications (northern or southern concentration of the new urban districts) from the viewpoint of the thermal environment.

49.4 Summary

In this chapter, future (a month of June in the 2030s) projections of the urban climate in Vinh City, Vietnam, using a regional atmospheric model, the WRF model, were introduced. In particular, the effects of future global warming and introducing a city master plan on the urban climate in the city were quantitatively discussed. Moreover, the impacts of concentrating and decentralizing urban districts in the city master plan on the urban climate were numerically clarified.

In actual urban planning, we should consider a lot of aspects, such as social welfare, disaster mitigation/prevention, nature and historic building conservation, and economic circumstances. In the inevitable future global warming era, the aspect of improving the thermal environment (or mitigation and adaptation to a hot environment) is also crucial in urban planning, especially in tropical cities.

References

1. Skamarock WC et al (2008) A description of the advanced research WRF version 3. NCAR/TN-475+STR, NCAR technical note
2. Kimura F, Kitoh A (2007) Downscaling by pseudo global warming method. The Final Report of ICCAP: 43–46
3. Iizuka S et al (2015) Impacts of disaster mitigation/prevention urban structure models on future urban thermal environment. Sustain Cities Soc 19:414–420
4. Kusaka H et al (2001) A simple single-layer urban canopy model for atmospheric models: comparison with multi-layer and slab models. Bound-Layer Meteorol 101:329–358
5. van Vuuren DP et al (2011) The representative concentration pathways: an overview. Clim Chang 109:5–31
6. IPCC (2014) Climate change 2014: synthesis report. IPCC Fifth Assessment report

Chapter 50
Urban Heat Island of Putrajaya City in Malaysia

Dilshan Remaz Ossen, Adeb Qaid, and Hasanuddin Bin Lamit

Abstract The administrative capital city of Malaysia, Putrajaya, is built on the garden city concept. The Putrajaya planning guidelines for sustainable cities suggested that 37% of the development area should be dedicated to green and open spaces. Further, they stated that 50% of the area for car parks and open spaces must be provided with shade trees and grass. Although the garden city concept was applied, the city still experiences high air temperatures. The purpose of this work is to highlight the recent trends in studies of the heat island and surface temperature conducted in Putrajaya city. The study further highlights findings and suggested solutions for mitigating of the heat island in the city.

Keywords Urban Heat Island · Putrajaya City · Tropical Planned City

50.1 Urban Heat Island Phenomenon

The urban heat island phenomenon is a common issue of urban environments. The temperature within urban areas is higher than that in surrounding rural areas, causing an uncomfortable environment for people in terms of thermal comfort. The factors that cause urban heat islands are also a result of the intense increase in building and infrastructure development to cater to the rising population in the cities, as well as the use of nonporous urban surfaces. These factors increase the heat absorption and release processes and the obstruction to air movement that will then create the urban heat island phenomenon.

Environment-friendly designs suggest integrating green areas, buildings, pedestrians, public spaces, urban corridors, and streets to produce an urban environment that is conducive to use. Studies have emphasized several ways to mitigate the urban heat island effect: the use of green spaces to lower the temperatures and thus creating cool

D. R. Ossen (✉) · A. Qaid
Department of Architecture Engineering, Kingdom University, Riffa, Bahrain

H. B. Lamit
Faculty of Built Environment, Universiti Teknologi Malaysia, Skudai, Johor, Malaysia

© Springer Nature Singapore Pte Ltd. 2018
T. Kubota et al. (eds.), *Sustainable Houses and Living in the Hot-Humid Climates of Asia*, https://doi.org/10.1007/978-981-10-8465-2_50

549

islands in urban areas, the use of urban geometry to create cool island effects, the utilization of light-colored urban surfaces and permeable urban surfaces, the preparation of guidelines to increase efficiencies, and better transportation and traffic management.

In recent years, environment-friendly design methods have been actively discussed and researched based on microclimate analyses, especially in Europe. However, it is quite difficult to directly apply these accumulated findings and skills to an urban development or landscape design in tropical countries, as each country has different climates and obviously differences in cultural expressions and aspirations. This section reviews and identifies the criterion that influenced the urban microclimate analysis of Putrajaya, a planned city in the tropics.

50.2 Description of Putrajaya City

Putrajaya, the federal administrative capital of Malaysia, covers 4931 hectares. A total of 40% of the city area is designated open space, and 552 hectares are designated for a man-made lake. The city is located in the Klang Valley between Kuala Lumpur, the capital city of Malaysia, and Kuala Lumpur International airport, within Lat. 2_55 °N and Long. 101_42 °E. Putrajaya is a newly planned city designed with the theme "a city in a garden." The master plan divided the city into 2 main areas and 20 precincts: the core (5 precincts) and the peripheral area (15 precincts) (see Fig. 50.1). Putrajaya Boulevard, 4.2 km long and 100 m wide, divides the core area along the main path axis, extending from the northeast to the southwest and populated by symmetrical, large federal government buildings. Putrajaya Boulevard is classified as a dispersed urban form, with a somewhat lower density [11]. The peripheral area, located approximately 3–5 km from the central precinct, includes 14 residential neighborhoods [11, 18].

The climate is typically hot and humid, with a uniform air temperature during the year, an annual average maximum temperature of 27.5 °C, a minimum average temperature of 25 °C, an average humidity of 62.6%, and long hours of sunshine and solar radiation—an average of 6 h per day and 4.39 $kWhm^{-2}$ of annual average solar irradiation. Winds are generally light and variable, with speeds ranging from 0 to 7.5 m/s [17].

50.3 Putrajaya City Local Plan Area Guidelines and Policies

The city was constructed based on guidelines, according to the local planning policies and strategies (Official Portal of Perbadanan Putrajaya, http://www.ppj.gov.my/portal/page?_pageid=311,1&_dad=portal&_schema=PORTAL#1475   accessed   on

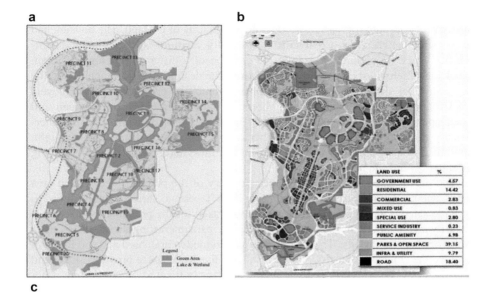

Fig. 50.1 (a) Master plan of Putrajaya (ppj), (b) the divided precincts [15], and (c) Putrajaya City (http://abckualalumpur.com/info_guide/putrajaya2.html)

4 February 2013) [5, 12]; the lake was designed to enhance the aesthetic of the city landscape, to present a good visual impact of the beauty of the area, and to provide continuous lake frontage that functions as a recreation, boating, fishing, water transport, and leisure activity resource.

The residential area was planned on a neighborhood planning concept with a systematic housing program, respecting the topography of the land. The housing had to be varied between low-density, medium-density, medium-high-density, and high-density types. This concept leads to the creation of a different typology of urban blocks in different urban neighborhood forms. The design concept, emphasizing the integration of different social communities, combines pedestrian paths, bicycle tracks, and walkways with the neighborhood, creating a hierarchy of green and open spaces. The low-density developments face the waterfront, while higher-density developments are located farther away, ensuring that adequate daylight, sunlight, air, and spaces provide a feeling of openness and privacy for each building oriented toward the public realm.

The green spaces are located on the periphery of the precinct, and shaded area landscape buffers are provided for adjacent land. Trees and shrubs were designed to provide screens and shade, and vegetation was used for the provision of a sense of security and a variety of interests. Urbanscapes create interesting views, vistas and a sense of place for the city. The landscape was designed in both hard and soft forms to achieve the concept of "a city in a garden," and the green space provides shade and cools the city. Policies on the environment are mostly focused on controlling water pollution, solid waste, and noise.

The street was designed to accommodate public utility services and a drainage system, providing an acceptable level of safety. The design had to minimize the negative impact of vehicular traffic. The neighborhood streets serve dual functions and have been used as a place for parking and as a space for social activity, interaction, and play. The road linkages are well connected to the expressway and the main distributor roads. Streetscapes were designed to enhance the city image.

The policies and strategies above were only centered on the urban design perspective and gave priority to accessibility, safety, image, aesthetics, amenities, society, and privacy— visually comforting and psychological aspects. The review of climate responsive policies was confined to buildings, while a general climate consideration of the city's or neighborhood's green areas, buildings, and street layouts that enhance the wind velocity to reduce the air temperature or avoid the creation of the "heat island" remained overlooked.

As Putrajaya is a new planned city, it became a subject of climate research in different fields, and the volume of research investigating urban heat islands in the city has been ever-increasing (see Table 50.1). Shahidan et al. [17] studied the cooling effect and the urban heat island in which the UHI was recorded at 2.6 °C. Ahmed et al. [3] found the heat island intensity to be 2 °C on average at nighttime and negligible during the daytime, using a fixed station measurement. In a field measurement study conducted by Thani et al. [19], the maximum air temperature reached 38 °C during the day. Morris et al. [8] computationally investigated the UHI and found that it ranges from 1.9 to 3.1 °C, and the average daily intensity was 0.79 °C hotter. Overall, the

Table 50.1 Urban heat island intensity, layers investigated, and approaches used

Author (s) and year	Approach and techniques	Type of urban heat island investigated	Layer	UHII
Ahmed et al. [3]	Filed measurement and fixed stations	Air temperature	UCL	2 °C at nighttime, negligible at daytime
Shahidan et al. [17]	Field measurement	Air temperature and surface temperature	UCL	Average monthly UHII 2.6 _C
Morris et al. [8]	Computational study WRF/Noah	Air temperature and surface temperature	UBL	Average daily intensity of 0.79 °C and the night hours, the UHII ranges from 1.9–3.1 °C
Morris et al. [9]	Computational study WRF/Noah/UCM coupled system	Air temperature and surface temperature of water body and vegetation	UBL	1.79 and 2.04 °C
Morris et al. [10]	Using the NCAR Weather and Research Forecasting (WRF) model and (NOAH LSM/UCM).	Air temperature	UBL	2.1 °C

urban heat island temperature range is between 2 and 3 °C. The increase in the urban heat island effect in the city is due to the wide, impervious surfaces and the urban construction. In contrast, Ahmad and Hashim [2] used a remote sensing thermal infrared band method and found that although the urban coverage of the city was increasing, the heat island phenomenon was observed to decrease, as the high moisture content due to antecedent precipitation had a removal effect on the urban heat island intensity. Therefore, the factors contributing to the formation of a UHI in Putrajaya were not only caused by urbanization but also by other climate change effects Salleh et al. [16]. However, studies of the urban heat island in Putrajaya have mostly focused on the urban surface temperature rather than the air temperature. Further research using different methods and considering the UCL and the UBL will clearly show the city heat island layer. Both layers were investigated in Putrajaya City by different methods and reflected a different urban heat island. The urban boundary layer (UBL) is a scale above the building roughness level affected by the presence of the land use or urban surface and reflects a variety of assemblages of processes—the so-called local or meso-scale—while the urban canopy layer (UCL) is a scale roughly between the ground and the roof level of the building and is known as a microscale, with site-specific characteristics and processes [4, 13].

Different areas in Putrajaya need to be investigated using a different methodology to extensively explore the urban heat island.

50.4   Urban Surface Temperature and the Heat Island

50.4.1   Landscape Design: Material, Material Properties, Water, and Greenery

According to Thani et al. [19], the temperature distribution is strongly influenced by the urban landscape morphology in which significant temperature differences are observed in various urban areas. In the Putrajaya area, according to Ho Chin Siong [6], a greenery of 420,000 trees was added (Broadleaf tree) and equally distributed to increase the evapotranspiration process and for cooling effects. All buildings' roofs were painted with high-albedo products; the albedo measurement of the urban area is between 0.15 and 0.22. The city streets and buildings were constructed using a variety of materials. The albedo "reflectivity" is the rating of the surface's material ability to reflect the solar radiation incident during the daytime. Emissivity is the ability of the material to reradiate back the absorbed heat to the atmosphere at nighttime. A material with a high albedo and a high emissivity is the best to maintain a cooling environment. High emissivity materials are important for reducing the urban heat island impact. The material, albedo, and emissivity of most ground-level and vertical facade buildings in the boulevard area are presented in Table 50.2.

50.4.2   Strategies Suggested to Mitigate the Urban Heat Island Intensity

Several studies were conducted in Putrajaya City, investigating the landscape setting, surface temperature, material textures, albedo effect on the air temperature of the city, and the urban heat island intensity, as shown in Table 50.3. Based on these studies, the researchers suggest several strategies to mitigate the heat intensity of the city. For example, Shahidan et al. [17] found that higher levels of tree canopy density (LAI 9.7) (e.g., *F. benjamina* species) coupled with "cool" materials (albedo of 0.8) produced the largest urban air temperature reduction of 2.7 °C. Three major physical factors, namely, a larger quantity of trees, higher canopy density, and cool materials, produced the best improvement in the indoor and the outdoor climate. Qaid and Ossen [14] investigated the building height and found that aspect ratios of 0.8–2 reduce the morning microclimate and night heat islands. An aspect ratio of 2–0.8 reduces the temperature of surfaces by 10–14 °C and that of the air by 4.7 °C (see Fig. 50.2). Ahmed et al. [3] found that a combination of cool material and trees can reduce the surface temperature by 7 °C. Morris et al. [9], conversely, found that a

Table 50.2 Surface material, albedo, and emissivity of the ground and the Putrajaya Boulevard buildings showing the common materials in city construction, adopted from Ahmed et al. [3]

Material in the ground and vertical surface	Locations	Albedo (α)	Emissivity (ε)
Polished rose granite	Ground and facade	0.4[**]	0.45[**]
Gray granite	Ground and facade	0.3[***]	0.45[**]
Dark granite	Ground and facade	0.2	0.72[*]
White granite	Ground and facade	0.43[**]	0.45[**]
Dark concrete	Ground and facade	0.2[***]	0.9[*]
Polished dark granite	Ground and facade	0.3[***]	0.72[*]
Polished white granite	Ground and facade	0.8[*]	0.45[**]
Embossed aluminum	Facade	0.85[*]	0.04[***]
White plaster	Facade	0.93[*]	0.91[*]
Galvanized metal	Facade	0.4	0.13[***]
Alucobond cladding	Facade	0.65[*]	0.77[*]
Smooth glass panel	Facade	0.7[*]	0.92~0.95[*]
Concrete, exposure stone	Ground	0.5[*]	0.65[**]
Concrete	Ground and facade	0.3[***]	0.94[*]
Concrete tiles	Ground	0.45[**]	0.63[**]
Dark concrete	Ground and facade	0.2[***]	0.9[*]
Gray concrete	Ground	0.3~0.4[**]	63[**]
Brown concrete			
Dark gray concrete			
Yellow concrete pigment			
Red concrete pigment			
Soil	Ground	0.45[**]	0.38[**]
Water body			
Vegetations			

* High (α and ε)
** Medium (α and ε)
*** Low (α and ε)

body of water and vegetation induced a daily temperature reduction of 0.14 and 0.39 °C, respectively. The vegetation's cooling effect was consistent during both mornings and nights, while that of water occurred only during the daytime. The daily mean temperature reductions were 0.047 °C and 0.024 °C per square kilometer of vegetation and body of water, respectively. However, water body and vegetation induced a daily reduction in temperature by 0.53 °C for Putrajaya City. The effect of the surface temperature and the albedo in Putrajaya City is well documented in the literature, but the effects of the building form and a body of water need further investigation.

Table 50.3 Physical and climatic parameters investigated and mitigation strategies

Parameter studied		Urban heat island strategies suggested	Authors
Physical landscape setting and design	Material and albedo of building facades and the ground level	Increase the albedo of the material used in the constructions	Abu Bakar [1], Ahmed et al. [3], Shahidan et al. [17] and Morris et al. [9]
		Use of granite with very light color and fine texture is the best	
		Use of cool material in general	
		Use of high- and low-albedo material on the street could modify street temperature and reduce glare during the day.	
	Vegetation and trees	Increase the canopy of the large trees and the cover of the small shrubs	Abu Bakar [1], Ahmed et al. [3], Shahidan et al. [17] and Morris et al. [9]
		Use of appropriate greenery in the landscape setting can create a thermally comforting environment	
		Increase the grass in the green area in urban streets and incorporate the grass with trees	
		Using of clustered trees is better than separated ones to reduce the heat	
		Increase the vegetation more than the water body for cooling effects during the day and the nighttime period	
	Water body	Increase the water body for a daytime cooling effect more than a nighttime effect	Morris et al. [9]
	Soil	Soil measure content reduce the heat island	Ahmad and Hashim [2], Salleh et al. [16]
Built form design	Buildings form and street orientation	Increase the buildings in the northeast to southwest direction	Qaid and Ossen [14]
		An aspect ratio of 2–0.8 reduces the temperature of the ground surfaces	
Climatic design	Wind velocity and direction	Increase the wind velocity by designing the building form in both street sides	Ahmed et al. [3], Qaid and Ossen [14]
		Street orientation should be facing wind directions	
		Asymmetrical streets are advised and are better than low symmetrical streets in	

(continued)

Table 50.3   (continued)

Parameter studied		Urban heat island strategies suggested	Authors
		enhancing wind when tall buildings confront the wind's direction	
	Solar radiation and shading	Providing shade to the exposed area can reduce heat	Abu Bakar [1], Qaid and Ossen [14]
		Asymmetrical streets are advised and are better than low symmetrical streets in blocking solar radiation, when tall buildings confront solar altitudes	

50.5   Conclusions

Heat islands in contemporary cities have been well documented, while the volume of research undertaken on smaller, growing cities is comparatively small. A planned city under construction is a fertile ground for climate research, offering the opportunity to formulate urban planning strategies to combat problems presented by urban heat islands and a rich source of knowledge for learning how temperature behavior changes throughout the year. There are still many gaps, and other areas and elements need to be investigated in Putrajaya City. The available knowledge is confined to climate zones and the urban heat island in a specific area of the boulevard. All of the factors, including thermal comfort, climate, microclimate, or bio-meteorological and human thermal comfort need to be investigated. There is a lack of strategies proposed to mitigate the heat island and a lack of studies that investigate the impacts of the heat island on outdoor and indoor thermal comfort. Bakar and Malek [5] found that most of the research attention in Malaysia was given to building design and its technological advancement in energy savings and conservation; very little research is being conducted on Malaysia's microclimate and the outdoor thermal environment, although these factors affect the energy consumption of buildings. Planners and designers should therefore consider policies and strategies presented that mitigate and control the UHI effect [7].

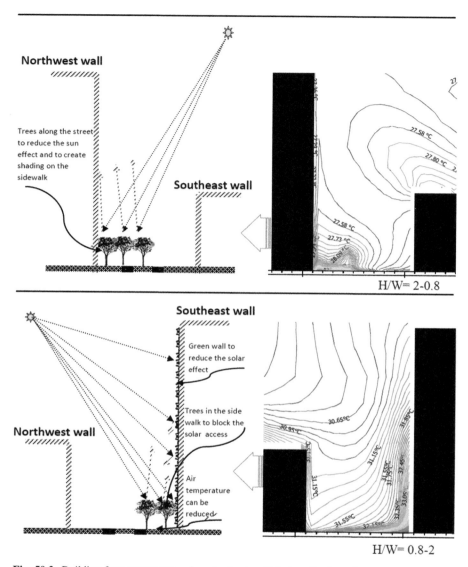

Fig. 50.2 Building form strategy to reduce the heat on Putrajaya Boulevard

References

1. Abu Bakar A (2012) The effects of ground surfaces–material, color & texture towards the adjacent thermal environment: a case study of plazas in Putrajaya, Malaysia
2. Ahmad S, Hashim NM (2007) Effects of soil moisture on urban heat island occurrences: case of Selangor, Malaysia. Humanit Soc Sci J 2(2):132–138

3. Ahmed A, Ossen D, Jamei E, Manaf N, Said I, Ahmad M (2015) Urban surface temperature behaviour and heat island effect in a tropical planned city. Theor Appl Climatol 119 (3–4):493–514. https://doi.org/10.1007/s00704-014-1122-2
4. Arnfield AJ (2003) Two decades of urban climate research: a review of turbulence, exchanges of energy and water, and the urban heat island. Int J Climatol 23(1):1–26. https://doi.org/10.1002/joc.859
5. Bakar AA, Malek NA (2015) Outdoor thermal performance investigations towards a sustainable tropical environment. Energy Sustain V Spec Contrib 206:23
6. Ho Chin Siong YM, Hashim OB (2011) Putrajaya Green City 2025: baseline and preliminary study
7. Li J-j, Wang X-r, Wang X-j, Ma W-c, Zhang H (2009) Remote sensing evaluation of urban heat island and its spatial pattern of the Shanghai metropolitan area, China. Ecol Complex 6 (4):413–420. https://doi.org/10.1016/j.ecocom.2009.02.002
8. Morris KI, Aekbal Salleh S, Chan A, Ooi MCG, Abakr YA, Oozeer MY, Duda M (2015) Computational study of urban heat island of Putrajaya, Malaysia. Sustain Cities Soc 19:359–372. https://doi.org/10.1016/j.scs.2015.04.010
9. Morris KI, Chan A, Ooi MC, Oozeer MY, Abakr YA, Morris KJK (2016a) Effect of vegetation and waterbody on the garden city concept: an evaluation study using a newly developed city, Putrajaya, Malaysia. Comput Environ Urban Syst 58:39–51. https://doi.org/10.1016/j.compenvurbsys.2016.03.005
10. Morris KI, Chan A, Salleh SA, Ooi MCG, Oozeer MY, Abakr YA (2016b) Numerical study on the urbanisation of Putrajaya and its interaction with the local climate, over a decade. Urban Clim 16:1–24. https://doi.org/10.1016/j.uclim.2016.02.001
11. Moser S (2010) Putrajaya: Malaysia's new federal administrative capital. Cities 27(4):285–297. https://doi.org/10.1016/j.cities.2009.11.002
12. Official Portal of Perbadanan Putrajaya. http://www.ppj.gov.my/portal/page?_pageid=311,1&_dad=portal&_schema=PORTAL#1475. Accessed 4 Feb 2013
13. Oke TR (1976) The distinction between canopy and boundary-layer urban heat islands. Atmos 14(4):268–277. https://doi.org/10.1080/00046973.1976.9648422
14. Qaid A, Ossen D (2015) Effect of asymmetrical street aspect ratios on microclimates in hot, humid regions. Int J Biometeorol 59(6):657–677. https://doi.org/10.1007/s00484-014-0878-5
15. Qureshi S, Ho C (2011) Towards Putrajaya green city 2025 implementing neighbourhood walkability in Putrajaya. Retrieved 5 Apr 2012
16. Salleh SA, Abd. Latif Z, Mohd WMNW, Chan A (2013) Factors Contributing to the Formation of an Urban Heat Island in Putrajaya, Malaysia. Procedia Soc Behav Sci 105:840–850. https://doi.org/10.1016/j.sbspro.2013.11.086
17. Shahidan MF, Jones PJ, Gwilliam J, Salleh E (2012) An evaluation of outdoor and building environment cooling achieved through combination modification of trees with ground materials. Build Environ 58:245–257. https://doi.org/10.1016/j.buildenv.2012.07.012
18. Siong HC (2006) Putrajaya – administrative Centre of Malaysia -planning concept and implementation. Paper presented at the Sustainable Urban Development and Governance Conference, SungKyunKwan University, Seoul
19. Thani SKSO, Mohamad NHN, Abdullah SMS (2013) The influence of urban landscape morphology on the temperature distribution of hot-humid Urban Centre. Procedia Soc Behav Sci 85:356–367. https://doi.org/10.1016/j.sbspro.2013.08.365